CRITICAL STABILITY CONSTANTS

Volume 2: Amines

CRITICAL STABILITY CONSTANTS

Volume 1 • Amino Acids

Volume 2 • Amines

CRITICAL STABILITY CONSTANTS

Volume 2: Amines

**by Robert M. Smith
and Arthur E. Martell**

Department of Chemistry
College of Science
Texas A & M University
College Station, Texas

PLENUM PRESS · NEW YORK AND LONDON

Library of Congress Cataloging in Publication Data

Martell, Arthur Earl, 1916-
  Critical stability constants.

  On vol. 2. Smith's name appears first on t.p.
  Includes bibliographical references.
  CONTENTS: v. 1. Amino acids—v. 2. Amines.
  1. Chemical equilibrium—Tables, etc. 2. Complex compounds—Tables, etc.
I. Smith, Robert Martin, 1927-   joint author. II. Title. [DNLM: 1. Amino acids.
2. Chemistry, Physical. QD503 M376c]
QD503.M37   541'.392   74-10610
ISBN 0-306-35212-5 (v. 2)

© 1975 Plenum Press, New York
A Division of Plenum Publishing Corporation
227 West 17th Street, New York, N.Y. 10011

United Kingdom edition published by Plenum Press, London
A Division of Plenum Publishing Company, Ltd.
4a Lower John Street, London, W1R 3PD, England

Printed in the United States of America

PREFACE

Over the past fifteen years the Commission on Equilibrium Data of the Analytical Division of the International Union of Pure and Applied Chemistry has been sponsoring a noncritical compilation of metal complex formation constants and related equilibrium constants. This work was extensive in scope and resulted in the publication of two large volumes of *Stability Constants* by the Chemical Society (London). The first volume, edited by L. G. Sillen (for inorganic ligands) and by A. E. Martell (for organic ligands), was published in 1964 and covered the literature through 1962. The second volume, subtitled Supplement No. 1, edited by L. G. Sillen and E. Hogfeldt (for inorganic ligands) and by A. E. Martell and R. M. Smith (for organic ligands), was published in 1971 and covered the literature up to 1969. These two large compilations attempted to cover all papers in the field related to metal complex equilibria (heats, entropies, and free energies). Since it was the policy of the Commission during that period to avoid decisions concerning the quality and reliability of the published work, the compilation would frequently contain from ten to twenty values for a single equilibrium constant. In many cases the values would differ by one or even two orders of magnitude, thus frustrating readers who wanted to use the data without doing the extensive literature study necessary to determine the correct value of the constant in question.

Because of difficulties of this nature, and because of the general lack of usefulness of a noncritical compilation for teaching purposes and for scientists who are not sufficiently expert in the field of equilibrium to carry out their own evaluation, we have decided to concentrate our efforts in this area toward the development of a critical and unique compilation of metal complex equilibrium constants. Although it would seem that decisions between available sets of data must sometimes be arbitrary and therefore possibly unfair, we have found the application of reasonable guidelines leads directly to the elimination of a considerable fraction of the published data of doubtful value. Additional criteria and procedures that were worked out to handle the remaining literature are described in the *Introduction* of this book. Many of these methods are quite similar to those used in other compilations of critical data.

In cases where a considerable amount of material has accumulated, it is felt that most of our critical constants will stand the test of time. Many of the data listed, however, are based on only one or a very few literature references and are subject to change when better data come along. It should be fully understood that this compilation is a continually changing and growing body of data, and will be revised from time to time as new results of these systems appear in the literature.

The scope of these tables includes the heats, entropies, and free energies of all reactions involving organic and inorganic ligands. The magnitude of the work is such that far more than a thousand book pages will be required. In order that the material be available in convenient form, the amino acid complexes, which amount to about a third of the total, are presented in Volume 1. Complexes of amines (which do not contain carboxylic acid functions) comprise Volume 2, while the remaining material should be published as Volume 3 within several months.

Robert M. Smith
Arthur E. Martell

Texas A&M University
College Station, Texas

CONTENTS

INTRODUCTION

Purpose

This compilation of metal complex equilibrium (formation) constants and the corresponding enthalpy and entropy values represent the authors' selection of the most reliable values among those available in the literature. In many cases wide variations in published constants for the same metal complex equilibrium indicate the presence of one or more errors in ligand purity, in the experimental measurements, or in calculations. Usually, the nature of these errors is not readily apparent in the publication, and the reader is frequently faced with uncertainties concerning the correct values. In the course of developing noncritical compilations of stability constants, the authors have long felt that these wide variations in published work constitute a serious impediment to the use of equilibrium data. Thus these critical tables were developed in order to satisfy what is believed to be an important need in the field of coordination chemistry.

Scope

These tables include all organic and inorganic ligands for which reliable values have been reported in the literature. The present volume, which includes approximately one third of the published data, is restricted to amines not containing a carboxylic acid function.

Values determined in nonaqueous solutions as well as values involving two or more different ligands (i.e., "mixed ligand" complexes) have not been included in this compilation but may be included in a subsequent volume. Hydrogen and hydroxide mixed complexes are included since these ions are always available in aqueous solution. In general, data were compiled for only those systems that involve metal ion equilibria. Data on potentially important ligands for which only acid–base equilibria are presently available are given in a separate table.

Selection Criteria

When several workers are in close agreement on a particular value, the average of their results has been selected for that value. Values showing considerable scatter have been eliminated. In cases where the agreement is poor and few results are available for comparison, more subtle methods were needed to select the best value. This selection was often guided by a comparison with values obtained for other metal ions with the same ligand and with values obtained for the same metal ion with similar ligands.

While established trends among similar metal ions and among similar ligands were valuable in deciding between widely varying values, such guidelines were used cautiously, so as not to overlook occasionally unexpected real examples of specificity or anomalous behavior.

When there was poor agreement between published values and comparison with other metal ions and ligands did not suggest the best value, the results of more experienced research groups who had supplied reliable values for other ligands were selected. When such assurances were lacking, it was sometimes possible to give preference to values reported by an investigator who had published other demonstrably reliable values obtained by the same experimental method.

In some cases the constants reported by several workers for a given group of metal ions would have similar relative values, but would differ considerably in the absolute magnitudes of the constants. Then a set of values from one worker near the median of all values reported were selected as the best constants. By this method it is believed that internal consistency was preserved to a greater extent than would be obtained by averaging reported values for each individual metal ion. When an important constant was missing from the selected set of values, but was available in another set of values not selected for this compilation, the missing constant was obtained by adjusting the nonselected values by a common factor, which was set so as to give the best agreement between the two groups of data.

Values reported by only one investigator are included in these tables unless there was some reason to doubt their validity. It is recognized that some of these values may be in error, and that such errors will probably not be detected until the work is repeated by other investigators, or until more data become available for analogous ligands or other closely related metal ions. Some values involving unusual metal ions have been omitted because of serious questions about the form of their complexes.

Papers deficient in specifying essential reaction conditions (e.g., temperature, ionic strength, nature of supporting electrolyte) were not employed in this compilation. Also used as a basis for disqualification of published data is lack of information on the purity of the ligand. Frequent deficiencies are lack of calibration of potentiometric apparatus, and failure to define the equilibrium quotients reported in the paper. In the case of optically active ligands, the omission of the nature of the ligand (D, L, or racemic mixture) resulted in serious ambiguities in the nature of all multiligand complexes formed. In such cases, values were omitted or indicated as being in doubt.

A bibliography for each ligand is included so that the reader may determine the completeness of the literature search employed in the determination of critical values. The reader may also employ these references to make his own evaluation if he has any questions or reservations concerning this compilation.

Arrangement

The arrangement of the tables is based on the placement of similar ligands together. These start with the simplest primary mono-amines and then proceed to those with additional coordinating groups and then to di- and other poly-amines. Similarly, secondary amines and tertiary amines are included and then followed by special groupings of imidazoles, pyridines, pyrimidines, purines, and amines containing phosphorus acid functions, in that order. Next there is a table of protonation constants for ligands for which no stability constants or only questionable metal stability constants are reported. Finally, there is a list of other ligands considered but not included in the tables for various reasons.

Metal Ions

The metal ions within each table are arranged in the following order: hydrogen, alkali metals, alkaline earth metals, lanthanides (including Sc and Y), actinides, transition metals, and posttransition metals. Within each group the arrangement is by increasing oxidation state of the metal, and within each oxidation state the arrangement follows the periodic table from top to bottom and from left to right. An exception is that Cu^+, Ag^+, Pd^{2+}, and Pt^{2+} are included with the posttransition metals.

Equilibrium

An abbreviated equilibrium quotient expression in the order products/reactants is included for each constant, and periods are used to separate distinct entities. Charges have been omitted as these can be determined from the charge of the metal ion and the abbreviated ligand formulas (such as HL) given after the name. Water has not been included in the equilibrium expressions since all of the values cited are for aqueous solutions. For example, $ML_2/M \cdot L^2$ for Ni^{2+} and ethylenediamine (en) would present the equilibrium: $Ni^{2+} + 2en \rightleftharpoons Ni(en)_2^{2+}$

The symbol H_{-1} (H_{-2}, etc.) is used to represent the ionization of one (two, etc.) amide protons or oxime protons from the metal complex. In the pyrimidine and purine sections, it is also used for the ionization of a proton from the ligand alone at high pH.

Equilibria involving protons are written as stability constants (protonation constants) rather than as ionization constants to be consistent with the metal complex formation constants. Consequently the $\triangle H$ and $\triangle S$ values have signs opposite to those describing ionization constants.

Log *K* Values

The log *K* values are the logarithms of the equilibrium quotients given in the second column at the specified conditions of temperature and ionic strength. The selected values are those considered to be the most reliable of the ones available. In some cases the value is the median of several values and in other cases it is the average of two or more values. The range of other values considered reliable is indicated by $+$ or $-$ quantities describing the algebraic difference between the other values and the selected value. Values considered to be of questionable validity are enclosed in parentheses. Such values are included when the evidence available is not strong enough to exclude them on the basis of the above criteria. Values concerning which there is considerable doubt have been omitted.

The log *K* values are given for the more commonly reported ionic strengths. Values for ionic strength 0.10 were given preference because this condition is most frequently used. Other ionic strengths often used are 0.5, 1.0, 3.0 and 0. Zero ionic strength is perhaps more important from a theoretical point of view, but several assumptions are involved in extrapolating or calculating from the measured values.

The temperature of 25°C was given preference in the tables because of its widespread use in equilibrium measurements and reporting other physical properties. When available, enthalphy changes ($\triangle H$) were used to calculate log *K* at 25°C when only measurements at other temperatures were available.

Other temperatures frequently employed are 20°C, 30°C, and 37°C. These are not included in the tables when there is a lack of column space and $\triangle H$ is available, since they may be calculated using the $\triangle H$ value. Values at other temperatures, especially those at 20°C and 30°C, were converted to 25°C to facilitate quantitative comparisons with the 25°C values listed.

The stereoisomerism of optically active ligands is indicated at the beginning of the name or in footnotes for 1:2 (metal:ligand) or higher ratios when different from that indicated in the name of the ligand given at the table heading. The 1:2 or higher complexes are placed in parentheses when the isomerism is not stated or involves a racemic mixture, since in such cases the nature of the complexes formed is not explicitly defined.

Equilibria involving protons have been expressed as concentration constants in order to be more consistent with the metal ion stability constants which involve only concentration terms. Concentration constants may be determined by calibrating the electrodes with solutions of known hydrogen ion concentrations or by conversion of pH values using the appropriate hydrogen ion activity coefficient. When standard buffers are used, mixed constants (also known as Bronsted or practical constants) are obtained which include both activity and concentration terms. Literature values expressed as mixed constants have been converted to concentration constants by using the hydrogen ion activity coefficients determined in KCl solution before inclusion in the tables. In some cases, papers were omitted because no indication was given as to the use of concentration or mixed constants. Some papers were retained despite this lack of information when it could be ascertained which constant was used by comparing to known values or by personal communication with the authors. For those desiring to convert the listed protonation constants to mixed constants, the following values should be added to the listed values at the appropriate ionic strength (the tabulation applies only to single proton association constants):

Ionic strength	Increase in log K
0.05	0.09
0.10	0.11
0.15	0.12
0.2	0.13
0.5	0.15
1.0	0.14
2.0	0.11
3.0	0.07

The values in the tables have not been corrected for complexation with medium ions for the most part. There are insufficient data to make corrections for most of the ligands, and in order to make values between ligands more comparable, the correction has not been made in the few cases where it could be made. In general the listed formation constants at constant ionic strength include competition by ions from KNO_3 and $NaClO_4$ and are somewhat smaller than they would be if measured in solutions of tetraalkylammonium salts.

Limited comparisons were made between values at different ionic strengths using observed trends. With simple amines there is a continuing increase of stability constants with an increase of ionic strength.

Enthalphy Values

The enthalphy of complexation values ($\triangle H$) listed in the tables have the units kcal/mole because of the widespread use of these units by workers in the field. These may be converted to SI units of kj/mole by multiplying the listed values by 4.184.

Calorimetrically determined values and temperature-variation-determined values from cells without liquid junction were considered of equal validity for the tables. Other temperature-variation-determined values were rounded off to the nearest kcal/mole and were enclosed in parentheses because of their reduced accuracy. Other values considered to be reliable but differing from the listed value were indicated by $+$ or $-$ quantities describing the algebraic difference between the other value and the selected value.

The magnitude of $\triangle H$ may vary with temperature and ionic strength, but usually this is less than the variation between different workers and no attempt has been made to show $\triangle H$ variation with changing conditions. These $\triangle H$ values may be used for estimating log K values at temperatures other than those listed, using the relationship

$$\frac{\triangle H}{2.303RT^2} = \frac{d \log K}{dT}$$

or, at 25°C,

$$\log K_2 = \log K_1 + \triangle H(T_2 - T_1)/0.00256.$$

This assumes that $\triangle C_p = 0$, which is not necessarily the case. The greater the temperature range employed, the greater the uncertainty of the calculated values.

Entropy Values

The entropy of complexation values ($\triangle S$) listed in the tables have the units cal/mole/degree and have been calculated from the listed log K and $\triangle H$ values, using the expression

$$\triangle G = \triangle H - T\triangle S$$

or, at 25°C,

$$\triangle S = 3.36\,(1.364 \log K + \triangle H).$$

These entropy values have been rounded off to the nearest cal/mole/degree, except in cases where $\triangle H$ values were quite accurate.

Bibliography

The references considered in preparing each table are given at the end of the table. The more reliable references are listed after the ions for which values are reported. In some tables groups of similar metal ions have been grouped together for the bibliography. The term "Other references" is used for those reporting questionable values, or values at conditions considerably different from those used in the tables, or values for metal ions not included in the tables because of questionable knowledge about the forms of their complexes. These additional references are cited to inform the reader of the extent of the literature search made in arriving at the selected values.

The biographical symbols used represent the year of the reference and the first letter of the surnames of the first two listed authors. In cases of duplication, letters a, b, c, etc., or the first letter of the third author's name are employed. The complete reference is given in the bibliography at the end of each volume.

In a work of this magnitude, there will certainly be errors and a few pertinent publications will have been overlooked by the compilers. We should like to request those who believe they have detected errors in the selection process, know of publications that were omitted, or have any suggestions for improvement of the tables, write to:

A. E. Martell, Head
Department of Chemistry
Texas A&M University
College Station, Texas 77843, U.S.A.

It is the intention of the authors to publish more complete and accurate revisions of these tables as demanded by the continually growing body of equilibrium data in the literature.

$$CH_3NH_2$$

CH_5N Methylamine   L

Metal ion	Equilibrium	Log K 25°, 0.2	Log K 25°, 0.5	Log K 25°, 0	ΔH 25°, 0	ΔS 25°, 0
H^+	$HL/H.L$	10.67	10.72	10.64 ±0.02	−13.2 ±0.1	4
Ag^+	$ML/M.L$		3.15	3.07	(−3)	(4)
	$ML_2/M.L^2$		6.68	6.89 −0.1	−11.7 +0.2	−8
Pt^{2+}	$ML_4/M.L^4$		40.1[a]			
Cd^{2+}	$ML/M.L$		2.75[b]			
	$ML_2/M.L^2$		4.81[b]		(−7)[c]	(−1)[b]
	$ML_3/M.L^3$		5.94[b]			
	$ML_4/M.L^4$		6.55[b]		(−14)[c]	(−17)[b]
Hg^{2+}	$ML/M.L$		8.66			
	$ML_2/M.L^2$		17.86			
	$ML_3/M.L^3$		18.2			
	$ML_4/M.L^4$		18.5			
	$MC1L/MC1_2.L$			2.40	−6.8	−12
	$ML_2/MC1_2.L^2$			4.61	−7.8	−5

_____

[a] 18°, 1.0;   [b] 25°, 2.1;   [c] 10-40°, 2.1

Bibliography:

H^+ 28HR,30HO,41EW,50BL,56CG,64WB,66PC, 71HT

Ag^+ 50B,55F,71HT

Pt^{2+} 61GG

Cd^{2+} 53SP

Hg^{2+} 66PC,72B

Other references: 03BE,03E,33AT,33T,35BW, 36BW,63PL,64SD,71SS

$$CH_3CH_2NH_2$$

C_2H_7N <u>Ethylamine</u>   L

Metal ion	Equilibrium	Log K 25°, 0.5	Log K 25°, 0	ΔH 25°, 0	ΔS 25°, 0
H^+	HL/H.L	10.66 ±0.00	10.636±0.000	-13.65-0.07	2.9
Cu^{2+}	$ML_4/M.L_4$	11.5[a]			
	$M(OH)_2L_2/M.(OH)^2.L^2$	15.9[a]			
Cu^+	$ML_2/M.L^2$	10.1[a]			
	MOHL/M.OH.L	10.8[a]			
Ag^+	ML/M.L	3.38 ±0.02	3.46	-5	-1
	$ML_2/M.L^2$	7.30 ±0.1	7.35 -0.03	-12.5 -0.5	-9
Pt^{2+}	$ML_4/M.L^4$	37.0[b]			
Zn^{2+}	ML/M.L	2.30			
	$ML_2/M.L^2$	4.33			
	$ML_3/M.L^3$	6.0			

_____

[a] 30°, 0.5; [b] 18°, 1.0

Bibliography:

H^+ 45CM,48BV,51EH,55F,68AL,69CI,71HT, 71LL

Cu^{2+},Cu^+ 67FH

Ag^+ 45CM,48BV,55F,68AL,71HT

Pt^{2+} 61GG

Zn^{2+} 71LL

Other references: 03BE,03E,28J,35BW,64SD, 72JJ

$$CH_2CH_2CH_2NH_2$$

| C_3H_9N | | Propylamine | | | L |

Metal ion	Equilibrium	Log K 25°, 0.5	Log K 25°, 0	ΔH 25°, 0	ΔS 25°, 0
H^+	HL/H.L	(10.74)	10.566±0.002	-13.75±0.09	2.2
Ag^+	ML/M.L	3.43	3.45	(-3)	(6)
	$ML_2/M.L^2$	7.35 ±0.03	7.44	-12.7	-9

Bibliography:

H^+	68AL,68CE,68OW,69CI,71HT	Other reference: 33T
Ag^+	68AL,71HT	

$$CH_3CH_2CH_2CH_2NH_2$$

| $C_4H_{11}N$ | | Butylamine | | | | L |

Metal ion	Equilibrium	Log K 25°, 1.0	Log K 25°, 0.5	Log K 25°, 0	ΔH 25°, 0	ΔS 25°, 0
H^+	HL/H.L	10.75	10.70 +0.05	10.640-0.001	-13.93+0.05	2.0
Ag^+	ML/M.L		3.44 ±0.01	3.42	(-4)	(2)
	$ML_2/M.L^2$		7.44 ±0.04	7.47	-12.6	-8
Hg^{2+}	ML/M.L		8.74			
	$ML_2/M.L^2$		18.14			
	$ML_3/M.L^3$		19.0			
	$ML_4/M.L^4$		20.0			

Bibliography:

H^+	50BL,62W,68AL,68CE,69CI,71HT,72CPa	Hg^{2+}	72B
Ag^+	50B,68AL,71HT	Other reference:	52S

$$CH_3CH_2CH_2CH_2CH_2NH_2$$

$C_5H_{13}N$ <u>Pentylamine</u>   L

Metal ion	Equilibrium	Log K 25°, 0.5	Log K 25°, 0	ΔH 25°, 0	ΔS 25°, 0
H^+	HL/H.L	(10.75)	10.597	−13.98	1.6
Ag^+	ML/M.L	3.49	3.57	−5	0
	$ML_2/M.L^2$	7.53 ±0.02	7.49	−12.3	−7

Bibliography:
H^+ 68AL,69CI,71HT Ag^+ 68AL,71HT

$$CH_3CH_2CH_2CH_2CH_2CH_2NH_2$$

$C_6H_{15}N$ <u>Hexylamine</u>   L

Metal ion	Equilibrium	Log K 25°, 0	ΔH 25°, 0	ΔS 25°, 0
H^+	HL/H.L	10.630	−13.93	1.9
Ag^+	ML/M.L	3.54	(−6)	(−4)
	$ML_2/M.L^2$	7.55	−12.9	−9

Bibliography: 71HT

$$\underset{\underset{\text{CH}_3\text{CHCH}_2\text{NH}_2}{|}}{\text{CH}_3}$$

$C_4H_{11}N$		2-Methylpropylamine (isobutylamine)			L

Metal ion	Equilibrium	Log K 25°, 0.5	Log K 25°, 0	ΔH 25°, 0	ΔS 25°, 0
H^+	HL/H.L	10.57	10.48	-13.92	1.3
Ag^+	ML/M.L	3.38			
	$ML_2/M.L^2$	7.24		$(-13)^a$	$(-10)^b$

^{a} 25-35°, 0.5; ^{b} 25°, 0.5

Bibliography:
H^+ 48BV,69CI Ag^+ 48BV

$$\text{s}\!\!\diagup\!\!\diagdown\!-\text{NH}_2$$

$C_5H_{11}N$		Cyclopentylamine			L

Metal ion	Equilibrium	Log K 37°, 0.15	Log K 25°, 0	ΔH 25°, 0	ΔS 25°, 0
H^+	HL/H.L	(10.03)	10.65	-14.3	1
Co^{2+}	ML/M.L	5.70			
Ni^{2+}	ML/M.L	6.82			
	$ML_2/M.L^2$	9.1			
Cu^{2+}	ML/M.L	8.01			
	$M_2(OH)_2L_2.H^2/M^2.L^2$	12.01			
Zn^{2+}	ML/M.L	4.2			
	$ML_2/M.L^2$	7.8			
	$M(OH)_2L_2.H^2/M.L^2$	4.5			

Bibliography:
H^+ 69CI,74MW Co^{2+}-Zn^{2+} 74MW

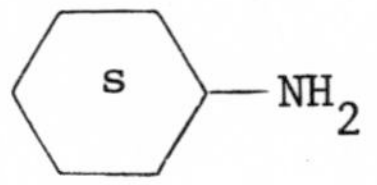

C$_6$H$_{13}$N		Cyclohexylamine			L

Metal ion	Equilibrium	Log K 37°, 0.15	Log K 25°, 0	ΔH 25°, 0	ΔS 25°, 0
H$^+$	HL/H.L	(9.93)	10.64	−14.4	0
Co^{2+}	ML/M.L	5.3			
Ni^{2+}	ML/M.L	5.9			
	ML$_2$/M.L^2	8.2			
Cu^{2+}	ML/M.L	7.67			
Zn^{2+}	ML/M.L	4.6			

Bibliography:

H$^+$ 32HS,69CI,74MW   Co^{2+}-Zn^{2+} 74MW

CH$_2$=CHCH$_2$NH$_2$

C$_3$H$_7$N		Prop-2-enylamine (allylamine)			L

Metal ion	Equilibrium	Log K 25°, 2.0	Log K 25°, 0	ΔH 25°, 0	ΔS 25°, 0
H$^+$	HL/H.L		9.52	−13.1	0
Ag$^+$	MHL/M.HL	0.114			

Bibliography:

H$^+$ 69CI   Ag$^+$ 67HV

$$CH_3CH=CHCH_2NH_2$$

C_4H_9N trans-But-2-enylamine   L

Metal ion	Equilibrium	Log K 25°, 2.0
Ag^+	MHL/M.HL	0.107

Bibliography: 67HV

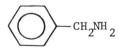

C_7H_9N Benzylamine   L

Metal ion	Equilibrium	Log K 25°, 1.0	Log K 25°, 0.5	Log K 25°, 0	ΔH 25°, 0	ΔS 25°, 0
H^+	HL/H.L	9.49	9.47	9.35	-13.04-0.06	-1.0
Ag^+	ML/M.L	3.02[a]	3.29			
	$ML_2/M.L^2$	6.80[a]	7.14			

_____

[a]30°, 1.0

Bibliography:
H^+ 48BV,49LM,54GFa,56RK Ag^+ 48BV,54GFa

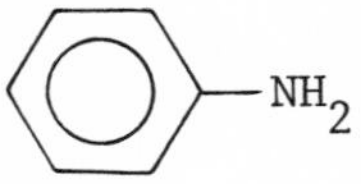

C$_6$H$_7$N   <u>Aminobenzene</u>   <u>(aniline)</u>   L

Metal ion	Equilibrium	Log K 25°, 0.1	Log K 25°, 1.0	Log K 25°, 0	ΔH 25°, 0	ΔS 25°, 0
H$^+$	HL/H.L	4.65 ±0.03	4.82	4.601±0.005	−7.31±0.07	−3.5
		4.71[a]			−7.64[a]	−4.1[a]
Hg$_2$$^{2+}$	ML/M.L		3.71[b]			
Cd^{2+}	ML/M.L	0.10[c]				
Hg^{2+}	ML/M.L		4.61[b]			
	ML$_2$/M.L^2		9.21[b]			

---

[a]25°, 0.5; [b]27°, 1.0; [c]25°, 0.3

Bibliography:

H$^+$ 37P, 47SW,49LM,61BR,64WD,67BH,68O, 70KP,70VS,73LP, 73BEM

Hg$_2$$^{2+}$,Hg^{2+} 64WD

Cd^{2+} 64NA

Other references: 24P, 36BW, 52G

C_7H_9N			2-Methylaniline	(o-toluidine)		L

Metal ion	Equilibrium	Log K 25°, 0.1	Log K 25°, 0.3	Log K 25°, 0	ΔH 25°, 0	ΔS 25°, 0
H^+	HL/H.L	4.57	(4.44)	4.447-0.01	-7.1 ±0.1	-3
Cd^{2+}	ML/M.L		-0.10			

Bibliography:
H^+ 61BR,64NA,66D,67BH,68O,71V   Other references: 24P,52G
Cd^{2+} 64NA

C_7H_9N			4-Methylaniline	(p-toluidine)		L

Metal ion	Equilibrium	Log K 25°, 0.1	Log K 25°, 0.3	Log K 25°, 0	ΔH 25°, 0	ΔS 25°, 0
H^+	HL/H.L	5.23	(4.79)	5.084±0.00	-7.6 ±0.0	-2
Cd^{2+}	ML/M.L		0.26			

Bibliography:
H^+ 61BR,64NA,66D,67BH,68O,70VS   Other reference: 52G
Cd^{2+} 64NA

$$HO_3SCH_2CH_2NH_2$$

| $C_2H_7O_3NS$ | 2-Aminoethanesulfonic acid | (taurine) | | | HL |

Metal ion	Equilibrium	Log K 25°, 0.5	Log K 25°, 0	ΔH 25°, 0	ΔS 25°, 0
H^+	HL/H.L	8.95	9.061	-10.00 ± 0.01	7.9
Ag^+	ML/M.L	2.97		-4.50[a]	-1.5[a]
	$ML_2/M.L^2$	6.15		-11.55[a]	-10.6[a]

---

[a] 25°, 0.5

Bibliography:

H^+ 53K,65HWH,72VT   Other references: 50A,59DG

Ag^+ 72VT

$$HO_3SCH_2CH_2CH_2NH_2$$

| $C_3H_9O_3NS$ | 3-Aminopropanesulfonic acid | | | HL |

Metal ion	Equilibrium	Log K 25°, 0.5	ΔH 25°, 0.5	ΔS 25°, 0.5
H^+	HL/H.L	9.89		
Ag^+	ML/M.L	3.17	-5.12	-2.7
	$ML_2/M.L^2$	6.75	-12.13	-9.8

Bibliography: 72VT

$$HO_3SCH_2CH_2CH_2CH_2NH_2$$

$C_4H_{11}O_3NS$		4-Aminobutanesulfonic acid		HL

Metal ion	Equilibrium	Log K 25°, 0.5	ΔH 25°, 0.5	ΔS 25°, 0.5
H^+	HL/H.L	10.39		
Ag^+	ML/M.L	3.38	-5.51	-3.0
	$ML_2/M.L^2$	7.08	-12.67	-10.1

Bibliography: 72VT

$$HO_3SCH_2CH_2CH_2CH_2CH_2NH_2$$

$C_5H_{13}O_3NS$		5-Aminopentanesulfonic acid		HL

Metal ion	Equilibrium	Log K 25°, 0.5	ΔH 25°, 0.5	ΔS 25°, 0.5
H^+	HL/H.L	10.61		
Ag^+	ML/M.L	3.49	-5.54	-2.6
	$ML_2/M.L^2$	7.40	-12.94	-9.6

Bibliography: 72VT

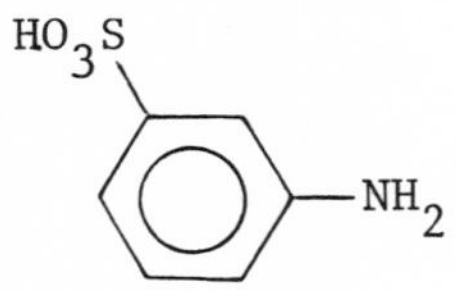

$C_6H_7O_3NS$		3-Aminobenzenesulfonic acid	(metanilic acid)		HL

Metal ion	Equilibrium	Log K 25°, 1.0	Log K 25°, 0	ΔH 25°, 0	ΔS 25°, 0
H^+	HL/H.L		3.738	−4.98±0.05	0.0
Ag^+	ML/M.L	1.23			
	$ML_2/M.L^2$	2.13			
	$ML_3/M.L^3$	2.31			
	$ML_4/M.L^4$	2.42			
Cd^{2+}	ML/M.L	0.26			
	$ML_2/M.L^2$	0.56			

Bibliography:

H^+ 65CS,68CWI,69CI Cd^{2+} 58ACa
Ag^+ 58AC

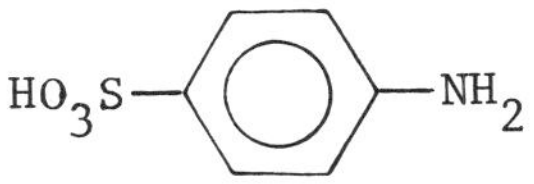

| $C_6H_7O_3NS$ | | 4-Aminobenzenesulfonic acid | (sulfanilic acid) | | | HL |

Metal ion	Equilibrium	Log K 25°, 0.1	Log K 25°, 1.0	Log K 25°, 0	ΔH 25°, 0	ΔS 25°, 0.1
H^+	HL/H.L	3.01		3.232	-4.25 ± 0.0	0.0
Ag^+	ML/M.L	1.14	1.03			
	$ML_2/M.L^2$	2.09	1.67			
	$ML_3/M.L^3$		2.07			
CH_3Hg^+	ML/M.L	2.60[a]				

_____

[a] 20°, 0.1

Bibliography:

H^+ 65CS,65HWH,65SS CH_3Hg^+ 65SS

Ag^+ 58AC

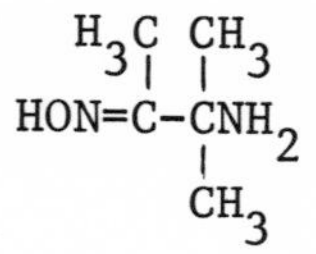

$$\begin{array}{cc} H_3C & CH_3 \\ | & | \\ HON=C-CNH_2 \\ | \\ CH_3 \end{array}$$

$C_5H_{12}ON_2$ 3-Amino-3-methylbutan-2-one oxime   L

Metal ion	Equilibrium	Log K 24°, 0.27	ΔH 25°, 0	ΔS 24°, 0.27
H^+	HL/H.L	8.99	-12.5	-1
Ni^{2+}	$ML_2/M.L^2$	8.6		
	$M(H_{-1}L)L.H/M.L^2$	6.2	-8.4	0
	$M(H_{-1}L)L/M(H_{-1}L)_2.H$	11.7		
Cu^{2+}	$ML_2/M.L^2$	11.9		
	$M(H_{-1}L)L.H/M.L^2$	7.9	-11.1	-1
	$M(H_{-1}L)L/M(H_{-1}L)_2.H$	9.8		

Bibliography: 58M,64WB   Other reference: 63LM

$$HOCH_2CH_2NH_2$$

C$_2$H$_7$ON   2-Aminoethanol (ethanolamine)   L

Metal ion	Equilibrium	Log K 25°, 0.1	Log K 25°, 0.5	Log K 25°, 0	ΔH 25°, 0	ΔS 25°, 0
H[+]	HL/H.L	9.52 ±0.02	9.62 ±0.03	9.498+0.001	-12.06 ±0.02	3.0
Ni[2+]	ML/M.L	2.98	3.06[a]		-3.4[a]	3[a]
	ML$_2$/M.L^2	5.33	5.52[a]		-7.3[a]	1[a]
	ML$_3$/M.L^3	7.33	6.95[a]		-9.9[a]	-1[a]
Cu[2+]	ML/M.L	5.7				
	ML$_2$/M.L^2	9.8				
	ML$_3$/M.L^3	13.0				
	ML$_4$/M.L^4	15.2 ±0.2				
Ag[+]	ML/M.L		3.13 ±0.02	3.2	(-6)[b]	(-6)
	ML$_2$M.L^2		6.68 ±0.00	6.76 ±0.0	(-12)[b]	(-9)
Zn[2+]	ML/M.L	3.7				
	ML$_2$/M.L^2	6.1				
	ML$_3$/M.L^3	9.4 ±0.2				
Cd[2+]	ML/M.L	2.77				
	ML$_2$/M.L^2	4.09				
	ML$_3$/M.L^3	5.6 ±0.1				
Hg[2+]	ML/M.L		8.51	8.56	-10.2	5
	ML$_2$M.L^2		17.32	17.33	-18.5	17
Pb[2+]	ML/M.L	7.56				

_____

[a] 25°, 0.4; [b] 10-40°, 0

Bibliography:

H[+]   48BV,49LM,51BP,56BR,59FO,62DG,68AL,
        68OW,69CI,73NK

Ni[2+]   62CW,62SG

Cu[2+]   59MP,65D

Ag[+]   48BV,56BR,58AS,59LB,68AL

Zn[2+]   59MP,65D

Cd[2+]   60MP

Hg[2+]   56BR,70E
Pb[2+]   59MP

Other references: 33T,53L,52S,54L,55FK,
        57S,59DG,61ALa,62FH,62MS,63MS,63Sa,
        64MS,65MP,66SKa,67FH, 69MI, 69MP,
        70UP,70UR,71BSa,71SSa,72JW,72VE,73UP

$$HOCHCH_2NH_2$$
$$|$$
$$CH_3$$

C_3H_9ON DL-1-Amino-2-propanol   L

Metal ion	Equilibrium	Log K 25°, 0.1	Log K 25°, 0.5	Log K 25°, 0	ΔH 25°, 0	ΔS 25°, 0
H^+	HL/H.L	9.49 ±0.00	9.61	9.464±0.006	−12.04	2.9
Ag^+	ML/M.L			3.23[a]		
	$ML_2/M.L^2$			(6.78)[a]		

---

[a] 20°, 0

Bibliography:

H^+ 68TE,73NKa   Other reference: 72VE

Ag^+ 64AK

$$HOCH_2CH_2CH_2NH_2$$

C_3H_9ON 3-Aminopropanol   L

Metal ion	Equilibrium	Log K 25°, 0.1	Log K 25°, 0.5	Log K 25°, 0	ΔH 25°, 0	ΔS 25°, 0
H^+	HL/H.L	10.11	10.22 ±0.01	10.088	−12.72±0.02	3.5
Ag^+	ML/M.L		3.25[a]			
	$ML_2/M.L^2$		7.04[a]			

---

[a] 20°, 0.5

Bibliography:

H^+ 68AL,68OW,69CI,73NKa Ag^+ 68AL

$$HOCH_2CH_2CH_2CH_2NH_2$$

$C_4H_{11}ON$ 4-Aminobutanol   L

Metal ion	Equilibrium	Log K 20°, 0.5	Log K 20°, 0
H^+	HL/H.L	10.61	10.38
Ag^+	ML/M.L	3.39	
	$ML_2/M.L^2$	7.30	

Bibliography: 68AL   Other reference: 72VE

$$HOCH_2CH_2CH_2CH_2CH_2NH_2$$

$C_5H_{13}ON$ 5-Aminopentanol   L

Metal ion	Equilibrium	Log K 20°, 0.5	Log K 20°, 0
H^+	HL/H.L	10.91	10.61
Ag^+	ML/M.L	3.42	
	$ML_2/M.L^2$	7.55	

Bibliography: 68AL

$$\text{C}_6\text{H}_5-\underset{\underset{\text{OH}}{|}}{\text{CH}}\text{CH}_2\text{NH}_2$$

$\text{C}_8\text{H}_{11}\text{ON}$ DL-2-Amino-1-phenylethanol   HL

Metal ion	Equilibrium	Log K 25°, 0.1
H^+	HL/H.L	11.90
	HL_2/HL.H	8.79
Cu^{2+}	ML/M.L	9.50
	ML_2/M.L^2	(15.5)

Bibliography:
H^+ 65JN Cu^{2+} 65JNa

$$\underset{\text{HOCH}_2\text{CHNH}_2}{\overset{\overset{\displaystyle O}{\overset{\|}{\text{CH}_3\text{OC}}}}{|}}$$

$\text{C}_4\text{H}_9\text{O}_3\text{N}$ DL-Serine methyl ester   L

Metal ion	Equilibrium	Log K 25°, 0.1	ΔH 25°, 0.1	ΔS 25°, 0.1
H^+	HL/H.L	7.03 ±0.01	(−11)[a]	(−5)
Ni^{2+}	ML/M.L	2.37[b]		
	ML_2/M.L^2	(4.35)[b]		

[a] 25-50°, 0.1;   [b] 25°, 0.15, optical isomerism not stated.

Bibliography:
H^+ 66HP,67HP,69LA,70HM Ni^{2+} 58LD

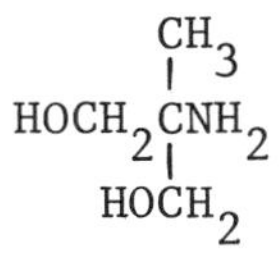

$$\begin{array}{c} \text{CH}_3 \\ | \\ \text{HOCH}_2\text{CNH}_2 \\ | \\ \text{HOCH}_2 \end{array}$$

$C_4H_{11}O_2N$ <u>2-Amino-2-methyl-1,3-propanediol</u>   L

Metal ion	Equilibrium	Log K 25°, 0.1	Log K 25°, 0	ΔH 25°, 0	ΔS 25°, 0
H^+	HL/H.L	8.82	8.797±0.005	−11.91±0.02	0.3
Ag^+	$ML_2/M.L^2$		6.90		

Bibliography:

H^+ 62HB,68OW,68TE   Other references: 59DG,71HS

Ag^+ 62HB

$$\begin{array}{c} HOCH_2 \\ | \\ HOCH_2CNH_2 \\ | \\ HOCH_2 \end{array}$$

| $C_4H_{11}O_3N$ | 2-Amino-2-(hydroxymethyl)-1,3-propanediol | | | | L |
| | (tris(hydroxymethyl)aminomethane, THAM) | | | | |

Metal ion	Equilibrium	Log K 25°, 0.1	Log K 25°, 0.5	Log K 25°, 0	ΔH 25°, 0	ΔS 25°, 0
H^+	HL/H.L	8.09	8.15	8.075	−11.36±0.03	−1.2
Ni^{2+}	ML/M.L	2.63				
	$ML_2/M.L^2$	4.5				
	$M_3(OH)_5L_3.H^5/M^3.L^3$	−13.4				
	$M_2(OH)_3L_3.H^3/M^2.L^3$	−27.0				
Cu^{2+}	ML/M.L	3.95				
	$ML_2/M.L^2$	7.63				
	$ML_3/M.L^3$	11.10				
	$ML_4/M.L^4$	14.1				
	ML/MOHL.H	6.0				
	$(MOHL)_2/(MOHL)^2$	2.2				
	$ML_2/MOHL_2.H$	6.32				
	$MOHL_2/M(OH)_2L_2.H$	7.90				
Ag^+	ML/M.L	3.14[a]	3.05		(−7)[b]	(−9)[a]
	$ML_2/M.L^2$	6.57[a]	6.53		(−15)[b]	(−20)[a]

[a] 25°, 0.05; [b] 0-60°, 0.05

Bibliography:
H^+ 49BP,61BH,63DG,66DGa,66HW,68CWI,68OW, 69BM,69HO,70GO,72O
Ni^{2+} 69BM
Cu^{2+} 69BM
Ag^+ 59S,66DGa
Other references: 63HS,71HS

$$CH_3OCH_2CH_2NH$$

| C_3H_9ON | | | 2-Methoxyethylamine | | | | L |

Metal ion	Equilibrium	Log K 25°, 0.5	Log K 30°, 1.0	Log K 25°, 0	ΔH 25°, 0	ΔS 25°, 0
H^+	HL/H.L	9.30	9.44	9.48	(−12)[a]	(4)
Ag^+	ML/M.L	2.95	3.15	3.23	(−6)[a]	(−2)
	$ML_2/M.L^2$	6.34	6.81	6.7	(−11)[a]	(−3)

[a] 10-40°, 0

Bibliography: 48BV,53L,59LB   Other reference: 54L

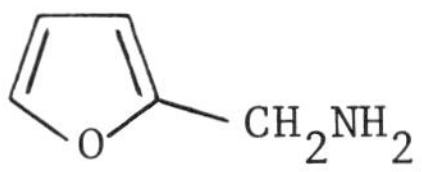

| C_5H_7ON | | 2-Furylmethylamine (2-(aminomethyl)furan) | L |

Metal ion	Equilibrium	Log K 30°, 1.0
H^+	HL/H.L	8.75
Ag^+	ML/M.L	2.64
	$ML_2/M.L^2$	5.98

Bibliography: 54GFa

| C_7H_9ON | 2-Methoxyaniline (o-anisidine) | | | | L |

Metal ion	Equilibrium	Log K 25°, 0.3	Log K 25°, 0	ΔH 25°, 0	ΔS 25°, 0
H^+	HL/H.L	(4.52)	4.527+0.001	-7.6 -0.1	-5
Cd^{2+}	ML/M.L	0.05			

Bibliography:

H^+ 61BR,64NA,69BHa,71V,72KP Cd^{2+} 64NA

| C_7H_9ON | 4-Methoxyaniline (p-anisidine) | | | | L |

Metal ion	Equilibrium	Log K 25°, 0.3	Log K 25°, 0	ΔH 25°, 0	ΔS 25°, 0
H^+	HL/H.L	(5.01)	5.357±0.00	-8.2 -0.1	-3
Cd^{2+}	ML/M.L	0.45			

Bibliography:

H^+ 64NA,66D, 69BH,70VS Cd^{2+} 64NA

$$\overset{\displaystyle O}{\underset{\displaystyle \|}{CH_3OCCH_2NH_2}}$$

$C_3H_7O_2N$		Glycine methyl ester		L

Metal ion	Equilibrium	Log K 25°, 0.1	ΔH 25°, 0.1	ΔS 25°, 0.1
H^+	HL/H.L	7.66 ±0.03	(−11)[a]	(−2)
Ni^{2+}	ML/M.L	2.45[b]		
Cu^{2+}	ML/M.L	3.83[b]		

---

[a] 20-30°, 0.04;   [b] 25°, 0.15

Bibliography:

H^+ 66HP,67HP,71AA Ni^{2+},Cu^{2+} 56WML

$$CH_3CH_2O\overset{\overset{\textstyle O}{\|}}{C}CH_2NH_2$$

C$_4$H$_9$O$_2$N   Glycine ethyl ester   L

Metal ion	Equilibrium	Log K 25°, 0.1	Log K 30°, 0.1	ΔH 25°, 0.1	ΔS 25°, 0.1
H$^+$	HL/H.L	7.69 ±0.02	7.55	(-11)[a]	(-2)
Co^{2+}	ML/M.L		1.43		
	ML$_2$/M.L^2		2.6		
Ni^{2+}	ML/M.L	2.49[b]	2.30		
	ML$_2$/M.L^2	4.58[b]	4.22		
Cu^{2+}	ML/M.L	4.04		(-8)[c]	(16)
	ML$_2$/M.L^2	(7.9)			
	ML$_3$/M.L^3	(12.7)			
Zn^{2+}	ML/M.L		1.79		
	ML$_2$/M.L^2		3.69		

---

[a] 0–50°, 0.1;   [b] 25°, 0.15;   [c] 20–45°, 0.1

Bibliography:

H$^+$ 41GH,65CJ,65R,66HJ,66HP,67HP,67W,68HA   Ni^{2+} 56WML,66HJ

Co^{2+},Zn^{2+} 66HJ   Cu^{2+} 65CJ

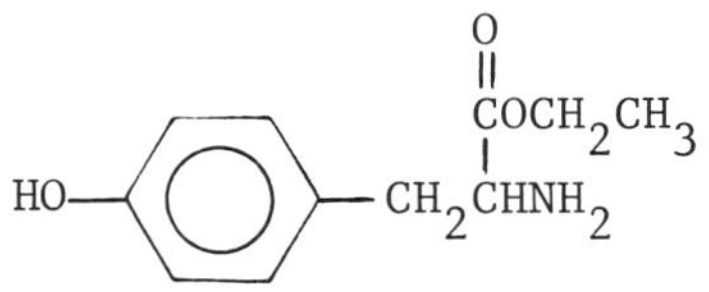

$$HO-\underset{}{\bigcirc}-CH_2CHNH_2 \ | \ \overset{O}{\overset{||}{C}}OCH_2CH_3$$

$C_{11}H_{15}O_3N$		L-Tyrosine ethyl ester	HL

Metal ion	Equilibrium	Log K 25°, 0.15	Log K 20°, 0.37
H^+	HL/H.L	9.68	9.71
	$H_2L/HL.H$	7.21	7.37
Cu^{2+}	ML/M.L		4.02
	$ML_2/M.L^2$		7.41
	$ML_3M.L^3$		9.72
	$ML_4/M.L^4$		11.5

Bibliography:

H^+ 58MEW,74W Cu^{2+} 74W

$$O$$
$$\parallel$$
$$H_2NCCH_2NH_2$$

$C_2H_6ON_2$		Glycinamide			L
Metal ion	Equilibrium	Log K 25°, 0.1	Log K 25°, 1.0	ΔH 25°, 0.1	ΔS 25°, 0.1
H^+	HL/H.L	7.93 ±0.01	8.19	−10.3 ±0.5	2
Ni^{2+}	ML/M.L	4.20[a]−0.02			
	$ML_2/M.L^2$	7.6[a] −0.3			
	$ML_3/M.L^3$	9.7[a]			
	$ML/M(H_{-1}L).H$	9.7[a]			
	$M(H_{-1}L)/MOH(H_{-1}L).H$	10.0[a]			
Cu^{2+}	ML/M.L	5.41 ±0.1	5.42	−6.0 ±0.5	4
	$ML_2/M.L^2$	9.63 ±0.1	9.83	−11.9 ±0.7	4
	$ML/M(H_{-1}L).H$	6.85 ±0.06			
	$M(H_{-1}L)/MOH(H_{-1}L).H$	7.96			
	$ML_2/M(H_{-1}L)L.H$	6.91			
	$ML_2/M(H_{-1}L)_2.H^2$	15.08		−16.6	13
Zn^{2+}	ML/M.L	3.28[a]			
Cd^{2+}	$ML_2/M.L^2$	5.2[a]			

[a] 25°, 0.16

Bibliography:
H^+ 57LD,60MC,67R,68Sa,71YM,72BB,72TS Zn^{2+}, Cd^{2+} 58LC
Ni^{2+} 57LD,60MC   Other reference: 56DR
Cu^{2+} 57LD,67R,68Sa,71YM,72BBL,72TS

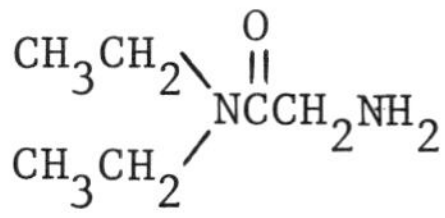

$$CH_3CH_2 \diagdown \overset{\displaystyle O}{\underset{\displaystyle \|}{N}}CCH_2NH_2$$
$$CH_3CH_2 \diagup$$

$C_6H_{14}ON_2$ Glycine-N,N-diethylamide   L

Metal ion	Equilibrium	Log K 25°, 0.1
H$^+$	HL/H.L	8.45
Cu^{2+}	ML/M.L	6.18
	ML$_2$M.L^2	11.30

Bibliography: 71YM

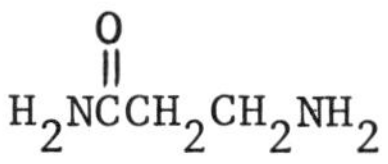

$$\overset{\displaystyle O}{\underset{\displaystyle \|}{H_2N}}CCH_2CH_2NH_2$$

$C_3H_8ON_2$ β-Alaninamide   L

Metal ion	Equilibrium	Log K 25°, 0.1
H$^+$	HL/H.L	9.19
Cu^{2+}	ML/M.L	5.1
	ML$_2$M.L^2	9.6

Bibliography: 71YM

$$CH_3CH_2 \diagdown \overset{\overset{O}{\parallel}}{N} CCH_2CH_2NH_2$$
$$CH_3CH_2 \diagup$$

C$_7$H$_{16}$ON$_2$ β-Alanine-N,N-diethylamide   L

Metal ion	Equilibrium	Log K 25°, 0.1
H$^+$	HL/H.L	9.44
Cu^{2+}	ML/M.L	5.51
	ML$_2$M.L^2	10.7

Bibliobraphy: 71YM

$$\overset{\overset{O}{\parallel}}{H_2NCCH_2N}\overset{\overset{O}{\parallel}}{HCCH_2NH_2}$$

C$_4$H$_9$O$_2$N$_3$ Glycylglycinamide   L

Metal ion	Equilibrium	Log K 25°, 0.1
H$^+$	HL/H.L	7.79 ±0.02
Cu^{2+}	ML/M.L	4.80 +0.2
	ML/M(H$_{-1}$L).H	5.00 ±0.01
	M(H$_{-1}$L)/M(H$_{-2}$L).H	7.92 -0.7
	M(H$_{-2}$L)/MOH(H$_{-2}$L).H	9.77

Bibliography: 72SGP,73YN   Other reference: 73KK

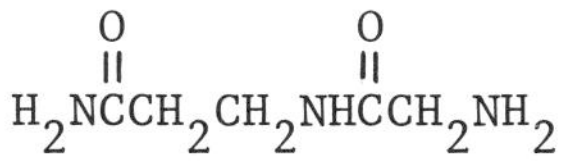

$$O \qquad O$$
$$\parallel \qquad \parallel$$
$$H_2NCCH_2CH_2NHCCH_2NH_2$$

$C_5H_{11}O_2N_3$ Glycyl-β-alaninamide   L

Metal ion	Equilibrium	Log K 25°, 0.1
H^+	HL/H.L	7.83
Cu^{2+}	ML/M.L	5.22
	ML/M(H_{-1}L).H	5.38
	M(H_{-1}L)/M(H_{-2}L).H	8.95

Bibliography: 73YN

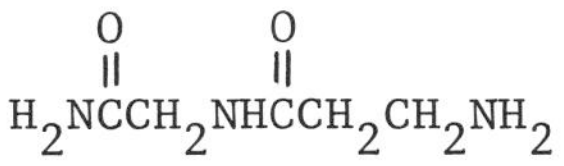

$$O \qquad O$$
$$\parallel \qquad \parallel$$
$$H_2NCCH_2NHCCH_2CH_2NH_2$$

$C_5H_{11}O_2N_3$ β-Alanylglycinamide   L

Metal ion	Equilibrium	Log K 25°, 0.1
H^+	HL/H.L	9.14
Cu^{2+}	ML/M.L	5.16
	ML/M(H_{-1}L).H	5.35

Bibliography: 73YN

$$\overset{\displaystyle O}{\overset{\displaystyle \|}{CH_3OC}}CH_2NH\overset{\displaystyle O}{\overset{\displaystyle \|}{C}}CH_2NH_2$$

| $C_5H_{10}O_3N_2$ | Glycylglycine methyl ester | | L |

Metal ion	Equilibrium	Log K 25°, 0.05
H^+	HL/H.L	7.78
Cu^{2+}	ML/M.L	4.11
	$ML/M(H_{-1}L).H$	5.23
	$M(H_{-1}L)L/M(H_{-1}L).L$	3.30
	$M(H_{-1}L)L/M(H_{-1}L)_2.H$	6.57
Zn^{2+}	ML/M.L	2.77
	$ML_2/M.L^2$	5.10

Bibliography:

H^+, Cu^{2+} 73NA Zn^{2+} 73NAa

$$CH_3CH_2O\overset{\displaystyle O}{\overset{\displaystyle \|}{C}}CH_2NH\overset{\displaystyle O}{\overset{\displaystyle \|}{C}}CH_2NH_2$$

| $C_6H_{12}O_3N_2$ | Glycylglycine ethyl ester | | L |

Metal ion	Equilibrium	Log K 25°, 0.16	Log K 25°, 0
H^+	HL/H.L	7.76 ±0.16	7.71
Ni^{2+}	ML/M.L	3.65	
	$ML_2/M.L^2$	6.60	
	$ML_3/M.L^3$	8.6	
	$ML/M(H_{-1}L).H$	(9.1)	
	$M(H_{-1}L)/MOH(H_{-1}l).H$	(9.7)	

Bibliography:

H^+ 37N,41GH,60MC,66HP,67HP   Other reference: 56DR

Ni^{2+} 60MC

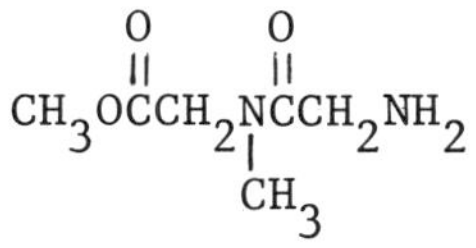

$$CH_3OCCH_2NCCH_2NH_2$$

with the structure showing two C=O groups (O above each carbon) and CH_3 below the N:

```
            O        O
            ||       ||
CH3OCCH2NCCH2NH2
            |
           CH3
```

$C_6H_{12}O_3N_2$ <u>Glycylsarcosine methyl ester</u>   L

Metal ion	Equilibrium	Log K 25°, 0.05
H^+	HL/H.L	8.01
Cu^{2+}	ML/M.L	5.18
	$ML_2/M.L^2$	9.09

Bibliography: 73NA

$$HSCH_2CH_2NH_2$$

C_2H_7NS	2-Aminoethanethiol			(2-mercaptoethylamine)		HL

Metal ion	Equilibrium	Log K 25°, 0.1	Log K 30°, 1.0	Log K 25°, 0	ΔH 25°, 0.01	ΔS 25°, 0
H^+	HL/H.L	10.71 ±0.02	10.69			
	H_2L/H.H	8.21 ±0.02	8.28	8.23	−7.43	12.7
Mg^{2+}	ML/M.L	2.30				
Ca^{2+}	ML/M.L	2.21				
Sr^{2+}	ML/M.L	1.55				
Ba^{2+}	ML/M.L	1.37				
Co^{2+}	ML/M.L		7.68			
	$ML_2/M.L^2$		14.71			
Ni^{2+}	ML/M.L		10.05			
	$ML_2/M.L^2$		19.81			
Cu^{2+} [a]						
Zn^{2+}	ML/M.L	9.90[b]	10.22			
	$ML_2/M.L^2$	18.74[b]	18.90			
Cd^{2+}	ML/M.L	10.97[b]				
	$ML_2/M.L^2$	19.75[b]				
Pb^{2+}	ML/M.L	11.10[b]				

---

[a] Reduced by ligand; [b] 25°, 0.15

Bibliography:

H^+ 51G,55LM,63TA,64IN

Mg^{2+}-Ba^{2+} 63TA

Co^{2+}-Cu^{2+} 51G

Zn^{2+} 51G,55LM

Cd^{2+},Pb^{2+} 55LM

Other references: 55FR,61KP,61KPa

$$\begin{array}{c} O \\ \parallel \\ CH_3OC \\ \mid \\ HSCH_2CHNH_2 \end{array}$$

$C_4H_9O_2NS$		L-Cysteine methyl ester		HL

Metal ion	Equilibrium	Log K 25°, 0.1	ΔH 25°, 0.1	ΔS 25°, 0.1
H^+	HL/H.L	8.93 ±0.01	(−5)[a]	(24)
	$H_2L/HL.H$	6.53 ±0.01	(−9)[a]	(−1)
Ni^{2+}	ML/M.L	7.6		
	$ML_2/M.L^2$	17.24		
	$ML_3/M.L^3$	24.3		
	$M_4L_6/M^4.L^6$	58.4		
	MHL/ML.H	3		
	$MHL_2/ML_2.H$	3		
Zn^{2+}	$ML_2/M.L^2$	15.91		
	MHL/ML.H	11.9		
	$MHL_2/ML_2.H$	4.85		
	ML/MOHL.H	−0.41		
Cd^{2+}	ML/M.L	8.89		
	$ML_2/M.L^2$	16.2		
	$M_2L_3/M^2.L^3$	29.73		
Hg^{2+}	$ML_2/ML.L$	6.33		
	$M_2L_3/(ML)^2.L$	9.9		
	$M_3L_4/(ML)^3.L$	12.9		
Pb^{2+}	ML/M.L	9.13		
	$ML_2/M.L^2$	15.29		
	$M(HL)_2M.H^2.L^2$	26.4		
	$MHL_2/ML_2.H$	6.62		
	MHL/ML.H	3		

[a] 25-50°, 0.1

Bibliography:
H^+ 65CM,66HP,69PPH
Ni^{2+}-Pb^{2+} 69PPH

Other reference: 55LM

$$CH_3SCH_2CH_2NH_2$$

C_3H_9NS <u>2-Methylthioethylamine</u>   L

Metal ion	Equilibrium	Log K 20°, 0.15	Log K 30°, 1.0	Log K 25°, 0	ΔH 25°, 0	ΔS 25°, 0
H^+	HL/H.L	9.44	9.31	9.34	$(-12)^a$	(2)
Ni^{2+}	ML/M.L		3.23		$(-5)^b$	$(-2)^c$
	$ML_2/M.L^2$		6.02		$(-11)^b$	$(-9)^c$
	$ML_3/M.L^3$		7.75		$(-15)^b$	$(-15)^c$
Cu^{2+}	ML/M.L	5.30	5.58	5.51	$(-7)^a$	(2)
	$ML_2/M.L^2$	9.68	10.68	10.68	$(-12)^a$	(9)
Cu^+	ML/M.L	5.65				
	$ML_2/M.L^2$	10.98				
Ag^+	ML/M.L		4.17			
	$ML_2/M.L^2$		6.88			
Cd^{2+}	ML/M.L		3.22			
	$ML_2/M.L^2$		5.5			

^{a}10-40°, 0; ^{b} 0-50°, 1.0; ^{c}30°, 1.0

Bibliography:
H^+, Cu^{2+} 54GF,59MB,62HP Ag^+ 51G
Ni^{2+} 54GF Cd^{2+} 56BF
Cu^+ 62HP

| C_5H_7NS | 2-Thienylmethylamine | (2-(aminomethyl)thiophene) | L |

Metal ion	Equilibrium	Log K 30°, 1.0
H^+	HL/H.L	8.78
Ag^+	ML/M.L	2.87
	$ML_2/M.L^2$	6.51

Bibliography: 54GFa

$HOCH_2CH_2SCH_2CH_2NH_2$

| $C_4H_{11}ONS$ | 2-(2-Aminoethylthio)ethanol | L |

Metal ion	Equilibrium	Log K 30°, 1.0	Log K 25°, 0	ΔH 25°, 0	ΔS 25°, 0
H^+	HL/H.L	(9.18)	9.26	(-12)[a]	(2)
Ni^{2+}	ML/M.L	3.28		(-4)[b]	(2)[d]
	$ML_2/M.L^2$	6.01		(-10)[b]	(-6)[d]
	$ML_3/M.L^3$	7.71		(-10)[b]	(2)[d]
Cu^{2+}	ML/M.L	5.44	5.32	(-5)[c]	(8)
	$ML_2/M.L^2$	10.41	10.23	(-11)[c]	(10)
Ag^+	ML/M.L	4.53	4.88	(-9)[a]	(-8)
	$ML_2/M.L^2$	7.46	8.90	(-15)[a]	(-10)

[a] 10-40°, 0;   [b] 30-50°, 1.0;   [c] 10-30°, 0;   [d] 30°, 1.0

Bibliography:
H^+,Cu^{2+},Ag^+ 53L,59LB   Other reference: 54L
Ni^{2+} 53L

$$H_2NCH_2CH_2NH_2$$

| $C_2H_8N_2$ | | Ethylenediamine | (en) | | | L |

Metal ion	Equilibrium	Log K 25°, 0.1	Log K 25°, 0.5	Log K 25°, 0	ΔH 25°, 1.0	ΔS 25°, 1.0
H^+	HL/H.L	9.89 ±0.07	10.04 ±0.05	9.928	-11.88[f]±0.06	5.6[f]
		10.04[a]	10.18[b]±0.06		-12.2	6
	H_2L/HL.H	7.08 ±0.03	7.31 ±0.02	6.848	-10.95[f]±0.08	5.4[f]
		7.22[a]	7.45[b]±0.02	7.93[d]±0.05	-10.6	-1
			7.69[c]±0.05	8.21[e]		
Mg^{2+}	ML/M.L		0.37[g]			
Cr^{2+}	ML/M.L		5.15[h]			
	ML_2/M.L^2		9.19[h]			
Mn^{2+}	ML/M.L	2.67	2.77[h]		-2.8	3[h]
	ML_2/M.L^2	(3.7)	4.87[h]		-6.0	2[h]
	ML_3/M.L^3		5.81[h]		-11.1	-11[h]
Fe^{2+}	ML/M.L		4.34[h]		-5.1	3[h]
	ML_2/M.L^2		7.66[h]		-10.4	0[h]
	ML_3/M.L^3		9.72[h]		-15.9	-9[h]
Co^{2+}	ML/M.L	(5.6)±0.3	5.96[h]±0.03		-6.9	4[h]
	ML_2/M.L^2	10.5 ±0.3	10.8[h] ±0.1		-14.0	2[h]
	ML_3/M.L^3	13.8	14.1[h] ±0.1		-22.2	-10[h]
Ni^{2+}	ML/M.L	7.35±0.1	7.47 ±0.1	7.32 +0.08	-9.0 ±0.1	4
			7.58[b]±0.04			
	ML_2/M.L^2	13.54 ±0.1	13.82 ±0.1	13.50 +0.1	-18.3 ±0.1	3
			14.02[b]±0.09			
	ML_3/M.L^3	17.71 ±0.1	18.13 ±0.1	17.61 +0.3	-28.0 +0.1	-10
			18.44[b]±0.1			

---

[a] 20°, 0.1;   [b] 25°, 1.0;   [c] 25°, 2.0;   [d] 25°, 3.0;   [e] 25°, 4.0;   [f] 25°, 0;   [g] 30°, 1.4;
[h] 25°, 1.4

Ethylenediamine (continued)

Metal ion	Equilibrium	Log K 25°, 0.1	Log K 25°, 0.5	Log K 25°, 0	ΔH 25°, 1.0	ΔS 25°, 1.0
Cu^{2+}	$ML/M.L$	10.54 ± 0.1	10.71 ± 0.09	$10.48 +0.04$	-13.1 ± 0.2	6
			$10.82^{b} \pm 0.1$	11.02^{i}	-12.6^{j}	
	$ML_2/M.L^2$	19.6 ± 0.2	20.04 ± 0.1	$19.55 +0.5$	-25.5 ± 0.2	7
			$20.2^{b} \pm 0.3$	20.61^{i}	-25.2^{j}	
	$MOHL/ML.OH$		0.73			
Cr^{3+}	$ML_2/MOHL_2.H$	4.86^{k}				
	$MOHL_2/M(OH)_2L_2.H$	7.34^{k}				
Co^{3+}	$ML_3/ML_2.L$		13.99^{b}			
	$ML_3/M.L^3$		48.68^{g}			
	$ML_2/MOHL_2.H$		5.80^{b}			
	$MOHL_2/M(OH)_2L_2.H$		8.10^{b}			
Cu^{+}	$ML_2/M.L^2$		11.2^{l}			
Ag^{+}	$ML/M.L$	4.70^{a}				
	$ML_2/M.L^2$	7.70^{a}				
	$MHL/ML.H$	7.68^{a}				
	$M_2L/ML.M$	1.8^{a}				
	$M_2L_2/(ML)^2$	3.8^{a}				
CH_3Hg^{+}	$ML/M.L$	8.25^{a}				
Pd^{2+}	$ML_2/ML.L$		18.4^{b}			
Pt^{2+}	$ML_2/M.L^2$		36.5^{m}			
Zn^{2+}	$ML/M.L$	5.7 ± 0.1	5.87 ± 0.1	5.66	-7.0	4^{h}
			5.92^{h}	6.15^{i}		
	$ML_2/M.L^2$	10.62 ± 0.02	10.97 ± 0.1	10.64	-11.9	11^{h}
			11.07^{h}	11.49^{i}		
	$ML_3/M.L^3$	13.23^{n}	13.03	13.89	-17.1	2^{h}
			12.93^{h}			

---

^{a} 20°, 0.1; ^{b} 25°, 1.0; ^{g} 30°, 1.4; ^{h} 25°, 1.4; ^{i} 25°, 2.15; ^{j} 25°, 0.5; ^{k} 4°, 0.1;

^{l} 25°, 0.3; ^{m} 18°, 1.0; ^{n} 25°, 0.15

Ethylenediamine   (continued)

Metal ion	Equilibrium	Log K 25°, 0.1	Log K 25°, 0.5	Log K 25°, 0	ΔH 25°, 1.0	ΔS 25°, 1.0
Cd^{2+}	ML/M.L	5.45 ±0.1	5.62[b]±0.06	5.41	(-6)[o]	(6)
				5.84[i]		
	$ML_2/M.L^2$	9.98[n]	10.21[b]±0.03	9.91	(-13.3)[pq]	(2)
				10.62[i]		
	$ML_3/M.L^3$	11.74	12.30[b]±0.04	12.69[i]	(-19.7)[pq]	(-10)
Hg^{2+}	ML/M.L	14.3				
	$ML_2/M.L^2$	23.24 ±0.07	23.42[h]		(-33)[r]	(-4)[h]
	MOHL/ML.OH	9.5				
	$MHL_2/ML_2.H$	5.2				
	$MH_2L_2/MHL_2.H$	4.2				
	$MH_2L_3/MHL_2.HL$	3.6				
	$MClL/MCl_2.L$			5.54	-8.7[f]	-4[f]
	$ML_2/MCl_2.L^2$			9.73	-17.7[f]	-15[f]
Pb^{2+}	ML/M.L	7.00				
	$ML_2/M.L^2$	8.45				

---

[b] 25°, 1.0; [f] 25°, 0; [h] 25°, 1.4; [i] 25°, 2.15; [n] 25°, 0.15; [o] 10-40°, 0; [p] 25°, 0.1; [q] not corrected for Cl^- complexes; [r] 10-40°, 0.1

Bibliography:

H^+ 45BA,48BN,52BM,52EP,52H,52SA,54DS, 56WM,57MC,61CP,63NMa,64FB,66PCI,67HW, 67VA,69HG,70FR,71HBM

Mg^{2+} 41B

Cr^{2+} 57PB

Mn^{2+} 41B,60CP,74MM

Fe^{2+} 41B,60CP

Co^{2+} 41B,50E,60CP,69PS,71GS,72NMa

Ni^{2+} 41B,45CM,50E,52BMa,52H,54DS,55PB, 59MB,60CP,63Ca,65NKK,67HWa,68FV,68PK, 68PS,70FR,71GS,74MM

Cu^{2+} 45CM,48BN,52BMa,53SPa,54DS,55CH,55JR, 55PB,57MC,59MB,63NMa,64SM,67HWa,67PS, 68PK,70FR,70GS,71HBM,71SSb,72BFa

Cr^{3+} 57H

Co^{3+} 41B,52BR

Cu^+ 61JW

Ag^+ 52SA

CH_3Hg^+ 65SS

Pd^{2+} 68RJ

Pt^{2+} 61GG

Zn^{2+} 45BA,53SPa,55NM,59MB,60CP,69PS,70FR, 71GS,74MM

Cd^{2+} 45BA,45CM,53SP,54DS,55CH,58BB,64SM, 73CV,73MB

Hg^{2+} 50B,55NR,56WM,61RM,66PCI

Pb^{2+} 69IM,74K

Other references: 28J,35BW,36BW,48MM,49LO, 50DL,54BM,56M,57B,57V,61ALa,61BS,61KPa, 63KV,64OM,66PS,67LK,67SP,67SS,68FD, 69MI,69VS,70MAc,71SSb,72HJ,73OI,73ZT

$$\overset{\displaystyle \underset{|}{CH_3}}{H_2NCHCH_2NH_2}$$

| $C_3H_{10}N_2$ | | DL-1-Methylethylenediamine | (1,2-propylenediamine, pn) | | | L |

Metal ion	Equilibrium	Log K 25°, 0.1	Log K 25°, 0.5	Log K 25°, 0	ΔH 25°, 0	ΔS 25°, 0
H[+]	HL/H.L	9.78 +0.09	9.90 ±0.09	9.72 +0.10	−11.9	5
					−12.3[a]	4[a]
	$H_2L/HL.H$	6.85 −0.01	7.06 −0.1	6.61 ±0.00	−9.7	−2
					−10.9[a]	−4[a]
Ni[2+]	ML/M.L	7.34 ±0.04	7.50 ±0.01	7.29 +0.01	−8.2	6
	$ML_2/M.L^2$	13.51[b]±0.08	13.86 ±0.00	13.43 +0.2	−16.4	6
	$ML_3/M.L^3$	17.82 +0.7	18.33 ±0.06	17.61 +1	−26.7	−9
					−27.0[c]	−9[c]
Cu[2+]	ML/M.L	10.58 ±0.1	10.76 ±0.09	10.44 ±0.1	−12.0	8
	$ML_2/M.L^2$	19.7[b] ±0.3	20.1 ±0.3	19.4 ±0.2	−23.9	10
					−26.0[c]	4[c]
Zn[2+]	ML/M.L	5.72 ±0.01	5.87 ±0.04	5.64		
	$ML_2/M.L^2$	(10.77)[e]	(11.20)[e]±0.3	(10.58)[e]		
	$ML_3/M.L^3$		(12.6)[de]			
Cd[2+]	ML/M.L		5.42[d]			
	$ML_2/M.L^2$		(9.97)[de]			
	$ML_3/M.L^3$		(12.12)[de]			
Hg[2+]	$ML_2/M.L^2$	(23.51)[e]			(−34)[f]	(−6)[c]
Pb[2+]	$ML_2/M.L^2$	8.62[g]				

---

[a] 25°, 0.5; [b] D-,L- and DL-isomers gave the same value; [c] 25°, 0.1; [d] 30°, 0.65; [e] optical isomerism not stated; [f] 0-40°, 0.1; [g] 25°, 0.2

Bibliography:

H[+]   45CM,53BM,61N,62NMK,63NMa,66PC,67VA   Cd[2+] 45CM
Ni[2+] 45CM,62NMH,70AB,72CHP,73PC   Hg[2+] 55NR,61RM
Cu[2+] 45CM,54BC,62NMK,67PC,70AB   Pb[2+] 74K
Zn[2+] 45CM,61NM,62NMB

Other references: 49LO,50DL, 50E,73OI

$$\begin{array}{c} CH_2CH_3 \\ | \\ H_2NCHCH_2NH_2 \end{array}$$

$C_4H_{12}N_2$ DL-1-Ethylethylenediamine   (1,2-butylenediamine)   L

Metal ion	Equilibrium	Log K 25°, 0.1		Log K 25°, 0	ΔH 25°, 0	ΔS 25°, 0.1
H[+]	HL/H.L	9.66		(9.39)	−11.5	6
	$H_2L/HL.H$	6.65		6.40	−9.9	−3
Ni[2+]	ML/M.L	7.24				
	$ML_2/M.L^2$	(14.09)[a]				
	$ML_3/M.L^3$	(19.32)[a]				
Cu[2+]	ML/M.L	10.37		(10.05)	−11.8	8
	$ML_2/M.L^2$	(19.47)[a]		(19.12)[a]	−23.8	9

[a] Optical isometism not stated.

Bibliography:

H[+]   66PC   Ni[2+]   72CHP   Cu[2+]   67PC

$$\begin{array}{c} CH_3 \\ | \\ H_2NCCH_2NH_2 \\ | \\ CH_3 \end{array}$$

$C_4H_{12}N_2$ 1,1-Dimethylethylenediamine   L

Metal ion	Equilibrium	Log K 25°, 0.1	Log K 25°, 0.65	Log K 25°, 0	ΔH 25°, 0	ΔS 25°, 0.1
H[+]	HL/H.L	9.66	9.85	(9.42)	−11.8	5
	$H_2L/HL.H$	6.43	6.64	6.18	−9.7	−3
Ni[2+]	ML/M.L	6.55	6.77	6.48	−7.0	6
	$ML_2/M.L^2$	(12.55)	12.17	(12.73)	−14.5	(9)
	$ML_3/M.L^3$	(15.61)	14.42	(16.4)	−18.5	(9)
Cu[2+]	ML/M.L	10.06	10.53	(9.77)	−11.3	8
	$ML_2/M.L^2$	18.93	19.58	18.69	−23.5	7

Bibliography:

H[+]   53BM,66PC   Cu[2+]   54BC,67PC

Ni[2+]   54BC,72CHP,73PC   Other reference: 64NK

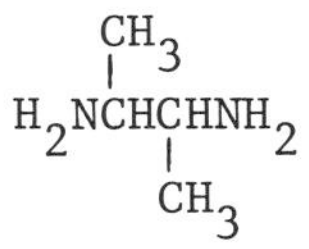

$$\begin{array}{c} CH_3 \\ | \\ H_2NCHCHNH_2 \\ | \\ CH_3 \end{array}$$

$C_4H_{12}N_2$ DL-1,2-Dimethylethylenediamine   (DL-2,3-butylenediamine)   L

Metal ion	Equilibrium	Log K 25°, 0.1	Log K 25°, 0.65
H^+	HL/H.L	(9.77)	9.85
	$H_2L/HL.H$	(6.74)	6.76
Ni^{2+}	ML/M.L	7.37	7.71
	$ML_2/M.L^2$	13.50[a]	14.19
	$ML_3/M.L^3$		18.50
Cu^{2+}	ML/M.L	10.87	11.39
	$ML_2/M.L^2$	20.16[a]	21.21

[a] D- and DL-isomers gave the same value.

Bibliography:

H^+ 53BM,70AB Ni^{2+},Cu^{2+} 54BC,70AB

$$H_2NCHCHNH_2$$
$$\begin{array}{cc} | & | \\ H_3C & CH_3 \end{array}$$

$C_4H_{12}N_2$ <u>meso-1,2-Dimethylethylenediamine</u> (<u>meso-2,3-butylenediamine</u>)   L

Metal ion	Equilibrium	Log K 25°, 0.1	Log K 25°, 0.46	Log K 25°, 0.65
H^+	HL/H.L	(9.80)	9.80	9.82
	$H_2L/HL.H$	(6.76)	6.80	6.77
Mn^{2+}	ML/M.L		2.72	
	$ML_2/M.L^2$		4.32	
Ni^{2+}	ML/M.L	6.71		7.04
	$ML_2/M.L^2$	12.39		12.74
	$ML_3/M.L^3$			15.63
Cu^{2+}	ML/M.L	10.41		10.72
	$ML_2/M.L^2$	19.44	.	20.06
Zn^{2+}	ML/M.L		6.06	
	$ML_2/M.L^2$		11.15	

Bibliography:

H^+ 53BM,63L,70AB Ni^{2+},Cu^{2+} 54BC,70AB

Mn^{2+},Zn^{2+} 63L

$H_2NCHCH_2NH_2$

$C_8H_{12}N_2$ DL-1-Phenylethylenediamine   L

Metal ion	Equilibrium	Log K 25°, 0.1	Log K 25°, 0	ΔH 25°, 0	ΔS 25°, 0
H^+	HL/H.L	8.85	(8.55)	-11.1	2
	H_2L/HL.H	6.01	5.76	-10.6	-9
Cu^{2+}	ML/M.L	8.36	8.33		
	$ML_2/M.L^2$	$(17.27)^a$	$(17.32)^a$		

-----

[a] Optical isomerism not stated.

Bibliography:

H^+ 66PC Cu^{2+} 67PC

$$H_3C\ CH_3$$
$$H_2N-C-C-NH_2$$
$$H_3C\ CH_3$$

$C_6H_{16}N_2$ 1,1,2,2-Tetramethylethylenediamine   L

Metal ion	Equilibrium	Log K 25°, 0.46	Log K 25°, 0.65
H^+	HL/H.L	9.96	9.98
	H_2L/HL.H	6.32	6.41
Ni^{2+}	$ML_2/M.L^2$	14.56	14.68
Cu^{2+}	ML/M.L		11.63
	$ML_2/M.L^2$		21.87
Zn^{2+}	ML/M.L	6.32	
	$ML_2/M.L^2$	11.92	

Bibliography:

H^+ 53BM,63L Zn^{2+} 63L

Ni^{2+} 54BC,63L   Other reference: 68Pb

Cu^{2+} 54BC

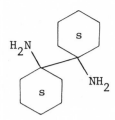

$C_{12}H_{24}N_2$ Bicyclohexyl-1,1'-diamine   L

Metal ion	Equilibrium	Log K 20°, 0.1
H^+	HL/H.L	10.41
	$H_2L/HL.H$	5.62
Co^{2+}	ML/M.L	5.3
	$ML_2/M.L^2$	10.1
	$ML_3/M.L^3$	(15.4)
Ni^{2+}	$ML_2/M.L^2$	14.9
Cu^{2+}	ML/M.L	12.20
	$ML_2/M.L^2$	23.15
Zn^{2+}	ML/M.L	6.35
	$ML_2/M.L^2$	(13.30)

Bibliography: 65TS

$C_6H_{14}N_2$ <u>meso(cis)-Cyclohexane-1,2-diamine</u>   L

Metal ion	Equilibrium	Log K 25°, 0.1	Log K 20°, 0.1	Log K 25°, 0	ΔH 25°, 0	ΔS 25°, 0
H^+	HL/H.L	(9.69)	9.99	9.80	(-11)[a]	(8)
	$H_2L/HL.H$	6.30	6.41	6.06	(-10)[a]	(-6)
Co^{2+}	ML/M.L		5.79			
	$ML_2/M.L^2$		10.34			
	$ML_3/M.L^3$		13.18			
Ni^{2+}	ML/M.L	7.12	7.41	7.21	(-9)[a]	(3)
	$ML_2/M.L^2$	13.83	13.54	13.09	(-15)[a]	(10)
	$ML_3/M.L^3$		16.48			
Cu^{2+}	ML/M.L	10.61	10.87	10.77	(-12)[a]	(9)
	$ML_2M.L^2$	19.97	20.54	20.32	(-24)[a]	(12)
Zn^{2+}	ML/M.L		6.08	5.81	(-7)[a]	(3)
	$ML_2M.L^2$		11.57	11.21	(-14)[a]	(4)
	$ML_2/MOHL_2.H$		10.6			
	$MOHL_2/M(OH)_2L.H$		11.5			
Cd^{2+}	ML/M.L		5.78	5.65	(-6)[a]	(6)
	$ML_2/M.L^2$		10.49	10.31	(-12)[a]	(7)
	$ML_2/MOHL_2.H$		10.9			

---

[a] 10-40°, 0

Bibliography:
H^+,Ni^{2+},Cu^{2+} 56SB,58BF,70AB Zn^{2+},Cd^{2+} 56SB,58BF
Co^{2+} 56SB

$C_6H_{14}N_2$ DL(trans)-Cyclohexane-1,2-diamine   L

Metal ion	Equilibrium	Log K 25°, 0.1	Log K 20°, 0.1	Log K 25°, 0	ΔH 25°, 0	ΔS 25°, 0
H[+]	HL/H.L	9.71 9.88[b]	9.89	(9.74)	(-12)[a]	(5)
	$H_2L/HL.H$	6.59 6.81[b]	6.72	6.29	(-10)[a]	(-5)
Mn[2+]	ML/M.L	2.94[b]				
	$ML_2/M.L^2$	(5.33)[bd]				
Co[2+]	ML/M.L		6.37			
	$ML_2/M.L^2$		(11.74)[d]			
	$ML_3/M.L^3$		(15.22)[d]			
Ni[2+]	ML/M.L	7.74	7.99	7.87	(-9)[a]	(6)
	$ML_2/M.L^2$	14.27[c]	14.98	14.64	(-17)[a]	(10)
	$ML_3/M.L^3$		20.07	19.38	(-26)[a]	(1)
Cu[2+]	ML/M.L	10.94	11.13	(11.03)	(-14)[a]	(4)
	$ML_2/M.L^2$	20.35[c]	20.93	(20.67)	(-26)[a]	(7)
Zn[2+]	ML/M.L	6.65[b]	6.37	6.19	(-5)[a]	(11)
	$Ml_2/M.L^2$	(12.12)[bd]	(11.98)[d]	(11.55)[d]	(-10)[a]	(19)
	$ML_3/M.L^3$	(14.1)[bd]				
	$ML_2/MOHL_2.H$		11.9			
	$MOHL_2/M(OH)_2L_2.H$		12.4			
Cd[2+]	ML/M.L		5.80	5.80	(-6)[a]	(6)
	$ML_2/M.L^2$		(10.51)[d]	(10.51)[d]	(-12)[a]	(8)
	$ML_2/MOHL_2.H$		12.4			

[a] 10-40°, 0; [b] 25°, 0.46; [c] D- and DL-isomers gave the same value;
[d] optical isomerism not stated.

Bibliography:
H[+]   56SB,58BF,63L,70AB   Ni[2+],Cu[2+] 56SB,58BF,70AB
Mn[2+]   63L   Zn[2+] 56SB,58BF,63L
Co[2+]   56SB   Cd[2+] 56SB,58BF

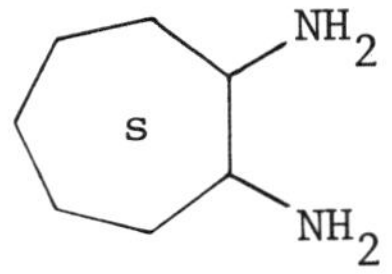

$C_7H_{16}N_2$

DL(trans)-Cycloheptane-1,2-diamine

L

Metal ion	Equilibrium	Log K 10°, 0
H^+	HL/H.L	10.48
	$H_2L/HL.H$	6.67
Ni^{2+}	ML/M.L	7.8
	$ML_2/M.L^2$	(14.4)
	$ML_3/M.L^3$	(18.2)
Cu^{2+}	ML/M.L	11.6
	$ML_2/M.L^2$	(21.7)
Zn^{2+}	ML/M.L	6.11
	$ML_2/M.L^2$	(11.64)
Cd^{2+}	ML/M.L	5.88
	$ML_2/M.L^2$	(10.88)

Bibliography: 58BF

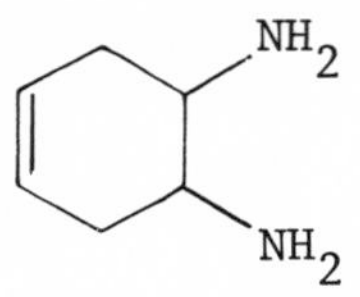

$C_6H_{12}N_2$ meso(cis)-1-Cyclohexene-4,5-diamine   L

Metal ion	Equilibrium	Log K 25°, 0.1
H[+]	HL/H.L	9.9
	H_2L/HL.H	6.6
Cu[2+]	ML/M.L	11.0
	MHL/ML.H	7.1

Bibliography: 57MC

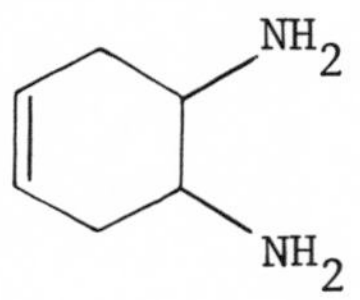

$C_6H_{12}N_2$ DL(trans)-1-Cyclohexene-4,5-diamine   L

Metal ion	Equilibrium	Log K 25°, 0.1
H[+]	HL/H.L	9.3
	H_2L/HL.H	6.2
Cu[2+]	ML/M.L	11.5
	MHL/ML.H	7.0

Bibliography: 57MC

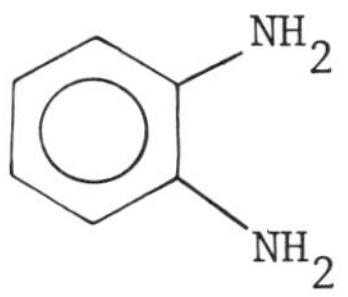

$C_6H_8N_2$ 1,2-Diaminobenzene   (1,2-phenylenediamine)   L

Metal ion	Equilibrium	Log K 25°, 0.1	Log K 25°, 0.3	ΔH 25°, 0.3	ΔS 25°, 0.3
H^+	HL/H.L	4.63	4.74	$(-6)^a$	(1)
Cu^{2+}	ML/M.L	4.55			
	$ML_2/M.L^2$	7.72			

---

[a] 15-40°, 0.3

Bibliography:

H^+ 71KT,71M Cu^{2+} 71KT

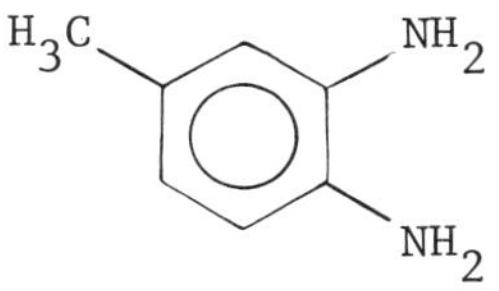

$C_7H_{10}N_2$ 1,2-Diamino-4-methylbenzene   (4-methyl-1,2-phenylenediamine)   L

Metal ion	Equilibrium	Log K 25°, 0.1	Log K 25°, 0.3	ΔH 25°, 0.3	ΔS 25°, 0.3
H^+	HL/H.L	4.79	4.83	$(-6)^a$	(2)
Cu^{2+}	ML/M.L	4.74			
	$ML_2/M.L^2$	8.50			

---

[a] 15-40°, 0.3

Bibliography:

H^+ 71KT,71M Cu^{2+} 71KT

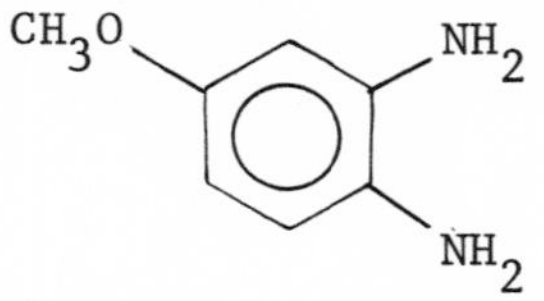

$C_7H_{10}ON_2$ <u>1,2-Diamino-4-methoxybenzene</u>   <u>(4-methoxyphenylenediamine)</u>   L

Metal ion	Equilibrium	Log K 25°, 0.1
H^+	HL/H.L	5.10
Cu^{2+}	ML/M.L	4.78
	$ML_2/M.L^2$	8.42

Bibliography: 71KT

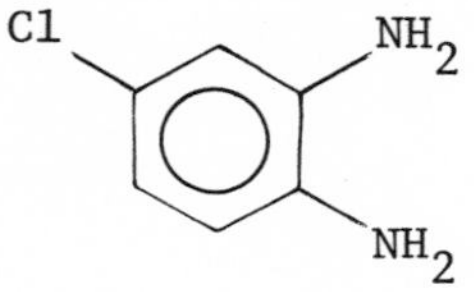

$C_6H_7N_2Cl$ <u>1,2-Diamino-4-chlorobenzene</u>   <u>(4-chlorophenylenediamine)</u>   L

Metal ion	Equilibrium	Log K 25°, 0.1
H^+	HL/H.L	3.94
	$H_2L/HL.H$	0.6
Cu^{2+}	ML/M.L	3.32
	$ML_2/M.L^2$	5.76

Bibliography: 71KT

$$H_2NCH_2CH_2CH_2NH_2$$

$C_3H_{10}N_2$ **Trimethylenediamine**   L

Metal ion	Equilibrium	Log K 25°, 0.1	Log K 25°, 1.0	Log K 25°, 0	ΔH 25°, 0.3	ΔS 25°, 0.5
H^+	HL/H.L	10.52 ±0.04	10.77 ±0.04	10.49 ±0.02	-13.2^d	4^d
		10.68^a	10.94^b	10.65^c+0.01	-13.1^c±0.0	5
	$H_2L/HL.H$	8.74 ±0.02	9.12 ±0.04	8.48 +0.01	-12.4^d	-3^d
		8.89^a	9.28^b	8.95^c±0.02	-12.7^c±0.1	-2
Ni^{2+}	ML/M.L	6.31 ±0.1	6.47 ±0.08	6.29 ±0.01	-7.2	5
				6.42^c	-7.8^e	
	$ML_2/M.L^2$	10.6 ±0.2	10.9 ±0.2	10.54 ±0.06	-14.6	0
				10.8^c	-15.0^e	
	$ML_3/M.L^3$	12.3^i	12.0			
Cu^{2+}	ML/M.L	9.75 ±0.07	10.00 ±0.02	9.61 ±0.03	-11.0	8
				9.84^c		
	$ML_2/M.L^2$	16.9 ±0.2	17.3 ±0.2	16.65	-22.0	4
				17.1^c	-22.8^e	
	ML/MOHL.H	7.66		7.42		
	$ML/M(OH)_2L.H^2$	19.36		19.10		
	$(MOHL)_2/(MOHL)^2$	2.41		2.17		
Ag^+	ML/M.L	5.85^a		5.71	$(-14)^f$	$(-21)^d$
	MHL/ML.H	7.34^a				
	$M_2L/ML.M$	0.6^a				
Cd^{2+}	ML/M.L	4.50 +0.2	4.72^b		$(-5)^g$	$(4)^h$
	$ML_2/M.L^2$	7.20 +0.5			$(-10)^g$	$(-1)^h$
	$ML_3/M.L^3$	8.0				
	MHL/ML.H		8.03^b			
Pb^{2+}	$ML_2/M.L^2$	8.16^j				

^{a} 20°, 0.1; ^{b} 20°, 1.0; ^{c} 25°, 0.5; ^{d} 25°, 0; ^{e} 25°, 1.2; ^{f} 10-40°, 0;
^{g} 0-49°, 0.15; ^{h} 25°, 0.1; ^{i} 25°, 0.15; ^{j} 25°, 0.2

Trimethylenediamine   (continued)

Bibliography:

H[+]   52SM,55PB,56HF,58BF,62SS,65NKS,67HW,   Ag[+]   52SM,58BF

          67VA,69CI,70BP,71SH,73FP   Cd[2+]   54IW,55CH,62SS

Ni[2+]   55CH,55PB,56HF,58BF,65NKS,67HWa,70MAa   Pb[2+]   74K

Cu[2+]   54IW,55CH,55PB,56HF,58BF,65NKS,67HWa,   Other reference: 59CG
          71SH

$$CH_3$$
$$H_2NCHCH_2CH_2NH_2$$

$C_4H_{12}N_2$	DL-1-Methyltrimethylenediamine		(1,3-diaminobutane)	L

Metal ion	Equilibrium	Log K 25°, 0.1	Log K 25°, 0.5	Log K 25°, 0
H[+]	HL/H.L	10.54	10.66	10.50
	$H_2L/HL.H$	8.67	8.90	8.42
Ni[2+]	ML/M.L	6.28	6.34	6.25
	$ML_2/M.L^2$	(10.49)[a]	(10.68)[a]	(10.43)[a]
Cu[2+]	ML/M.L	9.74	9.91	9.67
	$ML_2/M.L^2$	(17.08)[a]	(17.45)[a]	(16.92)[a]

[a] Optical isomerism not stated.

Bibliography:

H[+]   65NKS,68NT   Cu[2+]   65NKS

Ni[2+]   68NT

$$H_2NCH_2\overset{\overset{\displaystyle CH_3}{|}}{\underset{\underset{\displaystyle CH_3}{|}}{C}}CH_2NH_2$$

$C_5H_{14}N_2$ 2,2-Dimethyltrimethylenediamine   (2,2-dimethyl-1,3-diaminopropane)   L

Metal ion	Equilibrium	Log K 30°, 1.0	ΔH 30°, 1.0	ΔS 30°, 1.0
H^+	HL/H.L	10.22	$(-12)^a$	(7)
	H_2L/HL.H	8.18	$(-11)^a$	(1)
Co^{2+}	ML/M.L	4.88	$(-7)^a$	(-1)
	ML_2/M.L^2	7.95	$(-13)^a$	(-7)
Ni^{2+}	ML/M.L	6.59	$(-8)^a$	(3)
	ML_2/M.L^2	11.00	$(-15)^a$	(0)
Cu^{2+}	ML/M.L	9.94	$(-12)^a$	(5)
	ML_2/M.L^2	17.39	$(-24)^a$	(-1)
Ag^+	ML/M.L	4.66		
Zn^{2+}	ML/M.L	5.21	$(-5)^a$	(7)
	ML_2/M.L^2	10.41	$(-10)^a$	(14)

---

[a] 0-50°, 1.0

Bibliography:
H^+-Cu^{2+}, Zn^{2+} 56HF Ag^+ 52Ha

$$H_2NCH_2CH_2CH_2CH_2NH_2$$

$C_4H_{12}N_2$ Tetramethylenediamine   L

Metal ion	Equilibrium	Log K 25°, 0.1	Log K 25°, 0.5	Log K 25°, 0	ΔH 25°, 0.5	ΔS 25°, 0.5
H[+]	HL/H.L	10.72 ±0.07	10.87 ±0.07	10.65 +0.07	-13.6	4
		10.89[a]	11.28[b]±0.04		-13.6[e]	
	H₂L/HL.H	9.44 ±0.05	9.69 ±0.06	9.20 +0.04	-13.2	0
		9.61[a]	10.06[b]+0.05		-13.2[e]	
Ag[+]	ML/M.L	5.9[a]		5.48	(-14)[c]	(-22)[d]
	MHL/ML.H	8.0[a]				
Cd[2+]	ML/M.L		3.6[b]			
	MHL/ML.H		9.9[b]			
Hg[2+]	ML/M.L		17.96[b]			
	MHL/ML.H		4.29[b]			
	MH₂L₂/MHL.HL		6.83[b]			

---

[a] 20°, 0.1;   [b] 20°, 1.0;   [c] 10-40°, 0;   [d] 25°, 0:   [e] 25°,0.02

Bibliography:

H[+]   52SM,58BF,62SS,69CI,70BP,71OT,72KN   Cd[2+], Hg[2+] 62SS

Ag[+]   52SM,58BF

$$H_2NCH_2CH_2CH_2CH_2CH_2NH_2$$

$C_5H_{14}N_2$ Pentamethylenediamine   L

Metal ion	Equilibrium	Log K 25°, 0.1	Log K 25°, 0.5	Log K 20°, 1.0	ΔH 25°, 0.5	ΔS 25°, 0.5
H^+	HL/H.L	10.78	10.92	11.39	−13.9	3
		10.96[a]				
	H_2L/HL.H	9.85	10.15	10.59	−13.4	1
		10.02[a]				
Ag^+	ML/M.L	5.95[a]				
	MHL/ML.H	8[a]				
Hg^{2+}	ML/M.L			17.92		
	MHL/ML.H			4.83		
	MH_2L_2/MHL.HL			7.05		

---

[a] 20°, 0.1

Bibliography:
H^+ 52SM,62SS,70BP Hg^{2+} 62SS
Ag^+ 52SM

$$\begin{array}{cc} HON & NOH \\ \| & \| \\ H_2NC{-}CNH_2 \end{array}$$

$C_2H_6O_2N_4$ <u>Oxamide dioxime</u>   <u>(diaminoglyoxime)</u> H_2L

Metal ion	Equilibrium	Log K 20°, 0
H^+	HL/H.L	11.37
	$H_2L/HL.H$	2.95
Ni^{2+}	ML/M.L	2.7
	$ML_2/M.L^2$	4.7
	$ML_3/M.L^3$	7.3

Bibliography:

H^+ 57WM Ni^{2+} 58WM

$$\begin{array}{c} \text{OH} \\ | \\ H_2NCH_2CHCH_2NH_2 \end{array}$$

$C_3H_{10}ON_2$ 1,3-Diamino-2-propanol   (2-hydroxytrimethylenediamine)   L

Metal ion	Equilibrium	Log K 25°, 0.1	Log K 25°, 1.0	Log K 25°, 0	ΔH 25°, 0	ΔS 25°, 0
H[+]	HL/H.L	9.58 9.68[b]	9.81 +0.04	9.55 ±0.00	(−13)[a]	(0)
	H_2L/HL.H	7.98 8.19[b]	8.35 +0.03	7.75 +0.05	(−11)[a]	(−1)
Co[2+]	ML/M.L		3.90[c]			
	ML_2/M.L^2		7.14[c]			
Ni[2+]	ML/M.L	5.47[d]	5.64[c]	5.49	(−6)[a]	(5)
	ML_2/M.L^2	9.61[d]	10.02[c]	9.72	(−11)[a]	(8)
Cu[2+]	ML/M.L		(9.70)[c]			
	MOHL.H/M.L	(4.34)[d]		3.61	(−5)[a]	(0)
	$(MOHL)_2.H^2/M^2.L^2$	10.48	10.36	10.30		
Ag[+]	ML/M.L		5.80[c]	5.47	(−13)[a]	(−19)
Zn[2+]	ML/M.L		4.60[c]			
	ML_2/M.L^2		9.02[c]			

[a] 10-40°, 0;   [b] 25°, 0.5;   [c] 30°, 1.0;   [d] 30°, 0.16

Bibliography:

H[+]   55GF,58BF,70NT   Cu[2+]   51G,58BB,65MB,70NT,72NL

Co[2+], Zn[2+]   55GF   Ag[+]   55GF,58BF

Ni[2+]   55GF,58BF,65MB

$$H_2NCH_2CH_2OCH_2CH_2NH_2$$

$C_4H_{12}ON_2$ Oxybis(2-ethylamine)   (1,7-Diaza-4-oxaheptane)   L

Metal ion	Equilibrium	Log K 25°, 0	ΔH 25°, 0	ΔS 25°, 0
H^+	HL/H.L	9.75	(−12)[a]	(4)
	$H_2L/HL.H$	8.90	(−13)[a]	(−3)
Ni^{2+}	ML/M.L	5.62	(−7)[a]	(2)
	$ML_2/M.L^2$	9.01	(−14)[a]	(−6)
Cu^{2+}	ML/M.L	8.70	(−11)[a]	(3)
	$ML_2/M.L^2$	13.1	(−15)[a]	(10)
Ag^+	ML/M.L	5.45	(−14)[a]	(−22)

[a] 10–40°, 0

Bibliography: 59LB   Other reference: 54L

$$H_2NCH_2CH_2OCH_2CH_2OCH_2CH_2NH_2$$

$C_6H_{16}O_2N_2$ Ethylenebis(oxy-2-ethylamine)   (1,10-diaza-4,7-dioxadecane)   L

Metal ion	Equilibrium	Log K 25°, 0	ΔH 25°, 0	ΔS 25°, 0
H^+	HL/H.L	9.73	(−11)[a]	(8)
	$H_2L/HL.H$	8.75	(−11)[a]	(3)
Cu^{2+}	ML/M.L	7.89	(−8)[a]	(9)
Ag^+	ML/M.L	7.88	(−13)[a]	(−8)

[a] 10–40°, 0

Bibliography: 59LB   Other reference: 54L

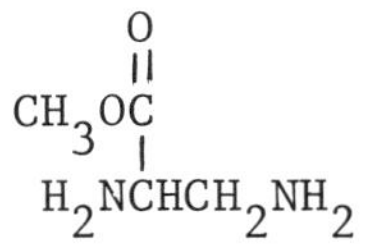

$$\begin{array}{c} O \\ \| \\ CH_3OC \\ | \\ H_2NCHCH_2NH_2 \end{array}$$

$C_4H_{10}O_2N_2$ DL-2,3-Diaminopropanoic acid methyl ester   L

Metal ion	Equilibrium	Log K 25°, 0.1	ΔH 25°, 0.1	ΔS 25°, 0.1
H^+	HL/H.L	8.25	$(-11)^a$	(1)
	$H_2L/HL.H$	4.64	$(-9)^a$	(-10)
Cu^{2+}	ML/M.L	8.99	$(-11)^a$	(4)
	$ML_2/M.L^2$	(16.75)	$(-22)^a$	(2)
	ML/MOHL.H	6.83		
Hg^{2+}	ML/M.L	6.38		
	$ML_2/M.L^2$	(11.48)		
	ML/MOHL.H	7.81		

---

[a] 25-50°, 0.1

Bibliography: 71HMa   Other reference: 61SS

$$H_2NCH_2\overset{\overset{O}{\|}}{C}NHCH_2CH_2NH\overset{\overset{O}{\|}}{C}CH_2NH_2$$

$C_6H_{14}O_2N_4$ Ethylenebis(iminocarbonylmethylamine)   (N,N'-diglycylethylenediamine,   DGEN)   L

Metal ion	Equilibrium	Log K 25°, 0.1	Log K 25°, 0.5	Log K 25°, 1.0
H^+	HL/H.L	8.22 +0.01	8.36[a]	8.39 −0.04
	H_2L/HL.H	7.48 ±0.00	7.66[a]	7.71 −0.08
Mg^{2+}	ML/M.L			0.54
Co^{2+}	ML/M.L	3.27 ±0.04		
Ni^{2+}	ML/M.L	5.38	(5.32)	5.65 −0.2
	$ML_2/M.L^2$	8.50		8.71
	$ML/M(H_{-2}L).H^2$	16.04	16.16	16.50
Cu^{2+}	ML/M.L	7.50	(8.26)[a]	8.13
	MHL/ML.H		5.34[a]	
	$ML/M(H_{-1}L).H$		6.36[a]	6.56
	$ML/M(H_{-2}L).H^2$	13.8	14.42[a]	15.01
	$(ML)^2/M_2(H_{-1}L)_2.H^2$	9.2		
	$M_2(H_{-1}L)_2/(M(H_{-2}L))^2.H^2$	18.40		
Zn^{2+}	ML/M.L	3.95		4.31
Cd^{2+}	ML/M.L			3.33

---

[a] 23°, 0.5

Bibliography:

H^+ 53CG,67ZF,69BMa,70BM,74SM Ni^{2+} 53CG,69BMa,70BM,71K

Mg^{2+}, Cd^{2+} 53CG Cu^{2+} 53CG,67ZF,69BMa

Co^{2+} 69BMa,74SM Zn^{2+} 53CG,69BMa

$$\underset{\text{H}_2\text{NCH}_2\overset{\overset{\text{O}}{\|}}{\text{C}}\text{NHCH}_2\text{CH}_2\text{CH}_2\text{NH}\overset{\overset{\text{O}}{\|}}{\text{C}}\text{CH}_2\text{NH}_2}{}$$

$C_7H_{16}O_2N_4$ **N,N'-Diglycyltrimethylenediamine**   L

Metal ion	Equilibrium	Log K 25°, 0.5
H^+	HL/H.L	8.38[a]
	H_2L/HL.H	7.69[a]
Ni^{2+}	ML/M.L	5.43
	ML/M(H_{-2}L).H^2	14.63
Cu^{2+}	ML/M.L	8.25[a]
	MHL/ML.H	5.46[a]
	ML/M(H_{-2}L).H^2	11.03[a]

[a] 23°, 0.5

Bibliography:

H^+, Cu^{2+} 67ZF Ni^{2+} 71K

$$\underset{\text{H}_2\text{NCH}_2\overset{\overset{\text{O}}{\|}}{\text{C}}\text{NHCH}_2\text{CH}_2\text{CH}_2\text{CH}_2\text{NH}\overset{\overset{\text{O}}{\|}}{\text{C}}\text{CH}_2\text{NH}_2}{}$$

$C_8H_{18}O_2N_4$ **N,N'-Diglycyltetramethylenediamine**   L

Metal ion	Equilibrium	Log K 23°, 0.5
H^+	HL/H.L	8.41
	H_2L/HL.H	7.78
Cu^{2+}	ML/M.L	8.64
	MHL/ML.H	5.12
	ML/M(H_{-1}L).H	7.62
	ML/M(H_{-2}1).H^2	14.99

Bibliography: 67ZF

$$H_2NCH_2\overset{\overset{O}{\|}}{C}NHCH_2CH_2CH_2CH_2CH_2NH\overset{\overset{O}{\|}}{C}CH_2NH_2$$

| $C_9H_{20}O_2N_4$ | N,N'-Diglycylpentamethylenediamine | L |

Metal ion	Equilibrium	Log K 23°, 0.5
H^+	HL/H.L	8.36
	H_2L/HL.H	7.73
Cu^{2+}	ML/M.L	8.46
	MHL/ML.H	5.34
	ML/M(H_{-1}L).H	7.32
	M(H_{-1}L)/MOH(H_{-1}L).H	10.15

Bibliography: 67ZF

$$H_2NCH_2\overset{\overset{O}{\|}}{C}NHCH_2CH_2CH_2CH_2CH_2NH\overset{\overset{O}{\|}}{C}CH_2NH_2$$

| $C_{10}H_{22}O_2N_4$ | N,N'-Diglycylhexamethylenediamine | L |

Metal ion	Equilibrium	Log K 23°, 0.5
H^+	HL/H.L	8.41
	H_2L/HL.H	7.78
Cu^{2+}	ML/M.L	8.93
	MHL/ML.H	5.03
	ML/M(H_{-1}L).H	7.70
	M(H_{-1}L)/MOH(H_{-1}L).H	10.47

Bibliography: 67ZF

$$O \qquad\qquad\qquad\qquad\qquad\qquad\qquad O$$
$$\parallel \qquad\qquad\qquad\qquad\qquad\qquad\qquad \parallel$$
$$H_2NCH_2CNHCH_2CH_2CH_2CH_2CH_2CH_2CH_2CH_2NHCCH_2NH_2$$

$C_{12}H_{26}O_2N_4$	N,N'-Diglycyloctamethylenediamine		L

Metal ion	Equilibrium	Log K 23°, 0.5
H^+	HL/H.L	8.41
	H_2L/HL.H	7.75
Cu^{2+}	ML/M.L	8.97
	MHL/ML.H	5.00

Bibliography: 67ZF

$$OO$$
$$\parallel\parallel$$
$$H_2NCH_2CH_2NHCCNHCH_2CH_2NH_2$$

$C_6H_{14}O_2N_4$	N,N'-Bis(2-aminoethyl)oxamide			L

Metal ion	Equilibrium	Log K 25°, 0.1	Log K 22°, 0.1	Log K 25°, 1.0
H^+	HL/H.L	9.23	9.31	9.54
	H_2L/HL.H	8.48	8.43	8.30
Co^{2+}	ML/M.L	2.78		
Cu^{2+}	ML/M.L		9.41	
	$ML/M(H_{-1}L).H$		7.51	
	$M(H_{-1}L)/M(H_{-2}L).H$		8.10	
	$M_2(H_{-2}L)/M(H_{-2}L).M$		7.36	

Bibliography:
H^+ 65WB,68GF,74SM Cu^{2+} 68GF
Co^{2+} 74SM

$$\underset{H_2NCH_2CH_2NHCCH_2CNHCH_2CH_2NH_2}{\overset{\displaystyle O \qquad O}{\overset{\displaystyle \| \quad \| }{}}}$$

$C_7H_{16}O_2N_4$ <u>N,N'-Bis(2-aminoethyl)malonamide</u>   L

Metal ion	Equilibrium	Log K 22°, 0.1
H^+	HL/H.L	9.40
	H_2L/HL.H	8.68
Cu^{2+}	ML/M.L	7.93
	ML/M(H_{-1}L).H	6.37
	M(H_{-1}L)/M(H_{-2}L).H	6.38

Bibliography: 68GF

$$H_2NCH_2CH_2SCH_2CH_2NH_2$$

$C_4H_{12}N_2S$	Thiobis(2-ethylamine)	(1,7-diaza-4-thiaheptane)		L

Metal ion	Equilibrium	Log K 30°, 1.0	ΔH 30°, 1.0	ΔS 30°, 1.0
H^+	HL/H.L	9.50	(-12)[a]	(3)
	H_2L/HL.H	8.70	(-12)[a]	(0)
Co^{2+}	ML/M.L	5.09	(-7)[a]	(0)
	ML_2/M.L^2	9.01	(-15)[a]	(−9)
Ni^{2+}	ML/M.L	7.27	(-9)[a]	(3)
	ML_2/M.L^2	13.37	(-20)[a]	(−6)
Cu^{2+}	ML/M.L	9.07	(-11)[a]	(5)
	ML_2/M.L^2	14.15	(-24)[a]	(−16)
Ag^+	ML/M.L	7.00	(-14)[a]	(−15)
Zn^{2+}	ML/M.L	5.31	(-5)[a]	(8)
	ML_2/M.L^2	8.88	(-10)[a]	(7)
Cd^{2+}	ML/M.L	5.47		
	ML_2/M.L^2	8.99		

[a] 0-50°, 1.0

Bibliography:
H^+,Co^{2+},Ni^{2+} 54GF Cd^{2+} 56BF
Cu^{2+}-Zn^{2+} 51G

$$H_2NCH_2CH_2SSCH_2CH_2NH_2$$

$C_4H_{12}N_2S_2$ Dithiobis(2-ethylamine)   (1,8-diaza-4,5-dithiaoctane)   L

Metal ion	Equilibrium	Log K 20°, 0.15	Log K 30°, 1.0
H+	HL/H.L	(9.04)	9.44
	H$_2$L/HL.H	8.70	8.68
Cu^{2+}	ML/M.L	6.70	
	MHL/ML.H	6.13	

Bibliography:
H+   51G,63HP   Cu^{2+} 63HP

$$H_2NCH_2CH_2OCH_2CH_2SCH_2CH_2NH_2$$

$C_6H_{16}ON_2S$ 2-[2-(2-Aminoethyloxy)ethylthio]ethylamine   L
                         (1,10-diaza-4-oxa-7-thiadecane)

Metal ion	Equilibrium	Log K 25°, 0	ΔH 25°, 0	ΔS 25°, 0
H+	HL/H.L	9.61	(−12)[a]	(4)
	H$_2$L/HL.H	8.68	(−12)[a]	(−1)
Ni^{2+}	ML/M.L	6.29	(−8)[a]	(2)
Cu^{2+}	ML/M.L	9.04	(−13)[a]	(−2)
Ag+	ML/M.L	8.27	(−14)[a]	(−9)

[a] 10–40°, 0

Bibliography: 59LB   Other reference: 54L

$$H_2NCH_2CH_2SCH_2CH_2SCH_2CH_2NH_2$$

$C_6H_{16}N_2S_2$ Ethylenebis(thio-2-ethylamine)   (1,10-diaza-4,7-dithiadecane)   L

Metal ion	Equilibrium	Log K 30°, 1.0	Log K 25°, 0	ΔH 25°, 0	ΔS 25°, 0
H^+	HL/H.L	9.47	9.46	(-12)[a]	(2)
	H_2L/HL.H	8.86	8.54	(-12)[a]	(-2)
Co^{2+}	ML/M.L	4.89			
Ni^{2+}	ML/M.L	7.90		(-11)[b]	(-1)[c]
Cu^{2+}	ML/M.L	11.32	10.63	(-15)[a]	(-2)
Cd^{2+}	ML/M.L	5.61			
	$ML_2/M.L^2$	8.05			

[a] 10-40°, 0;   [b] 0-50°, 1.0;   [c] 30°, 1.0

Bibliography:
H^+,Cu^{2+} 54GF,59MB Cd^{2+} 56BF
Co^{2+},Ni^{2+} 54GF   Other reference: 51G

$$\begin{array}{c} NH_2 \\ | \\ H_2NCH_2CHCH_2NH_2 \end{array}$$

$C_3H_{11}N_3$ 1,2,3-Triaminopropane   L

Metal ion	Equilibrium	Log K 20°, 0.1	Log K 20°, 0.5
H^+	HL/H.L	9.59	9.63
	$H_2L/HL.H$	7.95	8.08
	$H_3L/H_2L.H$	3.72	3.99
Co^{2+}	ML/M.L	6.80	
	MHL/ML.H	6.89	
Ni^{2+}	ML/M.L	9.30	
	MHL/ML.H	6.34	
Cu^{2+}	ML/M.L	11.1	
	$ML_2/M.L^2$	20.1	
	MHL/ML.H	7.30	
	$MHL_2/ML_2.H$	7.90	
	$MH_2L_2/MHL_2.H$	7.30	
Ag^+	ML/M.L	5.7	
	MHL/ML.H	7.56	
	$M_2L/ML.M$	1.2	
Zn^{2+}	ML/M.L	6.75	
	MHL/ML.H	7.09	
Cd^{2+}	ML/M.L	6.45	
	MHL/ML.H	7.89	
Hg^{2+}	ML/M.L		19.6
	MHL/ML.H		7.93

Bibliography: 50PSb

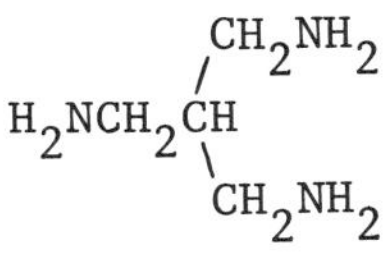

A. PRIMARY AMINES

$$\begin{array}{c} \quad\quad\quad CH_2NH_2 \\ \quad\quad\quad / \\ H_2NCH_2CH \\ \quad\quad\quad \backslash \\ \quad\quad\quad CH_2NH_2 \end{array}$$

$C_4H_{13}N_3$

Tris(aminomethyl)methane

L

Metal ion	Equilibrium	Log K 20°, 0.1	Log K 22°, 1.0
H^+	HL/H.L	10.39	10.51
	$H_2L/HL.H$	8.56	8.86
	$H_3L/H_2L.H$	6.44	6.90
Co^{2+}	ML/M.L	6.25	
	MHL/ML.H	7.89	
Ni^{2+}	ML/M.L	9.90	(9.23)
	MHL/ML.H	6.04	(7.18)
Cu^{2+}	ML/M.L	10.85	
	MHL/ML.H	8.24	
	$MH_2L/MHL.H$	3.46	
Ag^+	ML/M.L	8.70	
Zn^{2+}	MHL/M.HL	3.80	
	$MH_2L/MHL.H$	6.46	
Cd^{2+}	ML/M.L	5.40	
	MHL/ML.H	8.39	
	$MH_2L/MHL.H$	6.71	

Bibliography:

H^+,Ni^{2+} 61SB,62A Co^{2+},Cu^{2+}-Cd^{2+} 62A

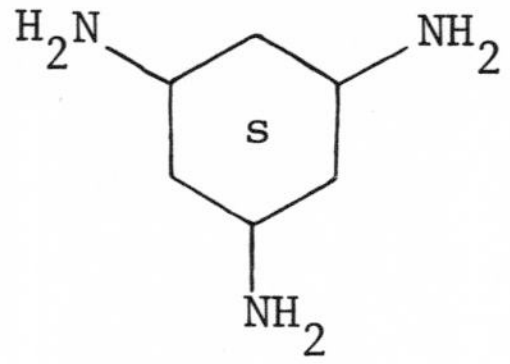

$C_6H_{15}N_3$ cis,cis-1,3,5-Triaminocyclohexane   L

Metal ion	Equilibrium	Log K 20°, 0.1
H^+	HL/H.L	10.4
	H_2L/HL.H	8.7
	H_3L/H_2L.H	6.9
Cu^{2+}	ML/M.L	10.1
	MHL/ML.H	7.0
	ML/MOHL.H	8.7
	MOHL/M(OH)$_2$L.H	11.3
Ag^+	ML/M.L	6.9
	MHL/ML.H	8.8
	M_2L/ML.M	2.4
Zn^{2+}	MHL/M.HL	3.7

Bibliography: 62BS

$$\begin{array}{ccc} H_2NCH_2 & & CH_2NH_2 \\ & C & \\ H_2NCH_2 & & CH_2NH_2 \end{array}$$

$C_5H_{16}N_4$ Tetrakis(aminomethyl)methane   L

Metal ion	Equilibrium	Log K 25°, 0.1
H^+	HL/H.L	9.89
	$H_2L/HL.H$	8.17
	$H_3L/H_2L.H$	5.67
	$H_4L/H_3L.H$	3.03
Co^{2+}	ML/M.L	7.59
	MHL/ML.H	7.82
	$MH_2L/MHL.H$	5.45
Ni^{2+}	ML/M.L	10.73
	$ML_2/M.L^2$	18.80
	MHL/ML.H	7.62
	$MHL_2/ML_2.H$	8.00
	$MH_2L_2/MHL_2.H$	7.57
Cu^{2+}	ML/M.L	11.0
	$ML_2/M.L^2$	19.43
	MHL/ML.H	7.5
	$MH_2L/MHL.H$	5.0
	$MHL_2/ML_2.H$	8.33
	$MH_2L_2/MHL_2.H$	7.55
	$M_2L/M^2.L$	19.59
Zn^{2+}	MHL/M.HL	5.00
	$MH_2L/MHL.H$	6.32
Cd^{2+}	ML/M.L	5.7
	MHL/ML.H	7.9

Bibliography:
H^+, Cu^{2+} 66ZB,68ZB Co^{2+}, Ni^{2+}, Zn^{2+}, Cd^{2+} 68ZB

$$CH_3NHCH_3$$

C$_2$H$_7$N   Dimethylamine   L

Metal ion	Equilibrium	Log K 25°, 0.2	Log K 25°, 0	ΔH 25°, 0	ΔS 25°, 0
H$^+$	HL/H.L	10.80	10.774-0.04	-12.0 ±0.1	9
Ag$^+$	ML$_2$/M.L^2		5.37	-9.7	-8

Bibliography:

H$^+$ 28HR,30HO,41EW,56CG,64WB,69CI   Ag$^+$ 55F

$$CH_3CH_2NHCH_2CH_3$$

C$_4$H$_{11}$N   Diethylamine   L

Metal ion	Equilibrium	Log K 25°, 0.5	Log K 25°, 0	ΔH 25°, 0	ΔS 25°, 0
H$^+$	HL/H.L	10.97	10.933	-12.75-0.02	7.3
Ag$^+$	ML/M.L	2.98[a]			
	ML$_2$/M.L^2	6.34	6.38	-10.7	-7

[a] 30°, 0.5

Bibliography:

H$^+$ 45CM,51EH,69CI

Ag$^+$ 45CM,55F   Other references: 68PM,72JJ

| C_4H_9N | | Pyrrolidine | (tetramethyleneimine) | | | L |

Metal ion	Equilibrium	Log K 25°, 0.2	Log K 25°, 0	ΔH 25°, 0	ΔS 25°, 0
H^+	HL/H.L	11.2	11.305	−12.9 ±0.1	8
Cu^{2+}	ML/M.L	6.4			
	$ML_2/M.L^2$	12.4			
	$ML_3/M.L^3$	17.8			
	$ML_4/M.L^4$	23.0			

Bibliography:

H^+ 61BM,63HB,71CC   Other reference: 60SP

Cu^{2+} 61BM

| $C_5H_{11}N$ | | Piperidine | (pentamethyleneimine) | | | | L |

Metal ion	Equilibrium	Log K 25°, 0.2	Log K 25°, 0.5	Log K 25°, 0	ΔH 25°, 0	ΔS 25°, 0
H^+	HL/H.L	11.01	11.12 ±0.01	11.123	−12.71±0.05	8.3
					−13.19[a]	6.6[a]
Ag^+	ML/M.L		3.10 ±0.07			
	$ML_2/M.L^2$		6.52 ±0.1			
Hg^{2+}	ML/M.L		8.70			
	$ML_2/M.L^2$		17.44			

---

[a] 25°, 0.5

Bibliography:

H^+ 48BV,50B,56BB,56CG,71CC,74BEB Hg^{2+} 50B

Ag^+ 48BV,50B,73BB   Other references: 60SP,63PL

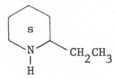

C$_6$H$_{13}$N		DL-2-Methylpiperidine		L

Metal ion	Equilibrium	Log K 25°, 0.5	ΔH 25°, 0.5	ΔS 25°, 0.5
H$^+$	HL/H.L	11.06	-14.03	3.5
Ag$^+$	ML/M.L	3.45		
	ML$_2$/M.L^2	(6.94)		

Bibliography:
H$^+$ 74BEB   Ag$^+$ 73BB

C$_7$H$_{15}$N		DL-2-Ethylpiperidine		L

Metal ion	Equilibrium	Log K 25°, 0.5	ΔH 25°, 0.5	ΔS 25°, 0.5
H$^+$	HL/H.L	11.10	-14.18	3.2
Ag$^+$	ML/M.L	3.84		
	ML$_2$/M.L^2	(7.36)		

Bibliography:
H$^+$ 74BEB   Ag$^+$ 73BB

C$_8$H$_{17}$N   <u>DL-2-Propylpiperidine</u>   L

Metal ion	Equilibrium	Log K 25°, 0.5	ΔH 25°, 0.5	ΔS 25°, 0.5
H$^+$	HL/H.L	11.09	−14.44	2.3
Ag$^+$	ML/M.L	4.0		
	ML$_2$/M.L^2	(7.52)		

Bibliography:

H$^+$ 74BEB   Ag$^+$ 73BB

C$_6$H$_{13}$N   <u>DL-3-Methylpiperidine</u>   L

Metal ion	Equilibrium	Log K 25°, 0.5	ΔH 25°, 0.5	ΔS 25°, 0.5
H$^+$	HL/H.L	11.05	−13.75	4.4
Ag$^+$	ML/M.L	3.04		
	ML$_2$/M.L^2	(6.43)		

Bibliography:

H$^+$ 74BEB   Ag$^+$ 73BB

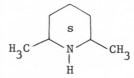

C$_6$H$_{13}$N   4-Methylpiperidine   L

Metal ion	Equilibrium	Log K 25°, 0.5	ΔH 25°, 0.5	ΔS 25°, 0.5
H$^+$	HL/H.L	11.08	-13.48	5.5
Ag$^+$	ML/M.L	3.20		
	ML$_2$/M.L^2	6.50		

Bibliography:

H$^+$ 74BEB   Ag$^+$ 73BB

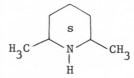

C$_7$H$_{15}$N   DL-2,6-Dimethylpiperidine   L

Metal ion	Equilibrium	Log K 25°, 0.5	ΔH 25°, 0.5	ΔS 25°, 0.5
H$^+$	HL/H.L	11.11	-14.53	2.1
Ag$^+$	ML/M.L	3.96		
	ML$_2$/M.L^2	(7.71)		

Bibliography:

H$^+$ 74BEB   Ag$^+$ 73BB

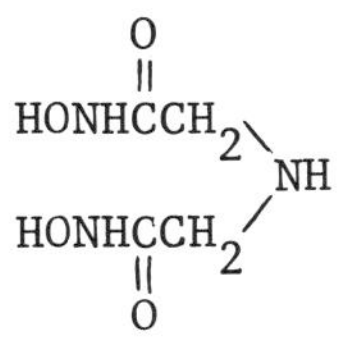

B. SECONDARY AMINES

$$\begin{array}{c} O \\ \| \\ HONHCCH_2 \\ \\ HONHCCH_2 \\ \| \\ O \end{array} \!\!\! NH$$

| $C_4H_9O_4N_3$ | Iminodiacethydroxamic acid | H_2L |

Metal ion	Equilibrium	Log K 20°, 0.1
H^+	$HL/H.L$	10.80
	$H_2L/HL.H$	7.23
	$H_3L/H_2L.H$	5.66
Cu^{2+}	$ML/M.L$	16.11
	$MHL/ML.H$	3.63
	$ML/MOHL.H$	7.30
	$MOHL/M(OH)_2L.H$	10.20
Fe^{3+}	$ML/M.L$	14.80
	$M_2L_3/M^2.L^3$	44.0

Bibliography: 72KMa

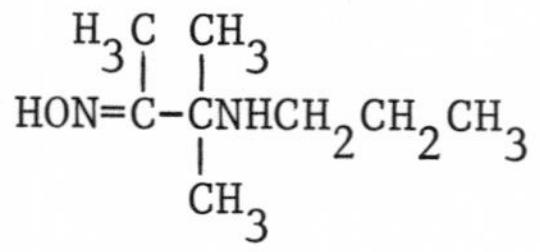

$$
\begin{array}{c}
H_3C \quad CH_3 \\
| \quad | \\
HON{=}C{-}CNHCH_2CH_2CH_3 \\
| \\
CH_3
\end{array}
$$

| $C_8H_{18}ON_2$ | | 3-Propylamino-3-methylbutan-2-one oxime | | | L |

Metal ion	Equilibrium	Log K 24°, 0.27	ΔH 25°, 0	ΔS 25°, 0.27
H^+	HL/H.L	9.01	-10.6	6
Ni^{2+}	$M(H_{-1}L)L.H/M.L^2$	1.2	2.0	12
	$M(H_{-1}L)L/M(H_{-1}L)_2.H$	8.4		

Bibliography: 58M,64WB

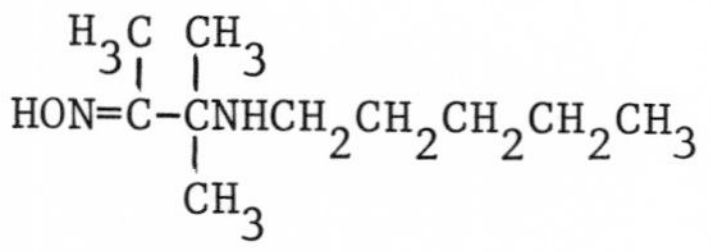

$$
\begin{array}{c}
H_3C \quad CH_3 \\
| \quad | \\
HON{=}C{-}CNHCH_2CH_2CH_2CH_2CH_3 \\
| \\
CH_3
\end{array}
$$

| $C_9H_{20}ON_2$ | | 3-Pentylamino-3-methylbutan-2-one oxime | L |

Metal ion	Equilibrium	Log K 24°, 0.27
H^+	HL/H.L	8.99
Ni^{2+}	$M(H_{-1}L)L.H/M.L^2$	1.1
	$M(H_{-1}L)L/M(H_{-1}L)_2.H$	8.4

Bibliography: 58M

$$HOCH_2CH_2NHCH_3$$

C_3H_8ON		2-(Methylamino)ethanol	(N-methylethanolamine)		L

Metal ion	Equilibrium	Log K 25°, 0.1	Log K 25°, 0	ΔH 25°, 0	ΔS 25°, 0
H^+	HL/H.L	9.9	9.88	-11.1	8
Cu^{2+}	ML/M.L	5.0			
	$ML_2/M.L^2$	9.0			
	$ML_3/M.L^3$	12.2			
	$ML_4/M.L^4$	14.7			
Zn^{2+}	ML/M.L	3.9			
	$ML_2/M.L^2$	6.8			
	$ML_3/M.L^3$	9.2			
	$ML_4/M.L^4$	11.4			

Bibliography:
H^+ 65D,69CI Cu^{2+}, Zn^{2+} 65D

$$HOCH_2CH_2NHCH_2CH_3$$

$C_4H_{11}ON$		2-(Ethylamino)ethanol	(N-ethylethanolamine)		L

Metal ion	Equilibrium	Log K 25°, 0.1	Log K 25°, 0	ΔH 25°, 0	ΔS 25°, 0
H^+	HL/H.L	10.00	9.961	-11.40	7.3
Cu^{2+}	ML/M.L	5.0			
	$ML_2/M.L^2$	9.1			
	$ML_3/M.L^3$	12.6			
	$ML_4/M.L^4$	15.5			

Bibliography:
H^+ 68TE   Other references: 55FK,67FH,70UP
Cu^{2+} 65D

$$\begin{array}{c} HOCH_2CH_2 \diagdown \\ NH \\ HOCH_2CH_2 \diagup \end{array}$$

$C_4H_{11}O_2N$ Iminodi-2-ethanol   (diethanolamine)   L

Metal ion	Equilibrium	Log K 25°, 0.1	Log K 25°, 0.5	Log K 25°, 0	ΔH 25°, 0	ΔS 25°, 0
H^+	HL/H.L	8.90	9.00 ±0.00	8.883	-10.07-0.6	6.9
Cu^{2+}	ML/M.L	5.4				
	$ML_2/M.L^2$	9.6				
	$ML_3/M.L^3$	12.8				
	$ML_4/M.L^4$	14.6				
Ag^+	ML/M.L		2.69			
	$ML_2/M.L^2$		5.48	5.80[a]		
Cd^{2+}	ML/M.L	2.40				
	$ML_2/M.L^2$	4.52				
Hg^{2+}	ML/M.L		7.84			
	$ML_2/M.L^2$		15.66			

[a] 20°, 0

Bibliography:
H^+ 56BR,62BR,65D,69CI,72VE
Cu^{2+} 65D
Ag^+ 56BR,58AS

Cd^{2+} 60MP
Hg^{2+} 56BR
Other references: 55FK,61ALa,62FH,63Sa, 64MSa,65MP,66SKa,67FH,69MI,69MP,70UR, 71SSa

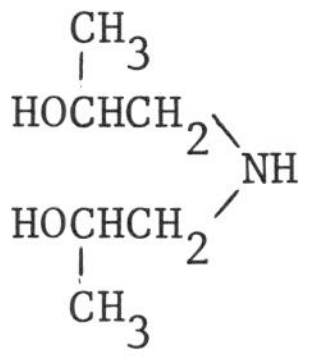

$$
\begin{array}{c}
\text{CH}_3 \\
| \\
\text{HOCHCH}_2 \\
\qquad\searrow \\
\qquad\qquad \text{NH} \\
\qquad\nearrow \\
\text{HOCHCH}_2 \\
| \\
\text{CH}_3
\end{array}
$$

$C_6H_{15}O_2N$ <u>DL-Iminodi-1-(2-propanol)</u>   <u>(bis(2-hydroxypropyl)amine)</u>   L

Metal ion	Equilibrium	Log K 20°, 0
H^+	HL/H.L	8.97
Ag^+	ML/M.L	2.95
	$ML_2/M.L^2$	(5.71)

Bibliography: 64AK

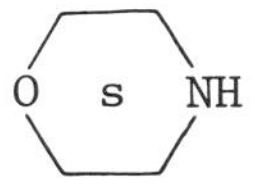

C_4H_9ON <u>Perhydro-1,4-oxazine</u>   <u>(morpholine)</u>   L

Metal ion	Equilibrium	Log K 25°, 0.5	Log K 25°, 1.0	Log K 25°, 0	ΔH 25°, 0	ΔS 25°, 0
H^+	HL/H.L	8.55	8.74	8.492	-9.3	7.7
Ag^+	ML/M.L	2.25				
	$ML_2/M.L^2$	4.92				

Bibliography:
H^+ 48BV,66HB,72KV Ag^+ 48BV

$$CH_3NHCH_2CH_2NH_2$$

| $C_3H_{10}N_2$ | N-Methylethylenediamine | | | | L |

Metal ion	Equilibrium	Log K 25°, 0.1	Log K 25°, 0.5	Log K 25°, 0	ΔH 25°, 0.5	ΔS 25°, 0.5
H^+	HL/H.L	10.11	10.21 ±0.08	10.04 ±0.02	−11.25	9.0
	$H_2L/HL.H$	7.04	7.27 ±0.01	6.76 ±0.02	−10.3	−1
Ni^{2+}	ML/M.L	7.17	7.33 −0.01	7.08 ±0.03	(−9)[a]	(2)[b]
	$ML_2/M.L^2$	12.79	13.09 ±0.09	12.55 ±0.04	(−17)[a]	(0)[b]
	$ML_3/M.L^3$		15.1[c]	15.2	(−21)[a]	(−1)[b]
Cu^{2+}	ML/M.L	10.33	10.49 ±0.07	10.21 ±0.03	−11.5	9
	$ML_2/M.L^2$	18.93	19.2 ±0.2	18.71 ±0.01	−23.7	8
Zn^{2+}	ML/M.L			5.34	(−5)	(8)[b]
	$ML_2/M.L^2$			9.71	(−9)[a]	(14)[b]
Cd^{2+}	ML/M.L		5.47[d]			
	$ML_2/M.L^2$		9.56[d]			
	$ML_3/M.L^3$		11.4[d]			
	MHL/M.HL		1.5[d]			
	MOHL/ML.OH		9.71[d]			

[a] 10-40°, 0;   [b] 25°, 0;   [c] 25°, 0.65;   [d] 25°, 1.0

Bibliography:

H^+	50E,52BM,59MB,67VA,70NK	Zn^{2+}	59MB
Ni^{2+}	52BMa,59MB,70NK	Cd^{2+}	73CV
Cu^{2+}	52BMa,59MB,70NK,72BFa		

$$CH_3CH_2NHCH_2CH_2NH_2$$

$C_4H_{12}N_2$ N-Ethylethylenediamine   L

Metal ion	Equilibrium	Log K 25°, 0.1	Log K 25°, 0.5	Log K 25°, 0
H^+	HL/H.L	10.16	10.27 +0.09	(10.15)
	$H_2L/HL.H$	7.10	7.33 −0.10	6.86
Ni^{2+}	ML/M.L	(6.52)	6.78[a]−0.1	6.51
	$ML_2/M.L^2$		12.08[a]	11.35
	$ML_3/M.L^3$		14.1[a]	
Cu^{2+}	ML/M.L	(9.69)	10.19[a]−0.3	9.67
	$ML_2/M.L^2$		18.57[a]	17.43

[a] 25°, 0.65

Bibliography:

H^+ 50E,52BM,69NT Ni^{2+},Cu^{2+} 52BMa,69NT

$$CH_3CH_2CH_2NHCH_2CH_2NH_2$$

$C_5H_{14}N_2$ N-Propylethylenediamine   L

Metal ion	Equilibrium	Log K 25°, 0.65
H^+	HL/H.L	10.19
	$H_2L/HL.H$	(7.39)
Ni^{2+}	ML/M.L	6.60
	$ML_2/M.L^2$	11.76
	$ML_3/M.L^3$	13.8
Cu^{2+}	ML/M.L	9.98
	$ML_2/M.L^2$	18.14

Bibliography:

H^+ 52BM Ni^{2+}, Cu^{2+} 52BMa

$$H_3C\diagdown$$
$$\qquad CHNHCH_2CH_2NH_2$$
$$H_3C\diagup$$

$C_5H_{14}N_2$ N-2-Propylethylenediamine   L

Metal ion	Equilibrium	Log K 25°, 0.65
H^+	HL/H.L	(10.47)
	$H_2L/HL.H$	(7.55)
Ni^{2+}	ML/M.L	5.17
	$ML_2/M.L^2$	8.64
Cu^{2+}	ML/M.L	9.07
	$ML_2/M.L^2$	16.52

Bibliography:

H^+ 52BM Ni^{2+}, Cu^{2+} 52BMa

$$H_3C \diagdown$$
$$CHNHCH_2CNH_2$$

(structure: 1-(2-Propylamino)-2-methyl-2-propylamine)

| $C_7H_{18}N_2$ | 1-(2-Propylamino)-2-methyl-2-propylamine | | | L |

Metal ion	Equilibrium	Log K 25°, 0.1	Log K 25°, 0.5	Log K 25°, 0
H^+	HL/H.L	10.07	10.21	10.03
	H_2L/HL.H	6.60	6.82	6.36
Cu^{2+}	ML/M.L	8.75	8.93	8.70
	$ML_2/M.L^2$			15.62

Bibliography:

H^+ 72NTE Cu^{2+} 72NTL

$$CH_3CH_2CH_2CH_2NHCH_2CH_2NH_2$$

| $C_6H_{16}N_2$ | N-Butylethylenediamine | L |

Metal ion	Equilibrium	Log K 25°, 0.65
H^+	HL/H.L	10.15
	H_2L/HL.H	(7.38)
Ni^{2+}	ML/M.L	6.73
	$ML_2/M.L^2$	12.29
	$ML_3/M.L^3$	14.5
Cu^{2+}	ML/M.L	9.94
	$ML_2/M.L^2$	18.21

Bibliography:

H^+ 52BM Ni^{2+}, Cu^{2+} 52BMa

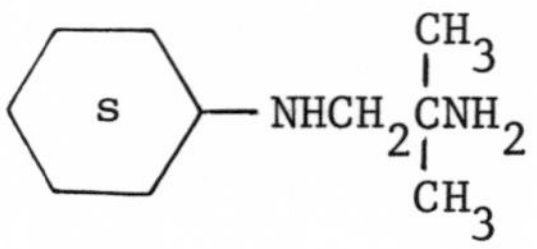

$$\text{[cyclohexyl ring]} - \text{NHCH}_2\underset{\underset{\text{CH}_3}{|}}{\overset{\overset{\text{CH}_3}{|}}{\text{C}}}\text{NH}_2$$

$C_{10}H_{22}N_2$ <u>1-(Cyclohexylamino)-2-methyl-2-propylamine</u>   L

Metal ion	Equilibrium	Log K 20°, 0.1
H^+	HL/H.L	10.18
	$H_2L/HL.H$	6.77
Ni^{2+}	$ML_2/M.L^2$	9.6
	$ML_3/M.L^3$	13.5
Cu^{2+}	ML/M.L	9.01
	$ML_2/M.L^2$	16.15

Bibliography: 65TS

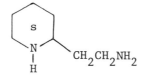

$C_6H_{14}N_2$ DL-2-(Aminomethyl)piperidine   L

Metal ion	Equilibrium	Log K 25°, 0	ΔH 25°, 0	ΔS 25°, 0
H^+	HL/H.L	9.58	$(-10)^a$	(10)
	$H_2L/HL.H$	6.23	$(-8)^a$	(2)
Co^{2+}	ML/M.L	4.86	$(-4)^a$	(9)
	$ML_2/M.L^2$	(9.2)	$(-10)^a$	(9)
Ni^{2+}	ML/M.L	6.29	$(-7)^a$	(5)
	$ML_2/M.L^2$	(11.36)	$(-15)^a$	(2)
	$ML_3/M.L^3$	(14.7)	$(-19)^a$	(4)
Cu^{2+}	ML/M.L	9.73	$(-10)^a$	(11)
	$ML_2/M.L^2$	(17.86)	$(-24)^a$	(1)

[a] 10-40°, 0

Bibliography: 63HG

$C_7H_{16}N_2$ DL-2-(2-Aminoethyl)piperidine   L

Metal ion	Equilibrium	Log K 20°, 0	Log K 25°, 0	ΔH 25°, 0	ΔS 25°, 0
H^+	HL/H.L	10.37	10.24	$(-11)^a$	(10)
	$H_2L/HL.H$	8.33	8.20	$(-10)^a$	(4)
Ni^{2+}	ML/M.L	5.28			
	$ML_2/M.L^2$	(9.59)			
Cu^{2+}	ML/M.L	8.76			
	$ML_2/M.L^2$	(14.7)			

[a] 10-40°, 0

Bibliography: 63HG

$$CH_3NHCH_2CH_2NHCH_2$$

$C_4H_{12}N_2$ N,N'-Dimethylethylenediamine   L

Metal ion	Equilibrium	Log K 25°, 0.1	Log K 25°, 0.5	Log K 25°, 0	ΔH 25°, 0.5	ΔS 25°, 0.5
H^+	HL/H.L	9.98 ±0.10	10.17 ±0.09	(10.03)	−10.74	10.5
	H_2L/HL.H	7.01 +0.06	7.30 ±0.06	6.80	−9.67	1.0
Ni^{2+}	ML/M.L	6.89 −0.2	7.05 ±0.02	6.84		
	ML_2/M.L^2	10.83 −0.3	11.19 +0.6	10.69		
	ML_3/M.L^3		13.3[a]			
Cu^{2+}	ML/M.L	10.02 −0.3	10.15 ±0.05	9.96	−11.10	9.2
	ML_2.M.L^2	17.05 −0.7	17.36 ±0.10	10.60	−20.77	9.8
	ML/MOHL.H	8.11 ±0.02		7.89		
	ML/M(OH)$_2$L.H^2	(18.2)	19.11	18.70	−19.4	22
	$(MOHL)_2$/$(MOHL)^2$	3.8		3.69		
	$(ML)^2$/$(MOHL)_2$.H^2	12.4	12.40	12.09	−15.6	4
Zn^{2+}	ML/M.L		5.51[b]			
	ML_2/M.L^2		9.73[b]			
Cd^{2+}	ML/M.L		5.20[c]			
	ML_2/M.L^2		8.74[c]			
	ML_3/M.L^3		10.59[c]			
	$MOHL_2$/ML_2.OH		10.94[c]			

---

[a] 25°, 0.65;   [b] 25°, 0.46;   [c] 25°, 1.0

Bibliography:

H^+ 53BM,54IG,59GM,63L,66NK,71PB Zn^{2+} 63L
Ni^{2+} 54BM,54IG,66NK Cd^{2+} 73CV
Cu^{2+} 54BM,54IG,59GM,66NK,72BFa

$$CH_3CH_2NHCH_2CH_2NHCH_2CH_3$$

$C_6H_{16}N_2$ <u>N,N'-Diethylethylenediamine</u>   L

Metal ion	Equilibrium	Log K 25°, 0.1	Log K 25°, 0.5	Log K 25°, 0
H^+	HL/H.L	10.10 +0.1	10.27 +0.1	(10.16)
	H_2L/HL.H	7.24 +0.08	7.52 ±0.03	6.96
Ni^{2+}	ML/M.L	5.58 +0.09	5.76 −0.2	5.50
	$ML_2/M.L^2$	8.02	8.37 +0.5	7.86
Cu^{2+}	ML/M.L	8.85	9.05 +0.2	8.79
	$ML_2/M.L^2$		(15.62)[a]	14.36
Zn^{2+}	ML/M.L		5.47[b]	
	$ML_2/M.L^2$		9.0[b]	

[a] 25°, 0.65;   [b] 25°, 0.46

Bibliography:

H^+	53BM,63L,63NMc,72TR	Cu^{2+}	54BM,63NMc
Ni^{2+}	54BM,65NKT,72TR	Zn^{2+}	63L

$$CH_3CH_2CH_2NHCH_2CH_2NHCH_2CH_2CH_3$$

$C_8H_{20}N_2$ N,N'-Dipropylethylenediamine   L

Metal ion	Equilibrium	Log K 25°, 0.65
H^+	HL/H.L	10.12
	$H_2L/HL.H$	7.38
Ni^{2+}	ML/M.L	5.52
	$ML_2/M.L^2$	8.0
Cu^{2+}	ML/M.L	8.79
	$ML_2/M.L^2$	14.34

Bibliography:

H^+ 53BM Ni^{2+}, Cu^{2+} 54BM

$$CH_3CH_2CH_2CH_2NHCH_2CH_2NHCH_2CH_2CH_2CH_3$$

$C_{10}H_{24}N_2$ N,N'-Dibutylethylenediamine   L

Metal ion	Equilibrium	Log K 25°, 0.65
H^+	HL/H.L	10.04
	$H_2L/HL.H$	7.31
Ni^{2+}	ML/M.L	5.43
Cu^{2+}	ML/M.L	8.67
	$ML_2/M.L^2$	13.51

Bibliography:

H^+ 53BM Ni^{2+}, Cu^{2+} 54BM

$C_4H_{10}N_2$		Piperazine	(perhydro-1,4-diazine)			L
Metal ion	Equilibrium	Log K 25°, 0.1	Log K 20°, 0.1	Log K 25°, 0	ΔH 25°, 0.1	ΔS 25°, 0.1
H^+	HL/H.L	9.71 ±0.02	9.84	9.731	-10.25[a]	10.2[a]
					-10.3 ±0.1	10
	$H_2L/HL.H$	5.59 +0.01	5.68	5.333	-7.43[a]	-0.5[a]
					-7.0 ±0.1	2
Ag^+	ML/M.L	3.33 -0.01	3.40		-6.41	-6.3
	$ML_2/M.L^2$	6.04			-10.45	-7.4
	MHL/ML.H		8.17			
	$M_2L/ML.M$		1.5			

[a] 25°, 0

Bibliography:

H^+ 52SM,63PC,68HR,72EH Ag^+ 52SM,73HB

$C_5H_{12}N_2$		2-Methylpiperazine		L
Metal ion	Equilibrium	Log K 25°, 0.1	ΔH 25°, 0.1	ΔS 25°, 0.1
H^+	HL/H.L	9.49	-10.0	10
	$H_2L/HL.H$	5.51	-6.7	3
Ag^+	ML/M.L	3.44	-6.29	-5.4
	$ML_2/M.L^2$	6.18	-10.10	-5.6

Bibliography:

H^+ 72EH Ag^+ 73HB

C$_5$H$_{12}$N$_2$ Perhydro-1,4-diazepine   (homopiperazine)   L

Metal ion	Equilibrium	Log K 25°, 0	ΔH 25°, 0	ΔS 25°, 0
H$^+$	HL/H.L	10.26	(-10)[a]	(13)
	H$_2$L/HL.H	6.6	(-8)[a]	(3)
Cu^{2+}	ML/M.L	7.8	(-6)[a]	(16)
	ML$_2$/M.L^2	13.8	(-11)[a]	(26)

[a] 10-40°, 0

Bibliography: 61PG

CH$_3$NHCH$_2$CH$_2$CH$_2$NHCH$_3$

C$_5$H$_{14}$N$_2$ N,N'-Dimethyltrimethylenediamine   L

Metal ion	Equilibrium	Log K 25°, 0.1	Log K 25°, 0.5	Log K 25°, 0
H$^+$	HL/H.L	10.64	10.80 +0.01	10.60
	H$_2$L/HL.H	8.81	9.10 +0.01	8.52
Ni^{2+}	ML/M.L		5.1	
Cu^{2+}	ML/M.L		8.38	

Bibliography:
H$^+$ 72KK,74KZ   Ni^{2+},Cu^{2+} 74KZ

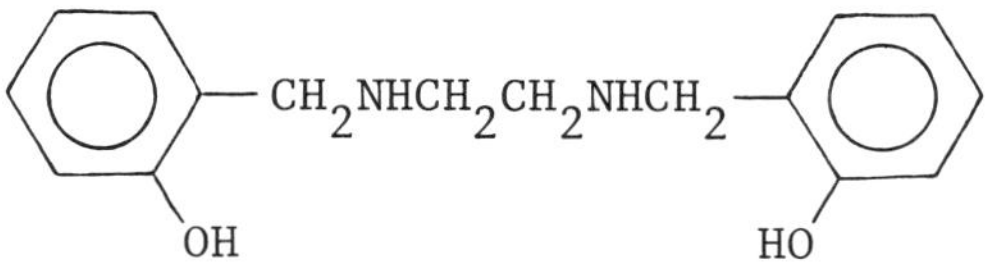

$C_{16}H_{20}O_2N_2$ Ethylenebis(iminomethylene-2-phenol) H_2L

                                  (N,N'-bis(2-hydroxybenzyl)ethylenediamine)

Metal ion	Equilibrium	Log K 25°, 0.1
H^+	HL/H.L	10.50
	$H_2L/HL.H$	9.80
	$H_3L/H_2L.H$	8.37
	$H_4L/H_3L.H$	6.17
Co^{2+}	ML/M.L	12.78
Cu^{2+}	ML/M.L	20.5
Zn^{2+}	ML/M.L	11.97

Bibliography: 68G

$$\begin{array}{ccc}
\text{H}_3\text{C}\;\;\text{CH}_3 & & \text{H}_3\text{C}\;\;\text{CH}_3 \\
| \quad | & & | \quad | \\
\text{HON=C-CNHCH}_2\text{CH}_2\text{NHC-C=NOH} \\
| & & | \\
\text{CH}_3 & & \text{H}_3\text{C}
\end{array}$$

$C_{12}H_{26}O_2N_4$ Ethylenebis[imino-3-(3-methylbutan-2-one oxime)]   L

(3,3,8,8-tetramethyl-4,7-diazadecane-2,9-dione dioxime)

Metal ion	Equilibrium	Log K 24°, 0.27	Log K 23°, 1.0
H^+	HL/H.L	8.21	8.31
	H_2L/HL.H	5.45	5.91
Ni^{2+}	ML/M.L	10.1	10.1
	ML/M(H_{-1}L).H	2.7	2.50
	M(H_{-1}L)/M(H_{-2}L).H	11.7	
Cu^{2+}	ML/M.L	13.0	
	ML/M(H_{-1}L).H	4.2	
	M(H_{-1}L)/M(H_{-2}L).H	9.7	

Bibliography:
H^+, Ni^{2+} 58M,62M Cu^{2+} 58M

$$\begin{array}{ccc}
\text{H}_3\text{C}\quad\;\text{CH}_3 & & \text{H}_3\text{C}\quad\;\text{CH}_3 \\
| \qquad | & & | \qquad | \\
\text{HON=CCH}_2\text{CNHCH}_2\text{CH}_2\text{NHCCH}_2\text{C=NOH} \\
| & & | \\
\text{CH}_3 & & \text{H}_3\text{C}
\end{array}$$

$C_{14}H_{30}O_2N_4$ Ethylenebis[imino-4-(4-methylpentan-2-one oxime)]   L

(4,4,9,9-tetramethyl-5,8-diazadodecane-2,11-dione dioxime)

Metal ion	Equilibrium	Log K 25°, 0.1	Log K 25°, 0.2	Log K 25°, 0	ΔH 25°, 0	ΔS 25°, 0
H^+	HL/H.L	9.41	9.47	9.35	-9.5	11
	H_2L/HL.H	6.35	6.46	5.89	-9.4	-5
Cu^{2+}	ML/M.L	13.24			-13.1[a]	17[a]
	ML/M(H_{-1}L).H	3.23			-5.7[a]	-4[a]

---

[a] 25°, 0.1

Bibliography: 72FH,74HP

$$HOCH_2CH_2NHCH_2CH_2NH_2$$

$C_4H_{12}ON_2$ 2-(2-Aminoethylamino)ethanol   (N-(2-hydroxyethyl)ethylenediamine)   L

Metal ion	Equilibrium	Log K 25°, 0.1	Log K 25°, 0.5	Log K 25°, 0	ΔH 25°, 0.1	ΔS 25°, 0.1
H+	HL/H.L	9.59 ±0.03	9.74 −0.01	9.56		
	H_2L/HL.H	6.60 −0.01	6.85 ±0.1	6.34		
Co2+	ML/M.L		(4.87)			
	ML_2/M.L^2		(9.9)			
	ML/MOHL.H		(9.07)			
Ni2+	ML/M.L	6.82	6.97 −0.3	6.76		
	ML_2/M.L^2	12.44	12.80 −0.3	12.28		
Cu2+	ML/M.L	10.09 ±0.02	10.33	10.02		
	ML_2/M.L^2	17.62 ±0.04	18.09 −0.1	17.45		
	ML/MOHL.H	7.32 ±0.02	7.51	7.09	(−6)[a]	(13)
	MOHL/M(OH)$_2$L.H	9.94		9.84	(−9)[a]	(15)
	$(MOHL)_2$/$(MOHL)^2$	2.15 ±0.04	2.29	1.87	(−4)[a]	(−3)
Zn2+	ML/M.L		(4.75)			
	ML_2/M.L^2		(10.2)			
	ML/MOHL.H		(8.05)			
Cd2+	ML/M.L		(4.93)			
	ML_2/M.L^2		(9.2)			

_____

[a] 0-43°, 0.1

Bibliography:
H+   62H,66NT,69CM,72HJ   Cu2+   57MC,59CG,59GM,66NT,72HJ
Co2+,Zn2+,Cd2+ 60HD   Other references: 50E,58HD
Ni2+   60HD,66NT

$$\begin{array}{c} CH_3 \\ | \\ HOCHCH_2NHCH_2CH_2NH_2 \end{array}$$

$C_5H_{14}ON_2$ DL-1-(2-Aminoethylamino)-2-propanol   L
                                    (N-(2-hydroxypropyl)ethylenediamine)

Metal ion	Equilibrium	Log K 25°, 0.5
H^+	HL/H.L	9.77
	H_2L/HL.H	6.93
Cu^{2+}	ML/M.L	10.40
	ML_2/M.L^2	(17.80)
	ML/MOHL.H	7.59
	MOHL/M(OH)$_2$L.H	10.08

Bibliography:

H^+	62H	Other reference: 50E
Cu^{2+}	60HD	

$$HOCH_2CH_2NHCH_2CH_2CH_2NH_2$$

$C_5H_{14}ON_2$ <u>2-(3-Aminopropylamino)ethanol</u>   L

(N-(2-hydroxyethyl)trimethylenediamine)

Metal ion	Equilibrium	Log K 25°, 0.1	Log K 25°, 0.5	Log K 25°, 0
H^+	HL/H.L	10.22	10.35 -0.09	10.19
	H_2L/HL.H	8.21	8.47 -0.04	7.95
Co^{2+}	ML/M.L		4.76	
	ML_2/M.L^2		7.98	
	ML_3/M.L^3		8.91	
Cu^{2+}	ML/M.L			10.42
	ML/MOHL.H			7.07
	MOHL/M(OH)$_2$L.H			10.17
	$(MOHL)_2$/$(MOHL)^2$			2.59

Bibliography:
H^+ 63NML,70ML Cu^{2+} 64NML

Co^{2+} 70ML

$$HOCH_2CH_2NHCH_2CH_2NHCH_2CH_2OH$$

$C_6H_{16}O_2N_2$ Ethylenediiminodi-2-ethanol   L

(N,N'-bis(2-hydroxyethyl)ethylenediamine)

Metal ion	Equilibrium	Log K 25°, 0.1	Log K 25°, 0.5	ΔH 25°, 0.1	ΔS 25°, 0.1
H^+	HL/H.L	9.24	9.32 −0.07		
	$H_2L/HL.H$	6.26	6.52 −0.01		
Co^{2+}	ML/M.L		5.13		
	$ML_2/M.L^2$		9.1		
Ni^{2+}	ML/M.L		6.67		
	$ML_2/M.L^2$		10.9		
Cu^{2+}	ML/M.L	9.68 −0.1	9.77		
	$ML_2/M.L^2$		15.5 ±0.1		
	ML/MOHL.H	7.15 −0.1	7.18	(−7)[a]	(9)
	$MOHL/M(OH)_2L.H$	9.37		(−6)[a]	(23)
	$(MOHL)_2/(MOHL)^2$	1.38		(−1)[a]	(3)
Zn^{2+}	ML/M.L		4.79		
	$ML_2/M.L^2$		9.1		
	ML/MOHL.H		8.23		
Cd^{2+}	ML/M.L		5.07		
	$ML_2/M.L^2$		8.9		

---

[a] 0-43°, 0.1

Bibliography:
H^+ 59GM,62H,72MP Cu^{2+} 57MC,59GM,60HD,72HJ
$Co^{2+},Ni^{2+},Zn^{2+},Cd^{2+}$ 60HD

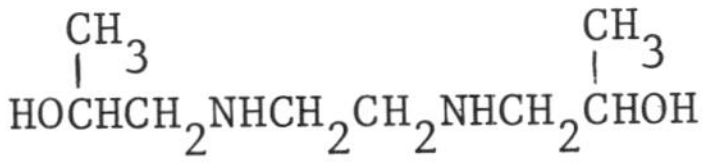

$$\begin{array}{cc} & CH_3 & & CH_3 \\ & | & & | \\ HOCHCH_2NHCH_2CH_2NHCH_2CHOH \end{array}$$

$C_8H_{20}O_2N_2$

<u>DL-Ethylenediiminodi-1-(2-propanol)</u>

<u>(N,N'-bis(2-hydroxypropyl)ethylenediamine)</u>

L

Metal ion	Equilibrium	Log K 25°, 0.5
H$^+$	HL/H.L	9.56
	H$_2$L/HL.H	6.26
Co^{2+}	ML/M.L	5.02
	ML$_2$/M.L^2	(9.5)
Ni^{2+}	ML/M.L	6.84
	ML$_2$/M.L^2	(10.9)
Cu^{2+}	ML/M.L	11.01
	ML$_2$/M.L^2	(16.64)
	ML/MOHL.H	7.34
	MOHL/M(OH)$_2$L.H	9.58
Zn^{2+}	ML/M.L	5.12
	ML$_2$/M.L^2	(9.57)
Cd^{2+}	ML/M.L	5.33
	ML$_2$/M.L^2	(8.6)

Bibliography:

H$^+$ 62H

Co^{2+}-Cd^{2+} 60HD

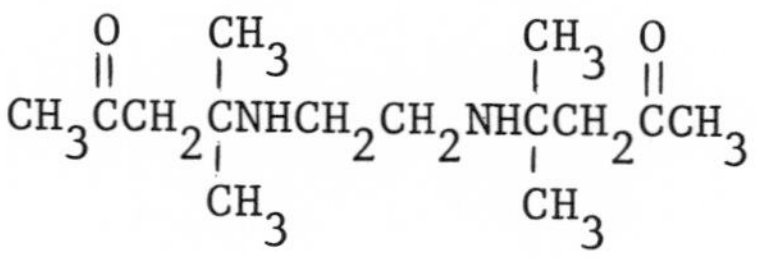

$$\underset{\substack{\text{CH}_3}}{\text{CH}_3}\overset{\substack{\text{O}\\||}}{\text{CCH}_2}\underset{\substack{|\\\text{CH}_3}}{\overset{\substack{\text{CH}_3\\|}}{\text{C}}}\text{NHCH}_2\text{CH}_2\text{NH}\underset{\substack{|\\\text{CH}_3}}{\overset{\substack{\text{CH}_3\\|}}{\text{C}}}\text{CH}_2\overset{\substack{\text{O}\\||}}{\text{C}}\text{CH}_3$$

$C_{14}H_{28}O_2N_2$ <u>Ethylenebis[imino-4-(4-methylpentan-2-one)]</u>   L

<u>(4,4,9,9-tetramethyl-5,8-diazadodecane-2,11-dione)</u>

Metal ion	Equilibrium	Log K 25°, 0.1
H^+	HL/H.L	9.41
	H_2L/HL.H	6.19
Cu^{2+}	ML/M.L	8.56

Bibliography: 74HP

B. SECONDARY AMINES

$$H_2NCH_2CH_2NHCH_2CH_2NH_2$$

$C_4H_{13}N_3$ <u>1,4,7-Triazaheptane</u>   (iminobis(2-ethylamine))   L

(diethylenetriamine, <u>dien</u>)

Metal ion	Equilibrium	Log K 25°, 0.1	Log K 20°, 0.1	Log K 25°, 0	ΔH 25°, 0.1	ΔS 25°, 0.1
H[+]	HL/H.L	9.84 ±0.05 9.88[a]−0.04	9.98	9.80 9.92[b]±0.03	−11.2	7
	H$_2$L/HL.H	9.02 ±0.06 9.09[a]−0.03	9.17	8.74 9.24[b]±0.00	−12.0	1
	H$_3$L/H$_2$L.H	4.23 ±0.03 4.47[a]±0.03	4.32	3.64 4.73[b]±0.00	−7.2	−5
Cr[2+]	ML/M.L		6.61			
	ML$_2$/M.L^2		9.34			
Mn[2+]	ML/M.L			3.99[c]		
	ML$_2$/M.L^2			6.91[b]	−7.0	8[b]
Fe[2+]	ML/M.L			6.23[c]		
	ML$_2$/M.L^2			10.53[b]	−13.0	5[b]
Co[2+]	ML/M.L	8.0 +0.2	8.1	8.57[b]	−8.2	8
	ML$_2$/M.L^2	13.9	14.1	14.77[b]	−18.4	2
Ni[2+]	ML/M.L	10.5 10.7[a]	10.7	10.96[b]	−11.9	8
	ML$_2$/M.L^2	18.6 18.9[a]	18.9	19.27[b]	−25.3	0
Cu[2+]	ML/M.L	15.9 ±0.1	16.1	15.6 16.34[b]−0.1	−18.0	12
	ML$_2$/M.L^2	20.9 +0.1	21.2		−26.2	8
	MHL/ML.H	3.2				
	MHL$_2$/ML$_2$.H	8.2				
	MOHL/ML.OH	4.6 ±0.1	4.6		−2.7	12

[a] 25°, 0.5;   [b] 25°, 1.0;   [c] 30°, 1.0

Diethylenetriamine   (continued)

Metal ion	Equilibrium	Log K 25°, 0.1	Log K 20°, 0.1	Log K 25°, 0	ΔH 25°, 0.1	ΔS 25°, 0.1
Ag^+	ML/M.L		6.1			
	MHL/ML.H		7.0			
	$M_2L/ML.M$		1.4			
Pd^{2+}	$MHL_2/ML_2.H$			6.1^b		
	$MH_2L_2/MHL_2.H$			2.5^b		
Zn^{2+}	ML/M.L	8.8	8.9	8.7	−6.5	18
				9.22^b		
	$ML_2/M.L^2$	14.3	14.5		−16.6	10
Cd^{2+}	ML/M.L	$(8.05)^a$	8.4			
	$ML_2/M.L^2$	13.84^a	13.8			
Hg^{2+}	ML/M.L			21.8^d		
	$ML_2/M.L^2$	25.02		29^d	$(−36)^e$	$(−6)$
	MHL/ML.H			3.9^d		
	MOHL/ML.OH			6.3^d		
Pb^{2+}	ML/M.L	8.50				
	$ML_2/M.L^2$	10.37				

---

[a] 25°, 0.5;   [b] 25°, 1.0;   [d] 20°, 0.5;   [e] 10−40°, 0.1

Bibliography:

H^+ 50JLM,50PSa,59MB,61CP,65PG,67SP,69ES, 69RJ,71AA,71HBM,72NMa

Cr^{2+} 65PG

Mn^{2+},Fe^{2+} 52JH,61CPS

Co^{2+} 50PSa,52JH,61CPS,72NMa

Ni^{2+} 50JLR,50PSa,61CPS,70MAb

Cu^{2+} 50JLR,50PSa,57MC,59MB,61CPS,69ES, 71AA,71HBM,73YB

Ag^+,Hg^{2+} 50PSa

Pd^{2+} 69RJ

Zn^{2+} 50PSa,52JH,59MB,61CPS

Cd^{2+} 50PSa,67SP

Hg^{2+} 50PSa,55NR,61RM

Pb^{2+} 69IM,74K

Other references: 49LO,50DL,69MI,70F

$$CH_3NHCH_2CH_2NHCH_2CH_2NH_2$$

$C_5H_{15}N_3$ 1,4,7-Triazaoctane   (N(1)-methyldiethylenetriamine)   L

Metal ion	Equilibrium	Log K 25°, 0.13
H⁺	HL/H.L	9.86
	$H_2L/HL.H$	9.18
	$H_3L/H_2L.H$	3.30
Cu²⁺	ML/M.L	15.32
	ML/MOHL.H	8.72

Bibliography: 71AA

$$H_2NCH_2CH_2NHCH_2CH_2CH_2NH$$

$C_5H_{15}N_3$ 1,4,8-Triazaoctane   (N-3-aminopropylethylenediamine)   L

Metal ion	Equilibrium	Log K 25°, 0.5
H⁺	HL/H.L	10.03
	$H_2L/HL.H$	9.11
	$H_3L/H_2L.H$	6.18
Ni²⁺	ML/M.L	10.7
	$ML_2/M.L^2$	17.8

Bibliography: 70WB   Other reference: 73AH

$$H_2NCH_2CH_2CH_2NHCH_2CH_2CH_2NH_2$$

$C_6H_{17}N_3$ 　　　　1,5,9-Triazanonane　　(iminobis(3-propylamine))　　　　　　L

Metal ion	Equilibrium	Log K 25°, 0.1	Log K 20°, 0.1	Log K 25°, 1.0	ΔH 25°, 0.1	ΔS 25°, 0.1
H^+	HL/H.L	10.65	10.81	10.76	−12.3	7
	$H_2L/HL.H$	9.57	9.73	9.80	−13.0	0
	$H_3L/H_2L.H$	7.72	7.85	8.19	−10.5	0
Co^{2+}	ML/M.L	6.92	7.02	7.01	−7.8	6
Ni^{2+}	ML/M.L	9.19	9.32	9.58	−10.6	7
	$ML_2/M.L^2$	12.74	12.96		−17.6	−1
Cu^{2+}	ML/M.L	14.20	14.40	14.73	−16.1	11
	ML/MOHL.H	9.7	9.8		−10.2	10
Zn^{2+}	ML/M.L	7.92	7.99		−5.4	18
	ML/MOHL.H	8.6	8.7		−8.7	10

Bibliography:
H^+-Cu^{2+} 56HF,66PN,66VA　　　　　Other reference: 73AH
Zn^{2+} 66PN,66VA

$$H_2NCH_2CH_2NHCH_2CH_2NHCH_2CH_2NH_2$$

$C_6H_{18}N_4$ <u>1,4,7,10-Tetraazadecane</u>   (<u>triethylenetetramine, trien</u>)   L

Metal ion	Equilibrium	Log K 25°, 0.1	Log K 20°, 0.1	Log K 25°, 1.0	ΔH 25°, 0.1	ΔS 25°, 0.1
H^+	HL/H.L	9.74 ±0.06 9.87[a]±0.09	9.92	10.02	-11.0	8
	H_2L/HL.H	9.08 ±0.02 9.21[a]±0.1	9.20	9.39	-11.3	4
	H_3L/H_2L.H	6.56 ±0.02 6.87[a]±0.01	6.67	7.00	-9.5	-2
	H_4L/H_3L.H	3.25 ±0.03 3.71[a]±0.05	3.32	4.00	-6.8	-8
Cr^{2+}	ML/M.L	7.9[b]	7.71			
Mn^{2+}	ML/M.L	4.90	4.93	5.46	-2.3	15
Fe^{2+}	ML/M.L	7.76	7.84	8.39	-6.1	15
Co^{2+}	ML/M.L	10.95 -0.5	11.09	11.35	-10.7	14
	MHL/ML.H		5.7			
Ni^{2+}	ML/M.L	13.8 +0.3 14.4[a]	14.0	(14.5)	-14.0 +0.1	16
	ML_2/M.L^2	18.6[a]				
	M_2L_3/M^2.L^3	36.9[a]				
	MHL/ML.H		4.8			
Cu^{2+}	ML/M.L	20.1 ±0.0	20.4	(20.9)	-21.6 +0.2	20
	MHL/ML.H		3.5			
	ML/MOHL.H	10.8	10.8		-14	2
Ag^+	ML/M.L		7.65			
	MHL/ML.H		8.0			
	MH_2L/MHL.H		6.2			
	M_2L/ML.M		2.4			

[a] 25°, 0.5;   [b] 26°, 0.1

Trien   (continued)

Metal ion	Equilibrium	Log K 25°, 0.1	Log K 20°, 0.1	Log K 25°, 1.0	$\Delta \bar{H}$ 25°, 0.1	ΔS 25°, 0.1
Zn^{2+}	ML/M.L	12.03 −0.1	12.14	(12.05)	−8.9 +0.6	25
	MHL/ML.H		5.1			
Cd^{2+}	ML/M.L	10.63 +0.2	10.75	11.04	−9.2	18
	MHL/ML.H		6.2			
Hg^{2+}	ML.M.L	25.0		25.3[c]		
	MHL/ML.H			5.5[c]		
Pb^{2+}	ML/M.L	10.4				

---

[c] 20°, 0.5

Bibliography:

H^+ 50JM,50S,63MR,63PCV,70WB,72NM

Cr^{2+} 65PG

Mn^{2+},Fe^{2+} 50S,52JH,61SP

Co^{2+} 50S,52JH,61SP,72NMa

Ni^{2+} 50S,51JM,57RS,61SP,63MR,65WH,70WB,
              71MA

Cu^{2+},Zn^{2+} 50S,51JM,57RS,59CG,61SP,63PC,
              65WH

Ag^+ 50S

Cd^{2+} 50S,52JH,57RS,65WH

Hg^{2+} 50S,57RS

Pb^{2+} 57RS,74K

Other references: 50DL,57H,57JB,68LC,69IM,
                    69MI,71SL

B. SECONDARY AMINES

$$H_2NCH_2CH_2NHCH_2CH_2CH_2NHCH_2CH_2NH_2$$

$C_7H_{20}N_4$	1,4,8,11-Tetraazaundecane	(N,N'-bis(2-aminoethyl)trimethylenediamine)	L

Metal ion	Equilibrium	Log K 25°, 0.5	ΔH 25°, 0.5	ΔS 25°, 0.5
H^+	HL/H.L	10.25	-11.0	10
	$H_2L/HL.H$	9.50	-11.3	5
	$H_3L/H_2L.H$	7.28	-10.0	0
	$H_4L/H_3L.H$	6.02	-9.2	-3
Ni^{2+}	ML/M.L	16.4	-17.9	15
	$ML_2/M.L^2$	20.1	-22.0	18
Cu^{2+}	ML/M.L	23.9	-27.7	16
Zn^{2+}	ML/M.L	12.8	-11.9	19

Bibliography: 70WB,72FB

$$H_2NCH_2CH_2CH_2NHCH_2CH_2NHCH_2CH_2CH_2NH_2$$

| $C_8H_{22}N_4$ | 1,5,8,12-Tetraazadodecane | (N,N'-bis(3-aminopropyl)ethylenediamine) | | | L |

Metal ion	Equilibrium	Log K 25°, 0.1	Log K 25°, 0.5	Log K 25°, 0	ΔH 25°, 0.1	ΔS 25°, 0.1
H^+	HL/H.L	10.53		10.46	−12.35	6.8
			10.66		−12.23[a]	7.8[a]
	$H_2L/HL.H$	9.77		9.51	−12.38	3.2
			9.96		−12.8[a]	3[a]
	$H_3L/H_2L.H$	8.30		7.81	−10.32	3.4
			8.53		−10.7[a]	3[a]
	$H_4L/H_3L.H$	5.59		4.86	−8.16	−1.8
			5.84		−9.8[a]	−6[a]
Ni^{2+}	ML/M.L		14.69		−19.2[a]	3[a]
	MHL/ML.H		5.72		−5.0[a]	9[a]
Cu^{2+}	ML/M.L	21.69		21.22	−24.8	16
			21.83		−25.9[a]	13[a]
	MHL/ML.H	3.53		3.49	−4.3	2
			3.57		−4.4[a]	2[a]
Zn^{2+}	ML/M.L		11.25		−10.6[a]	16[a]
	MHL/ML.H		6.59		−9.3[a]	−1[a]
	MOHL/ML.OH		4.0		−2.0[a]	12[a]

[a] 25°, 0.5

Bibliography:
H^+ 71HP,73BFa,73HP Cu^{2+} 73BFa,73HP
Ni^{2+},Zn^{2+} 73BFa

$$H_2NCH_2CH_2CH_2NHCH_2CH_2CH_2NHCH_2CH_2CH_2NH_2$$

$C_9H_{24}N_4$ 1,5,9,13-Tetraazatridecane   L

(N,N'-bis(3-aminopropyl)trimethylenediamine)

Metal ion	Equilibrium	Log K 25°, 0.1	Log K 20°, 0.1	ΔH 25°, 0.1	ΔS 25°, 0.1
H$^+$	HL/H.L	10.46	10.61	−12.20	6.9
	H$_2$L/HL.H	9.82	9.98	−12.5	3
	H$_3$L/H$_2$L.H	8.54	8.69	−11.7	0
	H$_4$L/H$_3$L.H	7.22	7.35	−10.9	−4
Co^{2+}	ML/M.L		7.81		
Ni^{2+}	ML/M.L	10.48	10.65	−13.2	4
Cu^{2+}	ML/M.L	17.05	17.30	−19.5	13
Zn^{2+}	ML/M.L	9.32	9.41	−7.35	18
Cd^{2+}	ML/M.L		8.04		
Hg^{2+}	ML/M.L		20.49		

Bibliography:

H$^+$,Ni^{2+},Zn^{2+} 62TA,72BF   Co^{2+},Cd^{2+},Hg^{2+} 62TA

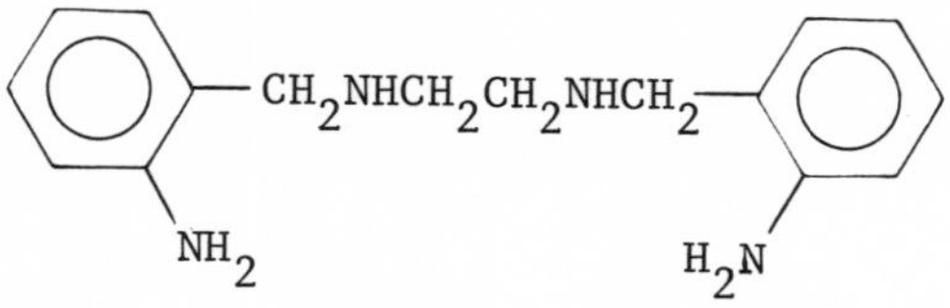

$$\text{CH}_2\text{NHCH}_2\text{CH}_2\text{NHCH}_2$$

$$\text{NH}_2 \qquad\qquad \text{H}_2\text{N}$$

$C_{16}H_{22}N_4$ <u>Ethylenebis(iminomethylene-2-phenylamine)</u>   L

<u>(N,N'-bis(2-aminobenzyl)ethylenediamine)</u>

Metal ion	Equilibrium	Log K 25°, 0.1
H^+	$HL/H.L$	9.00
	$H_2L/HL.H$	5.90
	$H_3L/H_2L.H$	2.32
	$H_4L/H_3L.H$	2.00
Co^{2+}	$ML/M.L$	(7.0)
Ni^{2+}	$ML/M.L$	10.00
Zn^{2+}	$ML/M.L$	7.17

Bibliography: 68G

$$H_2NCH_2CH_2NHCH_2CH_2NHCH_2CH_2NHCH_2CH_2NH_2$$

$C_8H_{23}N_5$ 1,4,7,10,13-Pentaazatridecane   L

(tetraethylenepentamine, tetren)

Metal ion	Equilibrium	Log K 25°, 0.1	Log K 20°, 0.1	ΔH 25°, 0.1	ΔS 25°, 0.1
H^+	HL/H.L	9.70 ±0.04	9.85	−10.8	8
	H_2L/HL.H	9.14 ±0.04	9.28	−11.3	4
	H_3L/H_2L.H	8.05 ±0.03	8.19	−10.7	1
	H_4L/H_3L.H	4.70 ±0.02	4.80	−7.9	−5
	H_5L/H_4L.H	2.97 ±0.07	3.06	−6.8	−9
Mn^{2+}	ML/M.L	6.55 +0.4	6.60	−3.7	18
Fe^{2+}	ML/M.L	9.85	9.96	−8.7	16
	MHL/ML.H	7.1			
Co^{2+}	ML/M.L	13.3 −0.1	13.5	−13.9	14
	MH_2L/ML.H	10.4			
Ni^{2+}	ML/M.L	17.4 ±0.4	17.6	−18.9 +0.5	16
	MHL/ML.H	4.1			
	MH_2L/MHL.H	4.0			
Cu^{2+}	ML/M.L	22.8 +0.1	23.1	−25.0 +1	20
	MHL/ML.H	5.2			
	MH_2L/MHL.H	3.8			
Ag^+	ML/M.L	7.4			
	MHL/ML.H	8.3			
	MH_2L/MHL.H	7.5			
	MH_3L/MH_2L.H	5.5			
Zn^{2+}	ML/M.L	15.1 +0.3	15.3	−13.9 −0.1	22
	MH_2L/ML.H^2	9.4			
Cd^{2+}	ML/M.L	14.0	14.2	−12.8	21
Hg^{2+}	ML/M.L	27.7			
Pb^{2+}	ML/M.L	10.5			

Tetraethylenepentamine (continued)

Bibliography:

H^+ 57JWa, 63PV,64PV,72NMa Cu^{2+},Zn^{2+} 58RH,63PV, 64PV,65WH

Mn^{2+} 58RH,63PV,64PV Ag^+ 73HT

Fe^{2+} 63PV,64PV Cd^{2+} 58RH,65WH

Co^{2+} 63PV,64PV,72NMa Hg^{2+},Pb^{2+} 58RH

Ni^{2+} 57JWa,58RH,63MR,63PV,64PV,65WH,71MA   Other references: 57JB,57JW,62JS

$$\begin{array}{c} CH_3CH_2 \\ \\ CH_3CH_2 \end{array} \!\!\! > \!\!\! NCH_2CH_3$$

| $C_6H_{15}N$ | | Triethylamine | | | | L |

Metal ion	Equilibrium	Log K 25°, 0.1	Log K 25°, 0.4	Log K 25°, 0	ΔH 25°, 0	ΔS 25°, 0
H^+	HL/H.L	10.75	10.77	10.715	-10.4 ±0.2	14
Ag^+	ML/M.L		2.6			
	$ML_2/M.L^2$		4.7			
Hg^{2+}	ML/M.L		7.8			
	$ML_2/M.L^2$		15			

Bibliography:

H^+ 50B,65PS,68CE,69CI Ag^+,Hg^{2+} 50B

C$_6$H$_{13}$N			N-Methylpiperidine				L

Metal ion	Equilibrium	Log K 23°, 0.2	Log K 25°, 0.5	Log K 25°, 0	ΔH 25°, 0	ΔS 25°, 0
H$^+$	HL/H.L	10.02	10.19	10.08	-9.44	14.5
					-9.91[a]	13.4[a]
Ag$^+$	ML/M.L		2.64			
	ML$_2$/M.L^2		3.8			

[a] 25°, 0.5

Bibliography:

H$^+$ 56CG,56ST,71CC,74BEB   Ag$^+$ 73BB

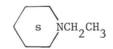

C$_7$H$_{15}$N		N-Ethylpiperidine			L

Metal ion	Equilibrium	Log K 25°, 0.5	ΔH 25°, 0.5	ΔS 25°, 0.5
H$^+$	HL/H.L	10.61	-10.39	13.7
Ag$^+$	ML/M.L	3.12		
	ML$_2$/M.L^2	5.20		

Bibliography:

H$^+$ 74BEB   Ag$^+$ 73BB

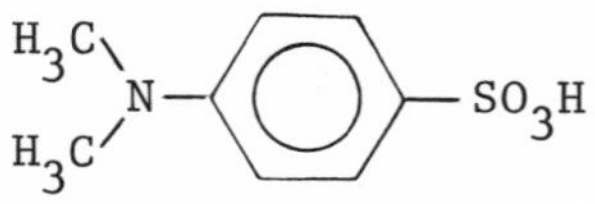

$$H_3C-N(CH_3)-C_6H_4-SO_3H$$

| $C_8H_{11}O_3NS$ | | 4-(Dimethylamino)benzenesulfonic acid | | HL |

Metal ion	Equilibrium	Log K 25°, 0.1	Log K 25°, 0
H^+	HL/H.L	3.37	3.60
Ag^+	ML/M.L	0.76	
	$ML_2/M.L^2$	1.36	

Bibliography: 58AC

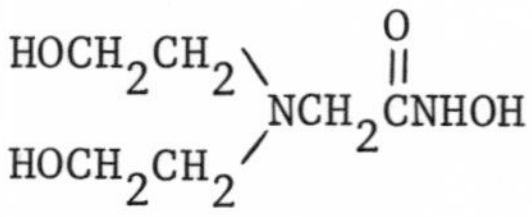

$$HOCH_2CH_2-N(CH_2CH_2OH)-NCH_2CNHOH$$

| $C_6H_{12}O_4N$ | | N,N-Bis(2-hydroxyethyl)aminoacethydroxamic acid | | HL |

Metal ion	Equilibrium	Log K 20°, 0.1
H^+	HL/H.L	8.97
	$H_2L/HL.H$	5.37
Cu^{2+}	ML/M.L	10.44
	$ML_2/M.L^2$	16.25
	$MOHL.L/ML_2.OH$	1.35
Fe^{3+}	MOHL/M.L.OH	26.17
	$M(OH)_2L/MOHL.OH$	9.50
	$M(OH)_3L/M(OH)_2L.OH$	4.70

Bibliography: 71KM

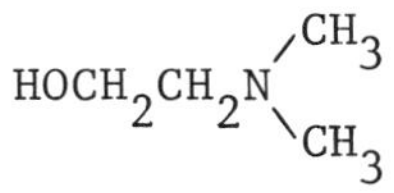

$$HOCH_2CH_2N{\nearrow}^{CH_3}_{\searrow CH_3}$$

$C_4H_{11}ON$ 2-(Dimethylamino)ethanol   L

Metal ion	Equilibrium	Log K 25°, 0.1	Log K 25°, 0.5	Log K 25°, 0	ΔH 25°, 0	ΔS 25°, 0
H^+	HL/H.L	9.29	9.49 −0.01	9.252 +0.01	−8.7	13
Cu^{2+}	ML/M.L	4.7				
	$ML_2/M.L^2$	8.7				
	$ML_3/M.L^3$	12.0				
	$ML_4/M.L^4$	14.9				
Ag^+	$ML_2/M.L^2$			3.80[a]±0.05		

[a] 20°, 0

Bibliography:
H^+ 69CI,70HH,73NK Ag^+ 61AL
Cu^{2+} 65D

$$HOCH_2CH_2N\begin{array}{c}CH_2CH_3\\CH_2CH_3\end{array}$$

$C_6H_{15}ON$ 2-(Diethylamino)ethanol   L

Metal ion	Equilibrium	Log K 25°, 0.1	Log K 25°, 0.5	Log K 25°, 0	ΔH 25°, 0	ΔS 25°, 0
H^+	HL.H.L	9.84	10.06	9.803+0.001	-9.69±0.03	12.4
Cu^{2+}	ML/M.L	4.9				
	$ML_2/M.L^2$	9.0				
	$ML_3/M.L^3$	12.2				
	$ML_4/M.L^4$	14.6				
Ag^+	$ML_2/M.L^2$			4.66^a±0.04		

---

[a] 20°, 0

Bibliography:
H^+ 68TE,69CI,73NK Ag^+ 61AL
Cu^{2+} 65D   Other references: 68HG,70UP

$$HOCH_2CH_2N\begin{array}{c}CH_3\\|\\CHCH_3\\\\CHCH_3\\|\\CH_3\end{array}$$

$C_8H_{19}ON$ 2-(Di-2-propylamino)ethanol   L

Metal ion	Equilibrium	Log K 20°, 0
H^+	HL/H.L	10.08
Ag^+	$ML_2/M.L^2$	4.0

Bibliography: 61AL

$$\begin{array}{c} HOCH_2CH_2 \\ \diagdown \\ NCH_3 \\ \diagup \\ HOCH_2CH_2 \end{array}$$

$C_5H_{13}O_2N$ N-Methyliminodi-2-ethanol   (N-methyldiethanolamine)   L

Metal ion	Equilibrium	Log K 25°, 0.1	Log K 25°, 0	ΔH 25°, 0	ΔS 25°, 0
H^+	HL/H.L	8.56	8.52	−9.2	8
Cu^{2+}	ML/M.L	4.9			
	$ML_2/M.L^2$	9.1			
	$ML_3/M.L^3$	12.5			
	$ML_4/M.L^4$	14.5			
Zn^{2+}	ML/M.L	4.3			
	$ML_2/M.L^2$	7.2			
	$ML_3/M.L^3$	9.1			
	$ML_4/M.L^4$	10.1			

Bibliography:
H^+ 59SG,65D Cu^{2+},Zn^{2+} 65D

$$HOCH_2CH_2 \diagdown$$
$$\qquad\qquad NCH_2CH_2OH$$
$$HOCH_2CH_2 \diagup$$

$C_6H_{15}O_3N$ Nitrilotri-2-ethanol   (triethanolamine)   L

Metal ion	Equilibrium	Log K 25°, 0.1	Log K 25°, 0.5	Log K 25°, 0	ΔH 25°, 0	ΔS 25°, 0
H^+	HL/H.L	7.8	7.90 −0.01	7.762	−8.1 ±0.1	8
			7.99^a	8.35^b		
Co^{2+}	ML/M.L		1.73			
Ni^{2+}	ML/M.L		2.27	2.92^c	-3.9^d	-3^d
	$ML_2/M.L^2$		3.09	4.74^c	-4.4^d	-1^d
Cu^{2+}	ML/M.L	3.9	4.23			
	$ML_2/M.L^2$	6.0				
Ag^+	ML/M.L		2.30			
	$ML_2/M.L^2$		3.64	4.23^e		
Cd^{2+}	ML/M.L	2.70				
	$ML_2/M.L^2$	4.60				
	$ML_3/M.L^3$	5.21				
Hg^{2+}	ML/M.L		6.90			
	$ML_2/M.L^2$		13.08			

[a] 25°, 1.0;   [b] 25°, 3.0;   [c] 30°, 1.0;   [d] 25°, 0.5;   [e] 20°, 0

Bibliography:

H^+ 47BR,60BA,65D,69CI,72BS,72VE Ag^+ 56BR,58AS

Co^{2+},Hg^{2+} 47BR Cd^{2+} 60MP

Ni^{2+} 63SG,72BS

Cu^{2+} 47BR,65D   Other references: 53BH,61ALa,62FH,63C,63Sa, 65MP,66SKa,69MP,70UR,71SSa,73K

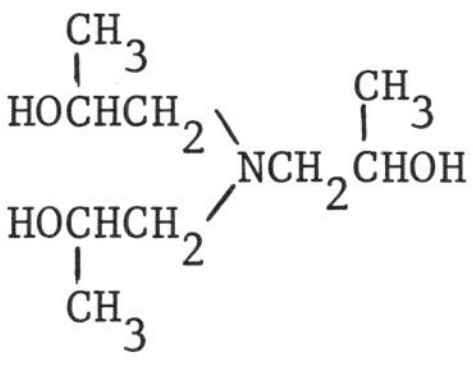

$$
\begin{array}{c}
CH_3 \\
| \\
HOCHCH_2 \searrow \qquad CH_3 \\
\qquad\qquad NCH_2CHOH \\
HOCHCH_2 \nearrow \\
| \\
CH_3
\end{array}
$$

$C_9H_{21}O_3N$ <u>DL-Nitrilotri-1-(2-propanol)</u>   (tris(2-hydroxypropyl)amine)   L

Metal ion	Equilibrium	Log K 25°, 0	ΔH 25°, 0	ΔS 25°, 0
H^+	HL/H.L	7.86 ±0.00	−8.9	6
Ag^+	ML/M.L	2.30[a]		
	$ML_2/M.L^2$	(4.27)[a]		

---

[a] 20°, 0

Bibliography:

H^+ 59SG,64AK Ag^+ 64AK

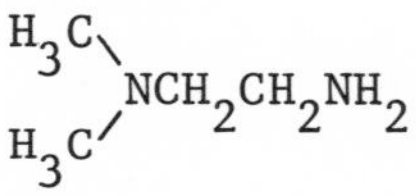

$$H_3C \diagdown$$
$$NCH_2CH_2NH_2$$
$$H_3C \diagup$$

$C_4H_{12}N_2$ N,N-Dimethylethylenediamine   L

Metal ion	Equilibrium	Log K 25°, 0.1	Log K 25°, 0.5	Log K 25°, 0	ΔH 25°, 0.5	ΔS 25°, 0.5
H^+	HL/H.L	9.50 ±0.08	9.69 +0.06	9.54	−10.42	9.4
	H_2L/HL.H	6.46 +0.06	6.69 +0.07	6.18	−8.42	2.4
Cu^{2+}	ML/M.L	9.19 ±0.04	9.33 ±0.05	9.08	−9.82	9.8
	$ML_2/M.L^2$	16.00 ±0.05	16.40 ±0.09	15.90	−19.3	10
	$ML/M(OH)_2L.H^2$		18.63		−20.9	15
	$(ML)^2/(MOHL)_2.H^2$		12.00		−15.6	3
Cd^{2+}	ML/M.L		4.81[a]			
	$ML_2/M.L^2$		8.11[a]			
	$ML_3/M.L^3$		9.41[a]			
	MOHL/ML.OH		8.69[a]			

---

[a] 25°, 1.0

Bibliography:
H^+ 54IG,67NKJ,71PB Cd^{2+} 73CV
Cu^{2+} 54IG,67NKJ,72BFa

$$CH_3CH_2\diagdown NCH_2CH_2NH_2$$
$$CH_3CH_2\diagup$$

$C_6H_{16}N_2$ <u>N,N-Diethylethylenediamine</u>   L

Metal ion	Equilibrium	Log K 25°, 0.1	Log K 25°, 0.5	Log K 25°, 0
H$^+$	HL/H.L	9.94 ±0.04	10.11 ±0.04	9.93
	H$_2$L/HL.H	6.81 ±0.04	7.05 ±0.03	6.51
Ni^{2+}	ML/M.L	4.57		
Cu^{2+}	ML/M.L	8.14 ±0.03	8.30	8.05
	ML$_2$/M.L^2	13.68 ±0.05	14.08	13.52

Bibliography:
H$^+$ 54IG,63NM,63NMb,72TR   Cu^{2+} 54IG,63NM,63NMb
Ni^{2+} 72TR

$C_6H_{14}N_2$ <u>N-(2-Aminoethyl)pyrrolidine</u>   L

Metal ion	Equilibrium	Log K 30°, 0
H$^+$	HL/H.L	9.74
	H$_2$L/HL.H	6.56
Ni^{2+}	ML/M.L	5.36
	ML$_2$/M.L^2	8.52
Cu^{2+}	ML/M.L	8.77
	ML$_2$/M.L^2	14.82

Bibliography: 61RF

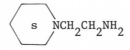

C$_7$H$_{16}$N$_2$ N-(2-Aminoethyl)piperidine   L

Metal ion	Equilibrium	Log K 30°, 0
H$^+$	HL/H.L	9.89
	H$_2$L/HL.H	6.38
Ni^{2+}	ML/M.L	4.30
Cu^{2+}	ML/M.L	7.77
	ML$_2$/M.L^2	13.60

Bibliography: 61RF

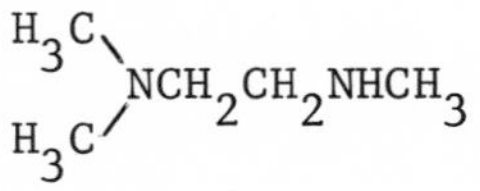

$$H_3C\diagdown NCH_2CH_2NHCH_3$$
$$H_3C\diagup$$

C$_8$H$_{20}$N$_2$ N,N,N'-Trimethylethylenediamine   L

Metal ion	Equilibrium	Log K 25°, 1.0
Cd^{2+}	ML/M.L	4.56
	ML$_2$/M.L^2	6.73
	ML$_3$/M.L^3	7.7
	MHL/M.HL	0.83
	M(OH)$_2$L/ML.(OH)^2	10.83

Bibliography: 73CV

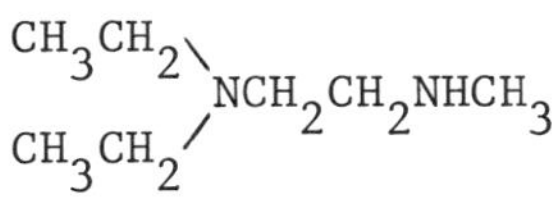

$$CH_3CH_2 \diagdown$$
$$ NCH_2CH_2NHCH_3$$
$$CH_3CH_2 \diagup$$

$C_7H_{18}N_2$ N,N-Diethyl-N'-methylethylenediamine   L

Metal ion	Equilibrium	Log K 25°, 0.1	Log K 25°, 0.5	Log K 25°, 0
H$^+$	HL/H.L	9.96	10.12	9.92
	H$_2$L/HL.H	6.76	7.07	6.48
Cu^{2+}	ML/M.L	7.69	7.87	7.64
	ML$_2$/M.L^2			12.48

Bibliography: 67NKA

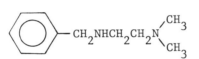

$C_{11}H_{18}N_2$ N-Benzyl-N',N'-dimethylethylenediamine   L

Metal ion	Equilibrium	Log K 25°, 0.1	Log K 25°, 0.5	Log K 25°, 0
H$^+$	HL/H.L	9.12	9.32	9.07
	H$_2$L/HL.H	5.83	6.11	5.57
Cu^{2+}	ML/M.L	7.67	7.86	7.63

Bibliography: 70T

H₃CN   s   NH

C$_5$H$_{12}$N$_2$ 1-Methylpiperazine   L

Metal ion	Equilibrium	Log K 25°, 0.1	ΔH 25°, 0.1	ΔS 25°, 0.1
H$^+$	HL/H.L	8.98	−8.4	13
	H$_2$L/HL.H	4.83	−4.0	9
Ag$^+$	ML/M.L	2.93	−5.58	−5.3
	ML$_2$/M.L^2	5.26	−9.55	−8.0

Bibliography:

H$^+$ 72EH   Ag$^+$ 73HB

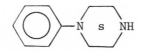

C$_{10}$H$_{14}$N$_2$ 1-Phenylpiperazine   L

Metal ion	Equilibrium	Log K 25°, 0.1	ΔH 25°, 0.1	ΔS 25°, 0.1
H$^+$	HL/H.L	8.60	−9.0	9
Ag$^+$	ML/M.L	2.97	−5.35	−4.4
	ML$_2$/M.L^2	5.74	−10.53	−9.1

Bibliography:

H$^+$ 72EH   Ag$^+$ 73HB

$$H_3C \diagdown \atop H_3C \diagup NCH_2CH_2N \diagup CH_3 \atop \diagdown CH_3$$

$C_6H_{16}N_2$	N,N,N',N'-Tetramethylethylenediamine				L

Metal ion	Equilibrium	Log K 25°, 0.1	Log K 25°, 0.5	Log K 25°, 0	ΔH 25°, 0.5	ΔS 25°, 0.5
H^+	HL/H.L	9.15 ±0.01	9.28 ±0.1	9.15	−7.40	17.7
	H_2L/HL.H	5.91 ±0.03	6.13 +0.1	5.58	−6.64	5.8
Ni^{2+}	ML/M.L	3.57	4.1^a			
Cu^{2+}	ML/M.L	7.20 +0.6	7.38 +0.8	(7.54)	−6.15	13.1
	$ML_2/M.L^2$	11.87				
	ML/MOHL.H	8.00	8.04		−9.5	5
	$ML/M(OH)_2L.H^2$	17.72	18.29		−21.2	13
	$(MOHL)_2/(MOHL)^2$	3.9	3.91		−2.0	11
	$M_3(OH)_4L_2/(M(OH)_2L)^2.M$	13.67			−12.3	21
	$M_2(OH)_2L/M(OH)_2L.M$		7.26			
Zn^{2+}	ML/M.L		3.62^a			
	$ML_2/M.L^2$		5.5^a			
Cd^{2+}	ML/M.L		3.87^b			
	$ML_2/M.L^2$		5.17^b			
	MHL/M.HL		1.04^b			

[a] 25°, 0.46;   [b] 25°, 1.0

Bibliography:

H^+	59GM,63L,67NKA,69CM,71PB,72TR	Zn^{2+}	63L
Ni^{2+}	63L,72TR	Cd^{2+}	73CV
Cu^{2+}	59GM,67NKA,69CM,72AP		

$C_6H_{12}N_2$	1,4-Diazabicyclo[2.2.2]octane (triethylenediamine)				L

Metal ion	Equilibrium	Log K 25°, 0.1	Log K 20°, 0.1	ΔH 25°, 0.1	ΔS 25°, 0.1
H+	HL/H.L	8.10 +0.7	8.19	−7.30 −7.21[a]	12.6
	H_2L/HL.H	4.14 −1	4.18	−3.01 −2.97[a]	8.9
Ag+	ML/M.L		1.65		

_____

[a] 25°, 0

Bibliography:
H+ 52SM,65PS,66LBH   Ag+   52SM

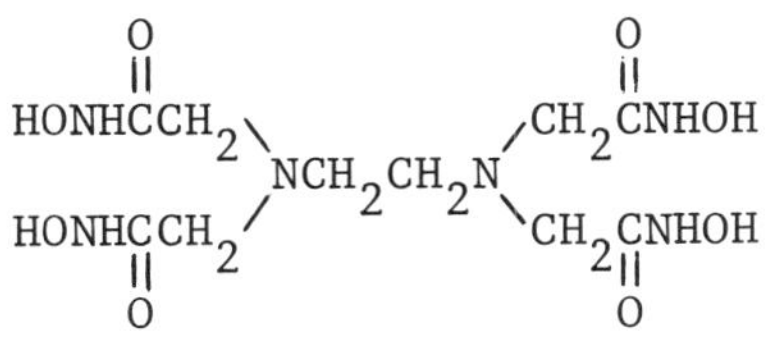

$$HONHCCH_2 \diagdown \quad \diagup CH_2CNHOH$$

(structure: Ethylenedinitrilotetraacethydroxamic acid)

$$HONHCCH_2 \diagup NCH_2CH_2N \diagdown CH_2CNHOH$$

$C_{10}H_{20}O_8N_6$ **Ethylenedinitrilotetraacethydroxamic acid** H_4L

Metal ion	Equilibrium	Log K 20°, 0.1
H^+	HL/H.L	11.1
	H_2L/HL.H	10.6
	H_3L/H_2L.H	7.23
	H_4L/H_3L.H	6.67
	H_5L/H_4L.H	6.05
	H_6L/H_5L.H	5.55
Cu^{2+}	ML/M.L	25.6
	MHL/ML.H	6.53
	MH_3L/MHL.H^2	10.68
	MH_4L/MH_3L.H	3.24
Fe^{3+}	ML/M.L	25.6
	MHL/ML.H	6.03
	MH_3L/MHL.H^2	10.00
	MH_4L/MH_3L.H	3.92
	MOHL/ML.OH	4.90
	$M(OH)_2L$/MOHL.OH	3.30

Bibliography: 72KM

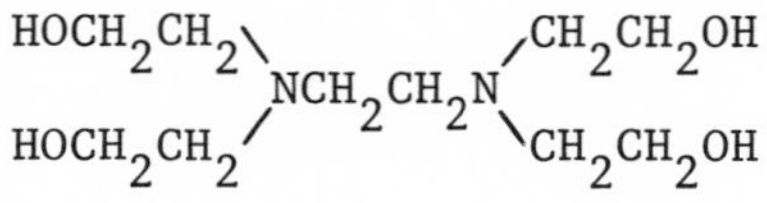

$$HOCH_2CH_2 \diagdown NCH_2CH_2N \diagup CH_2CH_2OH$$
$$HOCH_2CH_2 \diagup \qquad\qquad \diagdown CH_2CH_2OH$$

$C_{10}H_{24}O_4N_2$ <u>Ethylenedinitrilotetra-2-ethanol</u>   L

Metal ion	Equilibrium	Log K 25°, 0.1	Log K 25°, 0.5
H^+	HL/H.L	8.38	8.49 ±0.04
	H_2L/HL.H	4.37	4.47 ±0.03
Co^{2+}	ML/M.L		5.40 −0.3
	ML/MOHL.H		8.56
	MOHL/M(OH)$_2$L.H		9.76
Ni^{2+}	ML/M.L	6.5	6.50 −0.2
	ML/MOHL.H		9.07 +0.2
Cu^{2+}	ML/M.L		8.49 ±0.04
Zn^{2+}	ML/M.L		4.74
Cd^{2+}	ML/M.L		7.04

Bibliography:
H^+ 62H,64PG,65I,69RT Ni^{2+} 60HD,64PG,69RT
Co^{2+},Cu^{2+} 60HD,65I Zn^{2+},Cd^{2+} 60HD

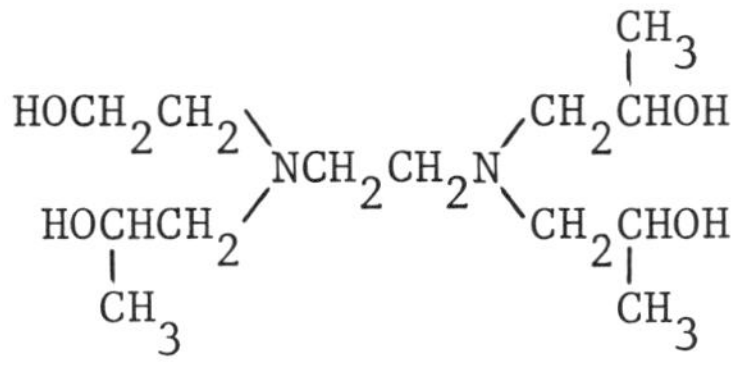

$$\begin{array}{c} \text{HOCH}_2\text{CH}_2 \\ \text{HOCHCH}_2 \\ | \\ \text{CH}_3 \end{array} \rangle \text{NCH}_2\text{CH}_2\text{N} \langle \begin{array}{c} \text{CH}_3 \\ | \\ \text{CH}_2\text{CHOH} \\ \text{CH}_2\text{CHOH} \\ | \\ \text{CH}_3 \end{array}$$

$C_{13}H_{30}O_4N_2$ <u>DL-N-(2-Hydroxyethyl)ethylenedinitrilotri-1-(2-propanol)</u>   L

Metal ion	Equilibrium	Log K 25°, 0.5
H^+	HL/H.L	8.80
	H_2L/HL.H	4.34
Co^{2+}	ML/M.L	5.96
	ML/MOHL.H	8.93
	MOHL/M(OH)$_2$L.H	10.12
Ni^{2+}	ML/M.L	7.25
Cu^{2+}	ML/M.L	9.36
Zn^{2+}	ML/M.L	5.67
Cd^{2+}	ML/M.L	7.73

Bibliography:
H^+ 62H Co^{2+}-Cd^{2+} 60HD

```
              CH3                    CH3
               |                      |
        HOCHCH2              CH2CHOH
               \NCH2CH2N/
        HOCHCH2              CH2CHOH
               |                      |
              CH3                    CH3
```

$C_{14}H_{32}O_4N_2$ DL-Ethylenedinitrilotetra-1-(2-propanol)   L

Metal ion	Equilibrium	Log K 25°, 0.1	Log K 25°, 0.5
H^+	HL/H.L	8.75	8.84
	H_2L/HL.H	4.24	4.35 ±0.02
Co^{2+}	ML/M.L		6.33
	ML/MOHL.H		8.99
	MOHL/M(OH)$_2$L.H		10.21
Ni^{2+}	ML/M.L	7.45	7.65
	ML/MOHL.H		9.76
	MOHL/M(OH)$_2$L.H		11.22
Cu^{2+}	ML/M.L		9.75
	ML/MOHL.H		6.77
	MOHL/M(OH)$_2$L.H		8.97
Zn^{2+}	ML/M.L		6.09
Cd^{2+}	ML/M.L		7.80

Bibliography:
H^+ 57HJ,62H,69RT Cu^{2+} 60HD,69RT
Co^{2+},Ni^{2+},Zn^{2+},Cd^{2+} 60HD   Other reference: 59K

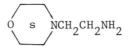

<u>N-(2-Aminoethyl)morpholine</u>   L

$C_6H_{14}ON_2$

Metal ion	Equilibrium	Log K 30°, 1.0
H^+	HL/H.L	9.31
	H_2L/HL.H	4.70
Ni^{2+}	ML/M.L	3.78
Cu^{2+}	ML/M.L	6.60
	ML_2/M.L^2	10.58

Bibliography: 56HF

$$CH_3$$
$$H_2NCH_2CH_2CH_2NCH_2CH_2CH_2NH_2$$

$C_7H_{19}N_3$ 5-Methyl-1,5,9-triazanonane   L

(N-methyliminobis(3-propylamine))

Metal ion	Equilibrium	Log K 25°, 0	ΔH 25°, 0	ΔS 25°, 0
H[+]	HL/H.L	10.59	(-11)[a]	(12)
	$H_2L/HL.H$	9.37	(-12)[a]	(3)
	$H_3L/H_2L.H$	6.43	(-10)[a]	(-4)
Co[2+]	ML/M.L	5.95[b]		
	$ML_2/M.L^2$	9.8[b]		
Ni[2+]	ML/M.L	7.30	(-8)[a]	(7)
	$ML_2/M.L^2$	10.9		
Cu[2+]	ML/M.L	12.5	(-14)[a]	(10)
	$ML_2/M.L^2$	16.9		
Cd[2+]	ML/M.L	5.92	(-5)[a]	(10)

[a] 10-40°, 0; [b] 10°, 0

Bibliography: 59GFa

$$CH_3$$
$$CH_3NHCH_2CH_2NCH_2CH_2NH_2$$

$C_6H_{17}N_3$ 4-Methyl-1,4,7-triazaoctane   L

(N,N'-dimethyldiethylenetriamine)

Metal ion	Equilibrium	Log K 25°, 0.13
H[+]	HL/H.L	10.03
	$H_2L/HL.H$	9.35
	$H_3L/H_2L.H$	2.82
Cu[2+]	ML/M.L	15.11
	ML/MOHL.H	9.00

Bibliography: 71AA

$$CH_3NHCH_2CH_2\overset{\overset{\displaystyle CH_3}{|}}{N}CH_2CH_2NHCH_3$$

$C_7H_{19}N_3$ 5-Methyl-2,5,8-triazanonane   L
                              (N,N',N''-trimethyldiethylenetriamine)

Metal ion	Equilibrium	Log K 25°, 0.13
H^+	HL/H.L	10.13
	H_2L/HL.H	9.37
	H_3L/H_2L.H	2.93
Cu^{2+}	ML/M.L	13.17
	ML/MOHL.H	8.58

Bibliography: 71AA

$$\overset{\displaystyle H_3C}{\underset{\displaystyle H_3C}{}}\!\!\!\!\diagdown\!\!\!\!\diagup N CH_2CH_2NHCH_2CH_2NH_2$$

$C_6H_{17}N_3$ 7-Methyl-1,4,7-triazaoctane   L
                              (N,N-dimethyldiethylenetriamine)

Metal ion	Equilibrium	Log K 25°, 0.13
H^+	HL/H.L	9.62
	H_2L/HL.H	8.63
	H_3L/H_2L.H	3.62
Cu^{2+}	ML/M.L	14.33
	ML/MOHL.H	8.47

Bibliography: 71AA

$$CH_3CH_2 \diagdown$$
$$NCH_2CH_2NHCH_2CH_2NH_2$$
$$CH_3CH_2 \diagup$$

$C_8H_{21}N_3$ 7-Ethyl-1,4,7-triazanonane   L

(N,N-diethyldiethylenetriamine)

Metal ion	Equilibrium	Log K 25°, 0.13
H$^+$	HL/H.L	9.90
	H$_2$L/HL.H	9.10
	H$_3$L/H$_2$L.H	3.93
Cu^{2+}	ML/M.L	13.2
	ML/MOHL.H	8.35

Bibliography: 71AA

$$\overset{\displaystyle CH_2CH_3}{\underset{\displaystyle |}{}}$$
$$CH_3CH_2NHCH_2CH_2NCH_2CH_2NHCH_2CH_3$$

$C_{10}H_{25}N_3$ 6-Ethyl-3,6,9-triazaundecane   L

(N,N',N''-triethyldiethylenetriamine)

Metal ion	Equilibrium	Log K 25°, 0.13
H$^+$	HL/H.L	10.13
	H$_2$L/HL.H	9.37
	H$_3$L/H$_2$L.H	2.93
Cu^{2+}	ML/M.L	13.17
	ML/MOHL.H	8.58

Bibliography: 71AA

$$CH_3CH_2 \diagdown NCH_2CH_2NHCH_2CH_2N \diagup CH_2CH_3$$
$$CH_3CH_2 \diagup \qquad\qquad\qquad\qquad \diagdown CH_2CH_3$$

$C_{12}H_{29}N_3$ 3,9-Diethyl-3,6,9-triazaundecane   L

(N,N,N'',N''-tetraethyldiethylenetriamine)

Metal ion	Equilibrium	Log K 25°, 0.13
H$^+$	HL/H.L	9.78
	H$_2$L/HL.H	9.03
	H$_3$L/H$_2$L.H	3.39
Cu^{2+}	ML/M.L	10.43
	ML/MOHL.H	7.61

Bibliography: 71AA   Other reference: 68MP

$$H_3C \diagdown \qquad\qquad CH_3 \qquad\qquad\qquad \diagup CH_3$$
$$\qquad\quad NCH_2CH_2NCH_2CH_2N$$
$$H_3C \diagup \qquad\qquad\qquad\qquad\qquad\qquad \diagdown CH_3$$

$C_9H_{23}N_3$ 2,5,8-Trimethyl-2,5,8-triazanonane   L

(N,N,N',N'',N''-pentamethyldiethylenetriamine)

Metal ion	Equilibrium	Log K 25°, 0.13
H$^+$	HL/H.L	9.22
	H$_2$L/HL.H	8.41
	H$_3$L/H$_2$L.H	2.09
Cu^{2+}	ML/M.L	12.16
	ML/MOHL.H	8.65

Bibliography: 71AA

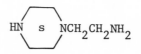

$C_6H_{15}N_3$ N-(2-Aminoethyl)piperazine   L

Metal ion	Equilibrium	Log K 25°, 0	ΔH 25°, 0	ΔS 25°, 0
H+	HL/H.L	9.55	(-12)[a]	(3)
	H₂L/HL.H	8.44	(-10)[a]	(5)
Cu²⁺	ML/M.L	5.5	(-5)[a]	(8)
	ML₂/M.L²	9.3	(-7)[a]	(19)

_____

[a] 10-40°, 0

Bibliography: 61PG

$$H_2NCH_2CH_2N \begin{array}{c} ^{CH_2CH_2NH_2} \\ ^{CH_2CH_2NH_2} \end{array}$$

$C_6H_{18}N_4$	Nitrilotris(2-ethylamine)	(tris(2-aminoethyl)amine, tren)				L

Metal ion	Equilibrium	Log K 25°, 0.1	Log K 20°, 0.1	Log K 25°, 0	ΔH 25°, 0.1	ΔS 25°, 0.1
H^+	HL/H.L	10.14	10.29 10.29[a]	10.03	−11.7	7
	H_2L/HL.H	9.43	9.59 9.84[a]	9.13	−12.8	0
	H_3L/H_2L.H	8.41	8.56 8.79[a]	7.85	−12.2	−2
Mn^{2+}	ML/M.L	5.8	5.8		−3.0	17
Fe^{2+}	ML/M.L	8.7	8.78 ±0.0		−6.3	19
Co^{2+}	ML/M.L	12.7	12.8		−10.7	22
Ni^{2+}	ML/M.L	14.6	14.8 −0.1	(14.8)[a]	−15.2	16
	MHL/M.HL			(9.85)[a]		
Cu^{2+}	ML/M.L	18.5	18.8 +0.2	18.4	−20.4	16
	MOHL/ML.OH	4.65	4.70		−3.7	9
Ag^+	ML/M.L		7.8			
	MHL/ML.H		8.1			
	MH_2L/MHL.H		7.3			
	M_2L/ML.M		2.4			
Zn^{2+}	ML/M.L	14.5	14.65 −0.05		−13.9	20
Cd^{2+}	ML/M.L		12.3			
Hg^{2+}	ML/M.L		25.8[a]			
	MHL/M.HL		4.5			

[a] 20°, 0.5

Bibliography:

H^+ 49AS,50PS,58BF,63PCa Ni^{2+} 49AS,50PS,63PCS,70MA

Mn^{2+},Co^{2+} 50PS,63PCS Cu^{2+} 49AS,50PS,58BF,63PCS

Fe^{2+},Ni^{2+},Zn^{2+} 49AS,50PS,63PCS Ag^+,Cd^{2+},Hg^{2+} 50PS

$$H_2NCH_2CH_2CH_2 \diagdown$$
$$NCH_2CH_2CH_2NH_2$$
$$H_2NCH_2CH_2CH_2 \diagup$$

| $C_9H_{24}N_4$ | Nitrilotris(3-propylamine) (tris(3-aminopropyl)amine) | | | | L |

Metal ion	Equilibrium	Log K 25°, 0.1	Log K 20°, 0.1	ΔH 25°, 0.1	ΔS 25°, 0.1
H^+	HL/H.L	10.51	10.65	−12.31	6.8
	H_2L/HL.H	9.82	9.98	−12.8	2
	H_3L/H_2L.H	9.13	9.28	−12.4	0
	H_4L/H_3L.H	5.62	5.71	−7.3	1
Co^{2+}	ML/M.L	6.36			
	ML/MOHL.H	10.79			
Ni^{2+}	ML/M.L	8.70	8.82	−9.2	9
	MHL/ML.H	7.08			
Cu^{2+}	ML/M.L	13.12	13.30	−14.7	11
	MHL/ML.H	8.15	8.28	−10.4	2
	ML/MOHL.H	9.79	9.90	−8.7	16
Zn^{2+}	ML/M.L	10.70	10.81	−8.9	19

Bibliography:

H^+,Ni^{2+}-Zn^{2+} 68DP,68VP Co^{2+} 68DP

$$H_2NCH_2CH_2CH_2N \quad s \quad NCH_2CH_2CH_2NH_2$$

$C_{10}H_{24}N_4$ <u>1,4-Bis(3-aminopropyl)piperazine</u>   L

Metal ion	Equilibrium	Log K 30°, 0
H^+	HL/H.L	10.22
	H_2L/HL.H	9.28
	H_3L/H_2L.H	6.15
Cu^{2+}	ML/M.L	13.60

Bibliography: 69WG

$$CH_3 \backslash NCH_2CH_2N / CH_2CH_2NH_2$$
$$H_2NCH_2CH_2 / \qquad \backslash CH_2CH_2NH_2$$

$C_9H_{25}N_5$ <u>N-Methyl-N,N',N'-tris(2-aminoethyl)ethylenediamine</u>   L
                           <u>(N-methylethylenedinitrilotris(2-ethylamine))</u>

Metal ion	Equilibrium	Log K 20°, 0.1
H^+	HL/H.L	10.00
	H_2L/HL.H	9.53
	H_3L/H_2L.H	8.69
	H_4L/H_3L.H	5.25
Cd^{2+}	ML/M.L	14.76
	MHL/ML.H	4.88
	MH_2L/MHL.H	5.08

Bibliography: 71SW

$$H_2NCH_2CH_2 \diagdown NCH_2CH_2N \diagup CH_2CH_2NH_2$$
$$H_2NCH_2CH_2 \diagup \qquad \diagdown CH_2CH_2NH_2$$

$C_{10}H_{28}N_6$ <u>Ethylenedinitrilotetrakis(2-ethylamine)</u>   L

(N,N,N',N'-tetrakis(2-aminoethyl)ethylenediamine, <u>penten</u>)

Metal ion	Equilibrium	Log K 25°, 0.1	Log K 20°, 0.1	Log K 20°, 0.5	ΔH 25°, 0.1	ΔS 25°, 0.1
H^+	HL/H.L	10.08 ±0.02	10.22	10.23	−11.3	8
	H_2L/HL.H	9.58 ±0.03	9.73	9.83	−11.5	5
	H_3L/H_2L.H	8.99 ±0.02	9.16	9.54	−11.2	−3
	H_4L/H_3L.H	8.42 ±0.02	8.58	8.85	−12.0	−2
	H_5L/H_4L.H	1.33	1.39		−4.5	−9
Mn^{2+}	ML/M.L	9.26	9.37		−8.9	13
Fe^{2+}	ML/M.L	11.1	11.2		−9.7	18
	MHL/ML.H		7.7			
Co^{2+}	ML/M.L	15.6	15.8		−14.8	22
	MHL/ML.H	6.82	6.95		−10.5	−4
Ni^{2+}	ML/M.L	19.1	19.3		−19.7	21
	MHL/ML.H	6.62	6.75		−10.0	−3
Cu^{2+}	ML/M.L	22.1	22.4		−24.5	19
	MHL/ML.H	8.01	8.16		−11.6	−2
	MH_2L/MHL.H		3.62			
	MH_3L/MH_2L.H		3.24			
Zn^{2+}	ML/M.L	16.06	16.24		−14.5	25
	MHL/ML.H	8.01	8.16		−11.5	−2
Cd^{2+}	ML/M.L		16.1			
	MHL/ML.H		6.49			
Hg^{2+}	ML/M.L			29.6		
	MHL/ML.H			8.62		
	MH_2L/MHL.H			4.7		
	MH_3L/MH_2L.H			2.6		

Bibliography:

H^+ 53SM,63PCa Cd^{2+}-Hg^{2+} 53SM

Mn^{2+}-Zn^{2+} 53SM,64SPC

$$H_2NCH_2CH_2 \diagdown$$
$$NCH_2CH_2CH_2N \diagup CH_2CH_2NH_2$$
$$H_2NCH_2CH_2 \diagup \diagdown CH_2CH_2NH_2$$

$C_{11}H_{30}N_6$ **Trimethylenedinitrilotetrakis(2-ethylamine)**   L

(N,N,N',N'-tetrakis(2-aminoethyl)trimethylenediamine, ptetraen)

Metal ion	Equilibrium	Log K 25°, 0.1	Log K 20°, 0.1	ΔH 25°, 0.1	ΔS 25°, 0.1
H^+	HL/H.L	10.24	10.39	−11.0	10
	$H_2L/HL.H$	9.56	9.70	−11.6	5
	$H_3L/H_2L.H$	9.18	9.34	−12.7	−1
	$H_4L/H_3L.H$	8.44	8.60	−12.4	−3
	$H_5L/H_4L.H$	2.5	2.5	−4.3	−3
Mn^{2+}	ML/M.L	5.30	5.33	−2.6	16
	$M_2L/ML.M$		2.2		
Co^{2+}	ML/M.L	13.29	13.45	−12.4	19
	MHL/ML.H		7.3		
	$M_2L/ML.M$		2.5		
Ni^{2+}	ML/M.L	18.46	18.69	−19.3	20
	MHL/ML.H	5.7	5.8	−7.2	2
	$M_2L/ML.M$		2.2		
Cu^{2+}	ML/M.L	21.10	21.40	−22.7	20
	MHL/ML.H	8.23	8.37	−11.4	−1
Zn^{2+}	ML/M.L	14.86	15.01	−12.0	28
	MHL/ML.H	7.7	7.8	−11.3	−3
	$M_2L/ML.M$		2.2		
Cd^{2+}	ML/M.L		12.84		
	MHL/ML.H		7.8		
	$MH_2L/MHL.H$		6.7		
	$M_2L/ML.M$		2.8		

Bibliography: 71PW

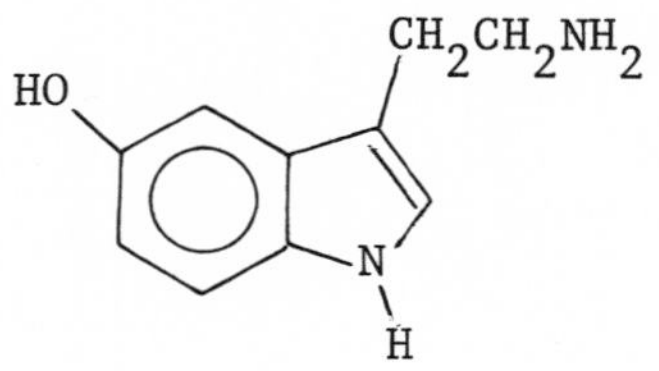

$$CH_2CH_2NH_2$$

HO

$C_{10}H_{12}ON_2$ <u>3-(2-Aminoethyl)-5-hydroxyindole</u>   <u>(5-hydroxytryptamine)</u>   HL

Metal ion	Equilibrium	Log K 20°, 0.37
H^+	HL/H.L	11.13
	H_2L/HL.H	9.85
Ni^{2+}	ML/M.L	5.1
	MHL/M.HL	3.4
Cu^{2+}	ML/M.L	9.9
Cd^{2+}	ML/M.L	3.6
	MHL/M.HL	3.14
Pb^{2+}	ML/M.L	8.04
	MHL/M.HL	5.02

Bibliography: 71WS

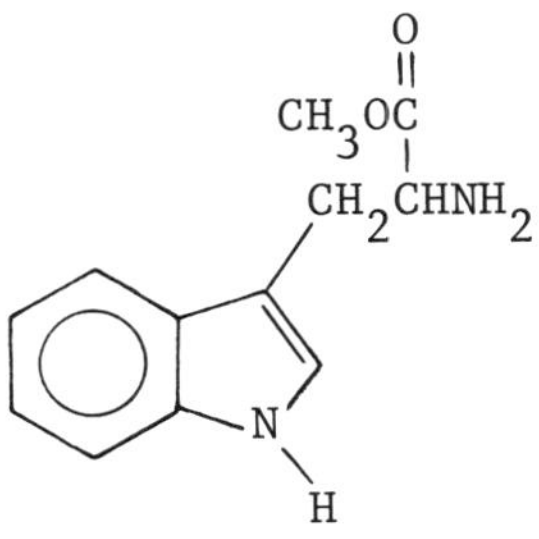

$$CH_3OC \overset{\displaystyle O}{\overset{\|}{\underset{|}{}}}$$

$$CH_2CHNH_2$$

$C_{12}H_{14}O_2N_2$ L-2-Amino-3-(3-indolyl)propanoic acid methyl ester   L

(tryptophan methyl ester)

Metal ion	Equilibrium	Log K 25°, 0.1	Log K 20°, 0.37
H^+	HL/H.L	7.3	7.18
Cu^{2+}	ML/M.L		3.95
	$ML_2/M.L^2$		7.14
	$ML_3/M.L^3$		9.3
	$ML_4/M.L^4$		10.8

Bibliography:

H^+ 67HP,74W Cu^{2+} 74W

$C_3H_4N_2$ 1,3-Diazole   (imidazole)   L

Metal ion	Equilibrium	Log K 25°, 0.16	Log K 25°, 1.0	Log K 25°, 0	ΔH 25°, 0.16	ΔS 25°, 0.16
H+	$L/H_{-1}L.H$		14.29[a]±0.04	14.44	(−18)[b]	(6)[c]
	HL/H.L	7.03 ±0.03	7.31 ±0.03	6.993	−8.71[c]±0.08	2.8
			7.90[d]±0.02		−8.80[e]±0.01	
Co2+	ML/M.L	2.40 ±0.07			(−4)[f]	(−2)
	$ML_2/M.L^2$	4.39 ±0.02			(−8)[f]	(−7)
	$ML_3/M.L^3$	5.92 ±0.03			(−12)[f]	(−13)
	$ML_4/M.L^4$	7.0 ±0.2			(−15)[f]	(−18)
	$ML_5/M.L^5$	7.4			(−18)[f]	(−27)
	$ML_6/M.L^6$	7.4			(−22)[f]	(−40)
Ni2+	ML/M.L	3.02 ±0.07			−5.8	−6
	$ML_2/M.L^2$	5.45 ±0.1			(−11)[f]	(−12)
	$ML_3/M.L^3$	7.5			(−15)[f]	(−16)
	$ML_4/M.L^4$	9.1			−18.4	−20
	$ML_5/M.L^5$	10.2			(−21)[f]	(−24)
	$ML_6/M.L^6$	10.7			(−24)[f]	(−32)
Cu2+	ML/M.L	4.18 ±0.03	4.66[d]		−7.6	−6
	$ML_2/M.L^2$	7.66 ±0.04	8.64[d]		(−14)[f]	(−12)
	$ML_3/M.L^3$	10.51 −0.01	11.94[d]		(−19)[f]	(−16)
	$ML_4/M.L^4$	12.6 ±0.1	14.60[d]		−23.0	−20
	$ML_6/M.L^6$		30.00[d]			
	$M_2(OH)_2L_3/M^2.(OH)^2.L^3$		21.22[d]			
Cu+	ML/M.L	6.83[h]	(5.78)[g]			
	$ML_2/M.L^2$	10.73[h]	10.98[g]			

---

[a] 25°, 0.5; [b] 15–35°, 0; [c] 25°, 0; [d] 25°, 3.0; [e] 25°, 0.2; [f] 10–50°, 0.16; [g] 20°, 0.15; [h] 25°, 0.1

Imidazole (continued)

Metal ion	Equilibrium	Log K 25°, 0.16	Log K 25°, 1.0	Log K 25°, 0	ΔH 25°, 0.16	ΔS 25°,0.16
Ag^+	$ML/M.L$	3.17^h	3.08 ± 0.03		-7.3^i	-10^h
	$ML_2/M.L^2$	6.94^h	6.90 ± 0.05		-15.7^i	-21^h
Zn^{2+}	$ML/M.L$	2.56 ± 0.04			-3.8	-1
	$ML_2/M.L^2$	4.89 ± 0.05				
	$ML_3/M.L^3$	$7.16 -0.01$				
	$ML_4/M.L^4$	9.19 ± 0.03			-16.2	-12
Cd^{2+}	$ML/M.L$	2.80	2.70		$(-4)^j$	(-1)
	$ML_2/M.L^2$	4.99 ± 0.09	5.08			
	$ML_3/M.L^3$	$6.46 -0.01$	6.65			
	$ML_4/M.L^4$	7.53 ± 0.05	7.60			
Hg^{2+}	$ML_2/M.L^2$	16.74^k				

-----

[h] 25°, 0.1; [i] 25°, 1.0; [j] 15-35°, 0.15; [k] 27°, 0.15

Bibliography:

H^+ 53TW,54EF,56WI,60BD,62W,64GHI,66DG,
         66KZ,67HW,68CWI,69LV,69NN,70WW,71S

Co^{2+} 58ME,66SK

Ni^{2+} 55MA,63CC,64BW,66SK

Cu^{2+} 54LW,55MA,58KD,61JW,62HP,63CC,64BW,
            66SK,71S

Cu^+ 54LW,61JW,62HP,66KZ

Ag^+ 60GG,64BW,66DGa,69NN

Zn^{2+} 54EF,57NG,58KD,64BW

Cd^{2+} 53TW,54LW,72J

Hg^{2+} 60BD

Other references: 54S,55LC,58SL,59ML,61S,
         65DF,69RW,70BL,71BL,71BLa,73S

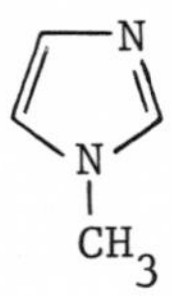

C$_4$H$_6$N$_2$ <u>1-Methylimidazole</u>   L

Metal ion	Equilibrium	Log K 25°, 0.16
H$^+$	HL/H.L	7.06
Cu^{2+}	ML/M.L	4.22
	ML$_2$/M.L^2	7.76
	ML$_3$/M.L^3	10.65
	ML$_4$/M.L^4	12.86
Ag$^+$	ML/M.L	3.00[a]
	ML$_2$/M.L^2	6.89[a]
Cd^{2+}	ML$_3$/M.L^3	6.49

_____

[a] 25°, 1.0

Bibliography:
H$^+$,Cu^{2+},Cd^{2+} 54LW   Ag$^+$ 64BW

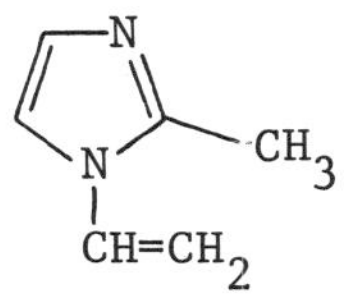

$C_6H_8N_2$ 1-Vinyl-2-methylimidazole   L

Metal ion	Equilibrium	Log K 25°, 1.0
H^+	HL/H.L	7.09
Ag^+	ML/M.L	3.48
	$ML_2/M.L^2$	7.01

Bibliography: 69NN

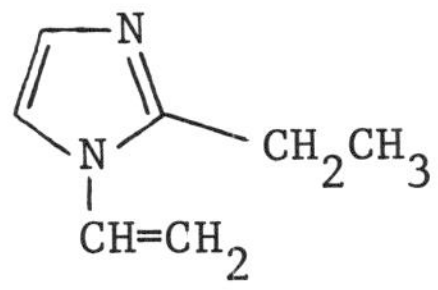

$C_7H_{10}N_2$ 1-Vinyl-2-ethylimidazole   L

Metal ion	Equilibrium	Log K 25°, 1.0
H^+	HL/H.L	6.88
Ag^+	ML/M.L	2.97
	$ML_2/M.L^2$	6.22

Bibliography: 69NN

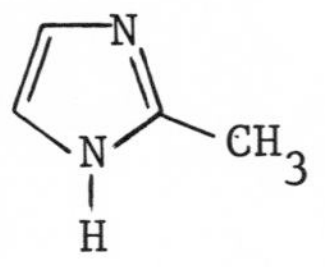

| $C_4H_6N_2$ | | 2-Methylimidazole | | L |

Metal ion	Equilibrium	Log K 25°, 1.0	Log K 25°, 0
H^+	HL/H.L	8.13	7.86
Ag^+	ML/M.L	3.11	
	$ML_2/M.L^2$	6.98	

Bibliography:

H^+ 38KN,69NN Ag^+ 69NN

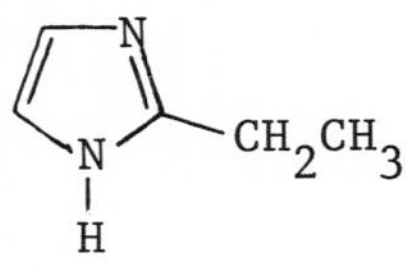

| $C_5H_8N_2$ | | 2-Ethylimidazole | | L |

Metal ion	Equilibrium	Log K 25°, 1.0
H^+	HL/H.L	8.02
Ag^+	ML/M.L	3.06
	$ML_2/M.L^2$	6.89

Bibliography: 69NN

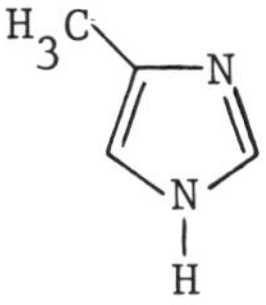

$C_4H_6N_2$ 4-Methylimidazole   L

Metal ion	Equilibrium	Log K 25°, 0.16	Log K 25°, 0	ΔH 25°, 0.16	ΔS 25°, 0.16
H^+	HL/H.L	7.57	7.54 ±0.02	$(-9)^a$	(4)
Cu^{2+}	ML/M.L	4.13			
	$ML_2/M.L^2$	7.62			
	$ML_3/M.L^3$	10.49			
	$ML_4/M.L^4$	12.45			
Zn^{2+}	ML/M.L	2.44			
	$ML_2/M.L^2$	4.97			
	$ML_3/M.L^3$	7.61			
	$ML_4/M.L^4$	9.99			

---

[a] 2.5-29°, 0.16

Bibliography:

H[+]   38KN,57NG Cu^{2+},Zn^{2+} 57NG

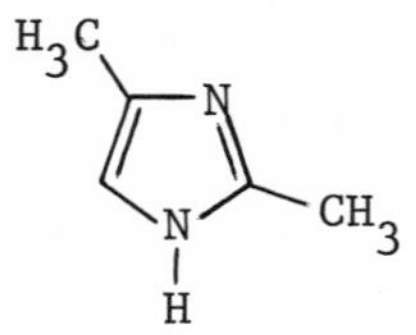

$C_5H_8N_2$ 2,4-Dimethylimidazole   L

Metal ion	Equilibrium	Log K 25°, 0.16	Log K 25°, 1.0	Log K 25°, 0	ΔH 25°, 0.16	ΔS 25°, 0.16
H^+	HL/H.L	8.40	8.57	8.37 ±0.01	(-9)[a]	8
Cu^{2+}	ML/M.L	3.8				
Ag^+	ML/M.L			3.44		
	$ML_2/M.L^2$			7.50		

---

[a] 2.5-29°, 0.16

Bibliography:

H^+ 38KN,57NG,69NN Ag^+ 69NN

Cu^{2+} 57NG

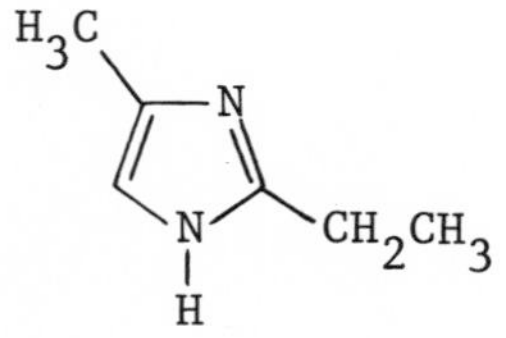

$C_6H_{10}N_2$ 2-Ethyl-4-methylimidazole   L

Metal ion	Equilibrium	Log K 25°, 1.0
H^+	HL/H.L	8.52
Ag^+	ML/M.L	3.64
	$ML_2/M.L^2$	7.74

Bibliography: 69NN

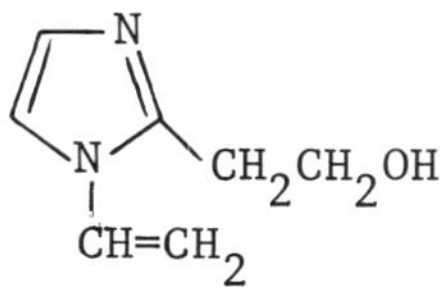

$$CH=CH_2$$ imidazole structure with CH_2CH_2OH

| $C_7H_{10}ON_2$ | 1-Vinyl-2-(2-hydroxyethyl)imidazole | L |

Metal ion	Equilibrium	Log K 25°, 1.0
H^+	HL/H.L	6.39
Ag^+	ML/M.L	2.86
	$ML_2/M.L^2$	6.24

Bibliography: 69NN

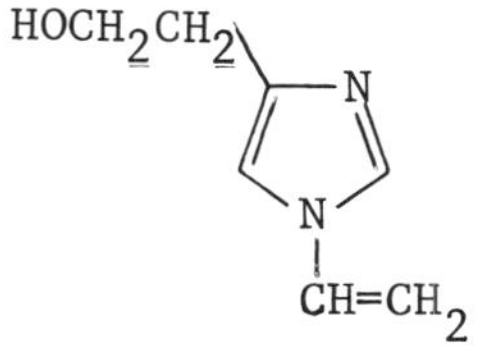

$HOCH_2CH_2$ imidazole structure with $CH=CH_2$

| $C_7H_{10}ON_2$ | 1-Vinyl-4-(2-hydroxyethyl)imidazole | L |

Metal ion	Equilibrium	Log K 25°, 1.0
H^+	HL/H.L	6.05
Ag^+	ML/M.L	3.03
	$ML_2/M.L^2$	6.40

Bibliography: 69NN

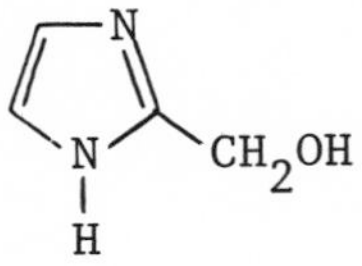

$C_4H_6ON_2$	2-Hydroxymethylimidazole			L

Metal ion	Equilibrium	Log K 25°, 3.0	ΔH 25°, 3.0	ΔS 25°, 3.0
H^+	HL/H.L	7.66	−10.0	2
Cu^{2+}	ML/M.L	4.27	−6.8	−3
	$ML_2/M.L^2$	8.19	−13.7	−8
	$ML_3/M.L^3$	11.32	−20.5	−17
	$ML_4/M.L^4$	13.4	−27.3	−30

Bibliography: 68W

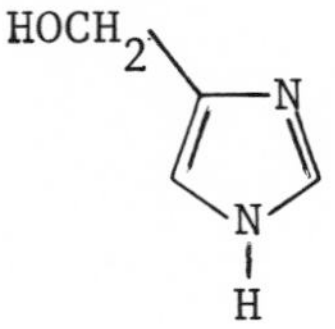

$C_4H_6ON_2$	4-Hydroxymethylimidazole				L

Metal ion	Equilibrium	Log K 25°, 3.0	Log K 25°, 0	ΔH 25°, 3.0	ΔS 25°, 3.0
H^+	HL/H.L	7.42	6.38	−9.3	3
Cu^{2+}	ML/M.L	4.25		−7.0	−4
	$ML_2/M.L^2$	8.15		−14.1	−10
	$ML_3/M.L^3$	11.26		−21.1	−19
	$ML_4/M.L^4$	13.3		−28.2	−34

Bibliography:
H^+ 38KN,68W Cu^{2+} 68W

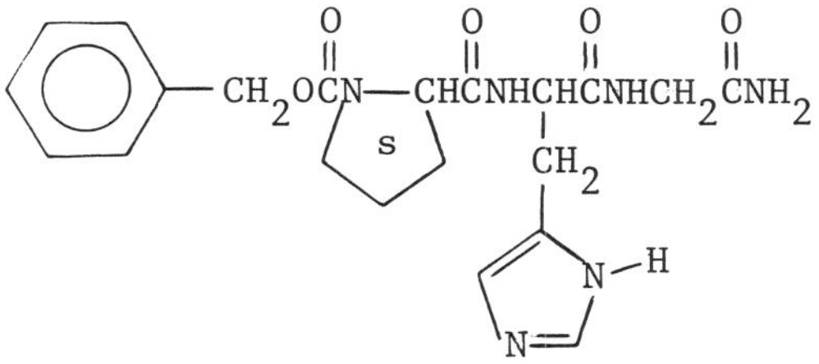

$C_{21}H_{26}O_5N_6$ <u>Benzyloxycarbonyl-L-prolyl-L-histidylglycinamide</u>   L

Metal ion	Equilibrium	Log K 25°, 0.16
H^+	HL/H.L	6.30
Cu^{2+}	ML/M.L	3.28
Zn^{2+}	ML/M.L	2.16

Bibliography: 59KC

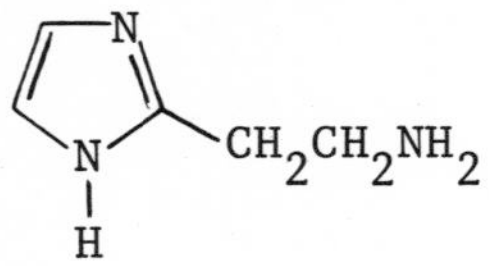

$C_5H_9N_3$		2-(2-Aminoethyl)imidazole (isohistamine)			L

Metal ion	Equilibrium	Log K 25°, 0.1		ΔH 25°, 0.1	ΔS 25°, 0.1
H^+	HL/H.L	9.27		-11.60	3.5
	$H_2L/HL.H$	6.04		-8.61	-1.3
Co^{2+}	ML/M.L	5.56			
	$ML_2/M.L^2$	9.58			
	$ML_3/M.L^3$	12.3			
Ni^{2+}	ML/M.L	7.12			
	$ML_2/M.L^2$	12.63			
	$ML_3/M.L^3$	16.16			
Cu^{2+}	ML/M.L	9.85		-11.51	6.5
	$ML_2/M.L^2$	16.98		-21.5	5

Bibliography:

H^+, Cu^{2+} 69EH,70EHT Co^{2+}, Ni^{2+} 69EH

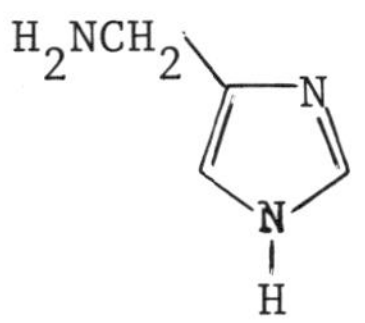

$C_4H_7N_3$ <u>4-Aminomethylimidazole</u>   L

Metal ion	Equilibrium	Log K 25°, 0.1	Log K 25°, 0.3	ΔH 25°, 0.3	ΔS 25°, 0.3
H^+	HL/H.L	9.24	9.15	−9.7	9
	$H_2L/HL.H$	4.62	4.77	−7.4	−3
Ni^{2+}	ML/M.L		5.85	−8.7	−2
	$ML_2/M.L^2$		10.67	−17.4	−10
	$ML_3/M.L^3$		13.79	−26.1	−24
Cu^{2+}	ML/M.L	(9.22)	8.73	−11.4	2
	$ML_2/M.L^2$	(17.17)	16.45	−22.8	−1

Bibliography:

H^+	67HW,71HG	Cu^{2+}	67HWa,71HG
Ni^{2+}	67HWa	Other reference:	60HJ

$$H_2NCH_2CH_2 \quad \text{(imidazole ring with N, N-H)}$$

$C_5H_9N_3$		4-(2-Aminoethyl)imidazole		(histamine)		L
Metal ion	Equilibrium	Log K 25°, 0.1	Log K 30°, 1.0	Log K 25°, 0	ΔH 25°, 0.1	ΔS 25°, 0.1
H^+	HL/H.L	9.83 ±0.03	9.72	9.86 −0.1	−11.95	4.9
					−13.01[a]	1.4[a]
	$H_2L/HL.H$	6.07 ±0.05	6.24	5.96 −0.2	−7.57	2.4
					−8.52[a]	−0.8[a]
Be^{2+}	ML/M.L	7.12				
	$ML_2/M.L^2$	12.47				
Mn^{2+}	ML/M.L	3.0[a]				
Co^{2+}	ML/M.L	5.03 ±0.1	5.34	5.12	(−7)[b]	(0)[c]
	$ML_2/M.L^2$	8.77 ±0.1	9.09	8.89	(−12)[b]	(0)[c]
	$ML_3/M.L^3$		10.97			
Ni^{2+}	ML/M.L	6.78 ±0.1	6.84	6.90	−7.8[d]	5
	$ML_2/M.L^2$	11.78 ±0.1	11.79	12.02	−14.8[d]	4
	$ML_3/M.L^3$	14.90 +0.1	14.87	15.30	−20.9[d]	−2
Cu^{2+}	ML/M.L	9.56 ±0.1	9.60	9.64	−11.4	6
					−13.4[a]	−1[a]
	$ML_2/M.L^2$	16.13 ±0.2	16.09	16.21	−21.0	3
					−23.4[a]	−5[a]
Cu^+	MHL/M.HL	8.87				
	$M(HL)_2/M.(HL)^2$	10.32				
Zn^{2+}	ML/M.L	5.4 ±0.2	5.77	5.03[e]		
	$ML_2/M.L^2$	10.0 ±0.1	10.50	9.81[e]		
	$ML_3/M.L^3$			12.09[e]		
Cd^{2+}	ML/M.L	4.75	4.83	4.82	(−7)[b]	(−3)[c]
	$ML_2/M.L^2$		8.2	8.22	(−12)[b]	(−3)[c]

[a] 25°, 0.2; [b] 10-40°, 0; [c] 25°, 0; [d] 25°, 0.3; [e] 37°, 0.15

Histamine   (continued)

Bibliography:

H[+]   55MA,56HF,58V,61NF,64DC,66KZ,66Z,
        67HW,67PS,69EH,69HG,70EHT,70MB,71RM,
        73GS

Be[2+] 70CA

MN[2+] 71RM

Co[2+] 55MA,56HF,61NF,69EH,69PS,71RM

Ni[2+] 55MA,56HF,59V,61NF,63CC,67HWa,68PS
        69EH,71RM

Cu[2+] 55MA,56HF,60V,61NF,63CC,64DC,67HWa,
        67PS,69EH,69HG,70MBa,71RM,73GS

Cu[+]   66KZ

Zn[2+] 52H,61S,63CC,69PS,71RM

Cd[2+] 52H,59V,61NF

Other references: 52A,53P,60HJ,62HJ,64AR,
        66PS,70Z

$$CH_3NHCH_2CH_2 \overset{\displaystyle}{\underset{\displaystyle}{\text{imidazole ring}}}$$

$C_6H_{11}N_3$ 4-(2-Methylaminoethyl)imidazole   L

Metal ion	Equilibrium	Log K 25°, 0.1
H[+]	HL/H.L	9.90
	$H_2L/HL.H$	5.87
Co[2+]	ML/M.L	4.45
	$ML_2/M.L^2$	7.25
Ni[2+]	ML/M.L	5.86
	$ML_2/M.L^2$	9.42
	$M_2L_3/M^2.L^3$	18.15
Cu[2+]	ML/M.L	8.35
	MHL/ML.H	4.63
	ML/MOHL.H	7.19
Zn[2+]	ML/M.L	4.83

Bibliography: 73BDa

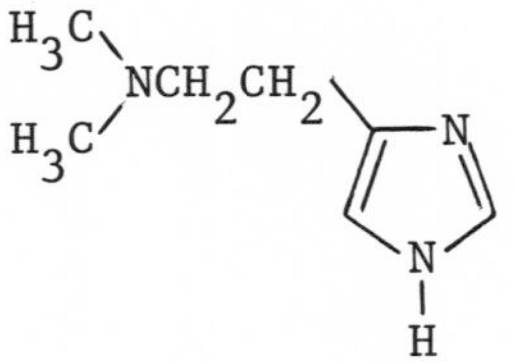

$C_7H_{13}N_3$ 4-(2-Dimethylaminoethyl)imidazole   L

Metal ion	Equilibrium	Log K 25°, 0.1
H^+	HL/H.L	9.33
	H_2L/HL.H	5.82
Co^{2+}	ML/M.L	2.82
Ni^{2+}	ML/M.L	3.88
Cu^{2+}	ML/M.L	6.56
	MHL/ML.H	5.80
	ML/MOHL.H	7.38
Zn^{2+}	ML/M.L	3.40

Bibliography: 73BDa

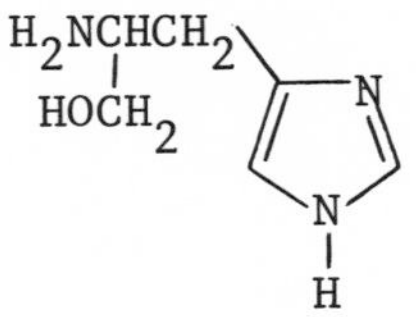

$C_6H_{11}ON_3$ DL-4-(2-Amino-3-hydroxypropyl)imidazole   (histidinol)   L

Metal ion	Equilibrium	Log K 25°, 0.15
H^+	HL/H.L	8.72
	H_2L/HL.H	5.81
Cu^{2+}	ML/M.L	9.35
	$ML_2/M.L^2$	(10.0)[a]

[a] Optical isomerism not stated.

Bibliography: 70WK

$$\begin{array}{c} \text{H}_2\text{NCHCH}_2 \\ | \\ \text{CH}_3\text{OC} \\ || \\ \text{O} \end{array}$$

$C_7H_{11}O_2N_3$ <u>L-Histidine methyl ester</u>   L

Metal ion	Equilibrium	Log K 25°, 0.1	ΔH 25°, 0.1	ΔS 25°, 0.1
H[+]	HL/H.L	7.22 ±0.03	-10.0	-1
	$H_2L/HL.H$	5.25 ±0.02	-7.6	-1
Be[2+]	ML/M.L	4.80		
	$ML_2/M.L^2$	(8.28)[a]		
Co[2+]	ML/M.L	4.24[b]		
	$ML_2/M.L^2$	(7.36)[b]		
Ni[2+]	ML/M.L	6.19[c]-0.1		
	$ML_2/M.L^2$	11.10[c]-0.2		
	$ML_3/M.L^3$	14.00[c]		
Cu[2+]	ML/M.L	8.49 ±0.03	-11.0	2
	$ML_2/M.L^2$	14.44 ±0.06	-19.2	2
	$ML_3/M.L^3$	16.0[c]		
Zn[2+]	ML/M.L	4.46[c]		
	$ML_2/M.L^2$	8.66[c]		
Cd[2+]	ML/M.L	3.98[c]		
	$ML_2/M.L^2$	6.79[c]		
	$ML_3/M.L^3$	8[c]		
Hg[2+]	ML/M.L	5.33		
	$ML_2/M.L^2$	9.47		
	$ML_3/M.L^3$	12.0		

---

[a] optical isomerism not stated;   [b] 25°, 0.25;   [c] 25°, 0.16

Bibliography:

H[+]   57LD,65AZ,65CMa,66HP,68HA,70MB,71HM   Zn[2+]   65AZ,65CMa

Be[2+]   70CA   Cd[2+]   65CMa

Co[2+]   65AZ   Ng[2+]   71HM

Ni[2+]   65CMa,71HM   Other references: 64AR,66PA

Cu[2+]   65CMa,67HM,70MBa,71HM

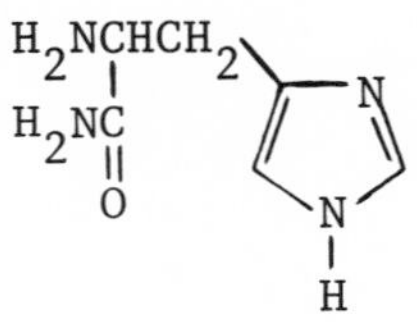

C$_6$H$_{10}$ON$_4$ L-Histidinamide   L

Metal ion	Equilibrium	Log K 25°, 0.5
H$^+$	HL/H.L	7.63
	H$_2$L/HL.H	5.70
Co^{2+}	ML/M.L	4.53
	ML$_2$/M.L^2	8.21

Bibliography: 69MM

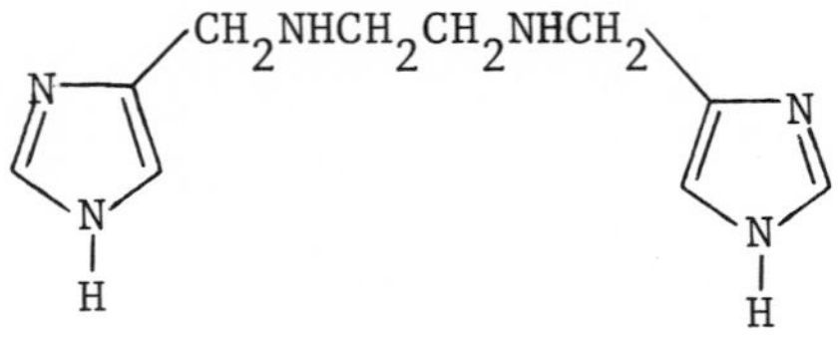

C$_{10}$H$_{16}$N$_6$ N,N'-Bis(4-imidazolylmethyl)ethylenediamine   L

Metal ion	Equilibrium	Log K 25°, 0.1
H$^+$	HL/H.L	9.05 -0.1
	H$_2$L/HL.H	6.56 +0.02
	H$_3$L/H$_2$L.H	4.26 +0.1
	H$_4$L/H$_3$L.H	3.21 +0.05
Co^{2+}	ML/M.L	11.43 -0.4
	MOHL/ML.OH	2.2[a]
Ni^{2+}	ML/M.L	14.02
Cu^{2+}	ML/M.L	16.5
Zn^{2+}	ML/M.L	10.39

[a] 25°, 0.2

Bibliography: H$^+$,Co^{2+} 68G,71ZK   Ni^{2+}-Zn^{2+} 68G

$C_8H_7N_3$ <u>2-(2-Pyridyl)imidazole</u>   L

Metal ion	Equilibrium	Log K 25°, 0.1	ΔH 25°, 0.1	ΔS 25°, 0.1
H^+	HL/H.L	5.47	−6.88	1.9
Fe^{2+}	ML/M.L	4.10		
	$ML_2/M.L^2$	7.90		
	$ML_3/M.L^3$	11.60	−19.4	−12
Co^{2+}	ML/M.L	5.26	−7.73	−1.9
	$ML_2/M.L^2$	10.05	−14.0	−1
	$ML_3/M.L^3$	13.87	−18.2	2
Ni^{2+}	ML/M.L	6.39	−9.25	−1.8
	$ML_2/M.L^2$	12.61	−17.9	−2
	$ML_3/M.L^3$	17.80	−23.0	4
Cu^{2+}	ML/M.L	7.94	−10.56	0.9
	$ML_2/M.L^2$	13.64	−17.2	5
	$ML_3/M.L^3$	16.92	−23.1	0
Zn^{2+}	ML/M.L	4.39	−6.37	−1.3
	$ML_2/M.L^2$	8.96	−13.2	−3
	$ML_3/M.L^3$	12.07	−14.9	5
Cd^{2+}	ML/M.L	4.70		
	$ML_2/M.L^2$	8.16		
	$ML_3/M.L^3$	10.74		
Hg^{2+}	ML/M.L	10.07		
	$ML_2/M.L^2$	18.28		

Bibliography:
H^+-Zn^{2+} 67EH,70EHP Cd^{2+},Hg^{2+} 67EH

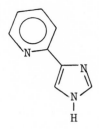

$C_8H_7N_3$ 4-(2-Pyridyl)imidazole   L

Metal ion	Equilibrium	Log K 25°, 0.1	ΔH 25°, 0.1	ΔS 25°, 0.1
H^+	HL/H.L	5.49	-5.76	5.8
Fe^{2+}	ML/M.L	4.93		
	$ML_2/M.L^2$	9.02		
	$ML_3/M.L^3$	13.76	-22.0	-11
Co^{2+}	ML/M.L	5.81		
	$ML_2/M.L^2$	11.32		
	$ML_3/M.L^3$	15.71		
Ni^{2+}	ML/M.L	7.20	-10.05	-0.7
	$ML_2/M.L^2$	13.95	-10.6	-2
	$ML_3/M.L^3$	19.82	-25.5	5
Cu^{2+}	ML/M.L	8.76	-11.71	0.8
	$ML_2/M.L^2$	15.16	-19.0	6
	$ML_3/M.L^3$	18.41	-23.5	5
Zn^{2+}	ML/M.L	5.42	-7.08	1.1
	$ML_2/M.L^2$	10.23	-12.9	4
	$ML_3/M.L^3$	13.84	-16.6	8

Bibliography:
H^+, Fe^{2+}, Ni^{2+}-Zn^{2+} 67EHP,70EHP Co^{2+} 67EHP

$C_7H_6N_2$ Benzo-1,3-diazole   (benzimidazole)   L

Metal ion	Equilibrium	Log K 25°, 0.16	Log K 20°, 0.15	ΔH 25°, 0.16	ΔS 25°, 0.16
H^+	HL/H.L	5.46	5.56	$(-9)^a$	(-5)
Cu^{2+}	ML/M.L	3.42	3.56	$(-9)^a$	(-15)
	$ML_2/M.L^2$	6.41	6.34		
	$ML_3/M.L^3$	8.92	9.00		
	$ML_4/M.L^4$	10.9	10.97		
Cu^+	ML/M.L		4.47		
	$ML_2/M.L^2$		9.73		

---
[a] 4-35°, 0.16

Bibliography:
H^+,Cu^{2+} 60LQ,62HP   Other reference: 69RW
Cu^+ 62HP

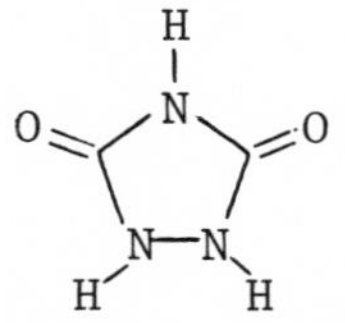

$C_2H_3O_2N_3$ 1,2,4-Triazolidin-3,5-dione   (urazole)   HL

Metal ion	Equilibrium	Log K 20°, 0.2
H^+	HL/H.L	5.69
Co^{2+}	ML/M.L	2.07
Ni^{2+}	ML/M.L	2.45
Zn^{2+}	ML/M.L	1.87

Bibliography: 63CO

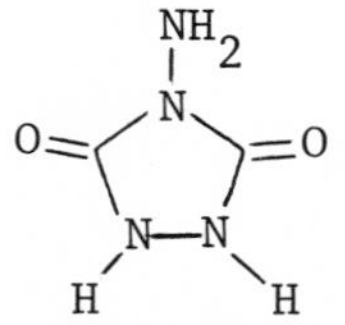

$C_2H_4O_2N_4$ 4-Amino-1,2,4-triazolidin-3,5-dione   (urazine)   HL

Metal ion	Equilibrium	Log K 20°, 0.2
H^+	HL/H.L	5.49
Co^{2+}	ML/M.L	2.34
Ni^{2+}	ML/M.L	2.65
	$ML_2/M.L^2$	4.80
Zn^{2+}	ML/M.L	2.17

Bibliography: 63CO

C_5H_5N			Pyridine (azine)			L
Metal ion	Equilibrium	Log K 25°, 0.1	Log K 25°, 0.5	Log K 25°, 0	ΔH 25°, 0	ΔS 25°, 0
H^+	HL/H.L	5.24 ±0.02	5.31 ±0.02	5.229	-4.92 ±0.1	7.4
			5.39[a]±0.02		-5.41[b]	6.2
Mn^{2+}	ML/M.L		0.14		-2^a	-6^b
Fe^{2+}	ML/M.L		0.6			
	$ML_2/M.L^2$		0.9			
Co^{2+}	ML/M.L	1.2	1.19 ±0.04			
	$ML_2/M.L^2$	1.8	1.70 ±0.00			
Ni^{2+}	ML/M.L	1.85	1.87 ±0.1		-3^a	-2^b
	$ML_2/M.L^2$		3.10 ±0.1			
	$ML_3/M.L^3$		3.71 ±0.0			
Cu^{2+}	ML/M.L	2.54	2.56 ±0.1	2.50	-4.0	-2
	$ML_2/M.L^2$		4.45 ±0.1	4.30	-8.9	-10
	$ML_3/M.L^3$		5.7 +0.1	5.16	-16.1	-30
	$ML_4/M.L^4$		6.5 +0.2	6.04	-22	-46
Cu^+	ML/M.L		4.84[c]			
	$Ml_2/M.L^2$		7.59[c]			
	$ML_3/M.L^3$		8.18[c]			
	$ML_4/M.L^4$		8.52[c]			
Ag^+	ML/M.L	(1.93)±0.03	2.06 ±0.06	2.05-0.05	-4.6	-6
					-4.78[b]±0.05	-6.6[b]
	$ML_2/M.L^2$	(4.22)±0.03	4.18 ±0.07	4.10+0.06	-11.2	-19
					-11.26[b]±0.08	-18.7[b]

[a] 25°, 1.0;   [b] 25°, 0.5;   [c] 20°, 1.0

Pyridine   (continued)

Metal ion	Equilibrium	Log K 25°, 0.1	Log K 25°, 0.5	Log K 25°, 0	ΔH 25°, 0	ΔS 25°, 0
Zn^{2+}	ML/M.L	1.0 ±0.1	0.99 ±0.1		-3^a	-6^b
	$ML_2/M.L^2$	1.6^d ±0.1	1.36 ±0.1			
	$ML_3/M.L^3$	1.9^d	1.55 ±0.05			
Cd^{2+}	ML/M.L	1.28 ±0.1	1.34 ±0.04			
	$ML_2/M.L^2$	2.02 −0.1	2.13 ±0.01			
	$ML_3/M.L^3$	2.3^d	2.41 ±0.09			
Hg^{2+}	ML/M.L		5.1			
	$ML_2/M.L^2$		10.0			
	$ML_3/M.L^3$		10.3			
	$ML_4/M.L^4$		10.6			

^{a} 25°, 1.0; ^{b} 25°, 0.5; ^{d} 30°, 0.1

Bibliography:

H^+ 48BV,49HJ,49LM,57LH,60SP,61DK,62HP,
64KS,65CC,67SB,68CWI,69G,69CI,70DT,
72MH,73BN,73CP,73YB

Mn^{2+} 63AB,73B

Fe^{2+} 73B

Co^{2+} 69NS,73B,73JV

Ni^{2+} 63AB,64KS,67SB,70FR, 70NB,71HB,73B

Cu^{2+} 48BV,61JW,63AB,64B,64KS,67SB,68IE,70FR

Cu^+ 73CP

Ag^+ 43VC,48BV,61CS,65N,66PV,67SB,68IE,
72B,72BE,72EB,72MH

Zn^{2+} 63AB,64KS,65SG,66DK,69NS,70FR,73B

Cd^{2+} 61DK,71BL,71HB,73B

Hg^{2+} 72B

Other references: 04E,30K,33AT,33T,35BW,
36BW,50DL,52F,53Nb,55MB,63BB,65N,
65PL,66DK,66FL,66GC,66LK,67FL,67N,
67RB,67SSa,67TK,69MB,69RJa,70NB,71SB

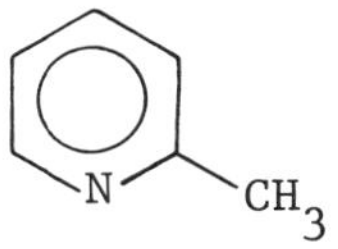

| C_6H_7N | | | 2-Methylpyridine | (2-picoline) | | L |

Metal ion	Equilibrium	Log K 25°, 0.1	Log K 25°, 0.5	Log K 25°, 0	ΔH 25°, 0	ΔS 25°, 0
H^+	HL/H.L	5.95	6.02 ±0.04	5.95 ±0.02	-6.22 +0.2	6.4
					-6.55[a]	5.6[a]
Cu^{2+}	ML/M.L	1.3	1.69[b]	1.75[c]		
	$ML_2/M.L^2$		2.8[b]	2.7[c]		
Cu^+	ML/M.L	5.40				
	$ML_2/M.L^2$	7.65				
	$ML_3/M.L^3$	8.5				
Ag^+	ML/M.L		2.33 ±0.06		-4.7[a]	-5[a]
	$ML_2/M.L^2$		4.66 +0.05		-11.2[a]	-16[a]

[a] 25°, 0.5;   [b] 25°, 0.6;   [c] 25°, 1.3

Bibliography:

H^+ 48BV,49LM,54AC,55BM,60SP,64KS,67SB, 72CS,72MH,73BEM

Cu^{2+} 64KS,64P,67SB

Cu^+ 64P

Ag^+ 48BV,67SB,72EB,72MH

Other references: 65PL,67RB,71SB

C_6H_7N 3-Methylpyridine   (3-picoline)   L

Metal ion	Equilibrium	Log K 25°, 0.1	Log K 25°, 0.5	Log K 25°, 0	ΔH 25°, 0	ΔS 25°, 0
H^+	HL/H.L	5.76 +0.1	5.87 −0.1	5.68 ±0.05	−5.71 +0.1	6.8
					−6.04[a]	6.6[a]
Ni^{2+}	ML/M.L	1.85	1.97[b]			
	$ML_2/M.L^2$		3.21[b]			
	$ML_3/M.L^3$		3.9[b]			
Cu^{2+}	ML/M.L	2.77	2.70[b]	2.76[c]		
	$ML_2/M.L^2$		4.72[b]	4.67[c]		
	$ML_3/M.L^3$		6.12[b]	6.13[c]		
	$ML_4/M.L^4$		6.9[b]	7.03[c]		
Cu^+	ML/M.L	5.60				
	$ML_2/M.L^2$	7.78				
	$ML_3/M.L^3$	8.6				
	$ML_4/M.L^4$	9.0				
Ag^+	ML/M.L		2.25 −0.1	2.00	−5.2[a]	−7[a]
	$ML_2/M.L^2$		4.48 −0.04	4.35	−11.9[a]	−19[a]
Zn^{2+}	ML/M.L	1.23[d]				
	$ML_2/M.L^2$	1.9[d]				
	$ML_3/M.L^3$	2.2[d]				
Cd^{2+}	ML/M.L	1.34[d]±0.07	1.28[e]			
	$ML_2/M.L^2$	2.3[d] ±0.1				
	$ML_3/M.L^3$	2.5[d]				

---

[a] 25°, 0.5;   [b] 25°, 0.6;   [c] 25°, 1.3;   [d] 30°, 0.1;   [e] 25°, 0.3

Bibliography:

H^+ 54AC,55BM,60SP,61DK,61ES,64KS,67SB, 72CS,72MH,73BEM

Ni^{2+} 64KS,67SB

Cu^{2+} 64KS,64P,67SB

Cu^+ 64P

Ag^+ 55MB,67SB,72EB,72MH

Zn^{2+} 66DK

Cd^{2+} 61DK,67N,68GS

Other references: 65PL,67RB,71SB

C$_6$H$_7$N   4-Methylpyridine   (4-picoline)   L

Metal ion	Equilibrium	Log K 25°, 0.1	Log K 25°, 0.5	Log K 25°, 0	ΔH 25°, 0	ΔS 25°, 0
H$^+$	HL/H.L	6.04 ±0.06	6.11 −0.09	6.03 ±0.05	−5.96 −0.1	7.6
					−6.44[a]	6.4[a]
Co^{2+}	ML/M.L		1.56	1.59[c]		
	ML$_2$/M.L^2		2.51	2.58[c]		
	ML$_3$/M.L^3		2.94			
	ML$_4$/M.L^4		3.17			
Ni^{2+}	ML/M.L	2.11	2.09	2.15[c]		
	ML$_2$/M.L^2		3.59	3.83[c]		
	ML$_3$/M.L^3		4.34	4.81		
	ML$_4$/M.L^4		4.70			
Cu^{2+}	ML/M.L	2.88	2.93[b]	2.99[d]		
	ML$_2$/M.L^2		5.16[b]	5.19[d]		
	ML$_3$/M.L^3		6.77[b]	6.82[d]		
	ML$_4$/M.L^4		8.08[b]	7.87[d]		
	ML$_5$/M.L^5			8.3[d]		
Cu$^+$	ML/M.L	5.65				
	ML$_2$/M.L^2	8.20				
	ML$_3$/M.L^3	8.8				
	ML$_4$/M.L^4	9.2				
Ag$^+$	ML/M.L		2.18 +0.06	2.03	−6.1[a]	−11[a]
	ML$_2$/M.L^2		4.64 +0.1	4.39	−12.8[a]	−22
Zn^{2+}	ML/M.L	1.40	1.30[e]	1.46[c]		
	ML$_2$/M.L^2		2.11[e]	2.45[c]		
	ML$_3$/M.L^3		2.85[e]	2.72[c]		
Cd^{2+}	ML/M.L		1.51[e] ±0.01			
	ML$_2$/M.L^2		2.47[e] −0.3			
	ML$_3$/M.L^3		2.90[e] ±0.07			

[a] 25°, 0.5; [b] 25°, 0.6; [c] 25°, 1.0; [d] 25°, 1.3; [e] 30°, 0.1

4-Methylpyridine   (continued)

Bibliography:

H+   48BV,54AC,55BM,55MB,60SP,61DK,62HP,   Cu+   64P
      64KS,67SB,72CS,72MH,73BEM,73BN   Ag+   48BV,67SB,72EB,72MH,73BN
Co2+   69LW,73NB   Zn2+   64KS,66DK,69LW
Ni2+   64KS,67SB,69LW,73NB   Cd2+   61DK,68GS
Cu2+   64KS,64P,67SB   Other references: 65PL,67RB,71HB,71SB,72PG

C$_7$H$_9$N   4-Ethylpyridine   L

Metal ion	Equilibrium	Log K 25°, 0.5	Log K 25°, 1.0	Log K 25°, 0	ΔH 25°, 0.5	ΔS 25°, 0.5
H+	HL/H.L	(6.33)	(5.89)	5.87	−6.67	6.6
Co^{2+}	ML/M.L		1.23			
	ML$_2$/M.L^2		2.06			
Ni^{2+}	ML/M.L		1.91			
	ML$_2$/M.L^2		3.34			
	ML$_3$/M.L^3		4.02			
Zn^{2+}	ML/M.L		1.32			
	ML$_2$/M.L^2		2.08			

Bibliography:

H+   54AC,71LW,73BEM   Co^{2+}-Zn^{2+} 71LW

C_7H_9N	2,3-Dimethylpyridine (2,3-lutidine)				L

Metal ion	Equilibrium	Log K 25°, 0.5	Log K 25°, 0	ΔH 25°, 0.5	ΔS 25°, 0.5
H^+	HL/H.L	6.75	6.57	-7.46	5.9
Ag^+	ML/M.L	2.45			
	$ML_2/M.L^2$	4.78			

Bibliography:

H^+ 54AC,72MH,73BEM Ag^+ 72MH

C_7H_9N	2,4-Dimethylpyridine (2,4-lutidine)				L

Metal ion	Equilibrium	Log K 25°, 0.5	Log K 25°, 0	ΔH 25°, ∿0	ΔS 25°, 0
H^+	HL/H.L	6.83 ±0.01	6.63	-7.2 -8.26[a]	6 3.6[a]
Co^{2+}	ML/M.L	0.51[b]			
Ni^{2+}	ML/M.L	0.56[b]			
Ag^+	ML/M.L	2.54 -0.07			
	$ML_2/M.L^2$	5.07 +0.1			
Zn^{2+}	ML/M.L	0.54[b]			

[a] 25°, 0.5;   [b] 25°, 1.0

Bibliography:

H^+ 48BV,54AC,60SP,72MH,73BEM Cu^{2+},Ni^{2+},Zn^{2+} 71WL

Ag^+ 48BV,72MH   Other reference: 73SB

C_7H_9N 2,5-Dimethylpyridine   (2,5-lutidine)   L

Metal ion	Equilibrium	Log K 25°, 0.5	Log K 25°, 0	ΔH 25°, ∼0	ΔS 25°, 0
H^+	HL/H.L	6.58 −0.1	6.40	−6.8 −7.12[a]	7 6.2[a]
Cu^{2+}	ML/M.L	1.78[b]			
	$ML_2/M.L^2$	2.8[b]			
Ag^+	ML/M.L	2.56 −0.1			
	$ML_2/M.L^2$	4.91 +0.04			

[a] 25°, 0.5;   [b] 25°, 0.6

Bibliography:
H^+ 54AC,60SP,67SB,72MH,73BEM Ag^+ 67SB,72MH
Cu^{2+} 67SB

C_7H_9N 2,6-Dimethylpyridine   (2,6-lutidine)   L

Metal ion	Equilibrium	Log K 25°, 0.5	Log K 25°, 0	ΔH 25°, ∼0	ΔS 25°, 0
H^+	HL/H.L	6.90 −0.2	6.72 +0.03	−7.2 −8.36[a]	7 3.5[a]
Ag^+	ML/M.L	2.68			
	$ML_2/M.L^2$	5.06			

Bibliography:
H^+ 54AC,55BM,60SP,67SB,72MH,73BEM   Other references: 65PL,73SB
Ag^+ 72MH

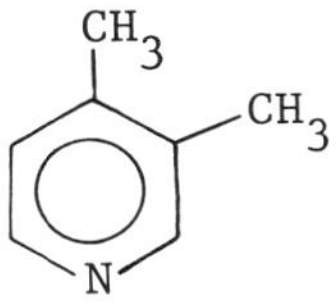

C_7H_9N

3,4-Dimethylpyridine   (3,4-lutidine)   L

Metal ion	Equilibrium	Log K 25°, 0.5	Log K 25°, 0	ΔH 25°, 0.5	ΔS 25°, 0.5
H^+	HL/H.L	6.65 -0.2	6.46	-7.48	5.3
Ni^{2+}	ML/M.L	2.26[a]			
	$ML_2/M.L^2$	3.2[a]			
	$ML_3/M.L^3$	5.21[a]			
Cu^{2+}	ML/M.L	3.11[a]			
	$ML_2/M.L^2$	5.38[a]			
	$ML_3/M.L^3$	7.46[a]			
	$ML_4/M.L^4$	8.7[a]			
Ag^+	ML/M.L	2.43			
	$ML_2/M.L^2$	4.85			

---

[a] 25°, 0.6

Bibliography:
H^+ 54AC,67SB,72MH,73BEM Ag^+ 72MH
Ni^{2+},Cu^{2+} 67SB

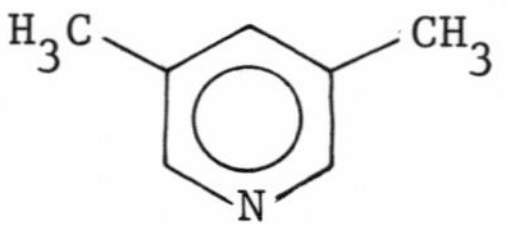

C_7H_9N 3,5-Dimethylpyridine   (3,5-lutidine)   L

Metal ion	Equilibrium	Log K 25°, 0.5	Log K 25°, 0	ΔH 25°, ~0	ΔS 25°, 0
H^+	HL/H.L	6.24 −0.08	6.17 ±0.00	−6.4	7
Ni^{2+}	ML/M.L	2.13[a]			
	$ML_2/M.L^2$	3.1[a]			
	$ML_3/M.L^3$	4.87[a]			
Cu^{2+}	ML/M.L	2.94[a]			
	$ML_2/M.L^2$	5.10[a]			
	$ML_3/M.L^3$	6.8[a]			
	$ML_4/M.L^4$	8.1[a]			
Ag^+	ML/M.L	2.36 +0.01			
	$ML_2/M.L^2$	4.66 −0.01			

---

[a] 25°, 0.6

Bibliography:

H^+ 54AC,60SP,61ES,67SB,72MH Ag^+ 67SB,72MH
Ni^{2+},Cu^{2+} 67SB

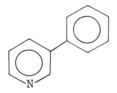

C$_{11}$H$_9$N   <u>2-Phenylpyridine</u>   L

Metal ion	Equilibrium	Log K 25°, 0.1
H$^+$	HL/H.L	4.64 ±0.02
Cu^{2+}	ML/M.L	1.0 ±0.3

Bibliography: 64KS,72SW

C$_{11}$H$_9$N   <u>3-Phenylpyridine</u>   L

Metal ion	Equilibrium	Log K 25°, 0.1
H$^+$	HL/H.L	4.81
Cu^{2+}	ML/M.L	2.25

Bibliography: 72SW

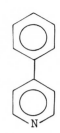

$C_{11}H_9N$ 4-Phenylpyridine   L

Metal ion	Equilibrium	Log K 25°, 0.1
H^+	HL/H.L	5.38
Cu^{2+}	ML/M.L	2.54

Bibliography: 72SW

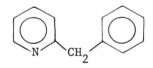

$C_{12}H_{11}N$ 2-Benzylpyridine   L

Metal ion	Equilibrium	Log K 25°, 0.1
H^+	HL/H.L	5.12
Cu^{2+}	ML/M.L	0.8

Bibliography: 72SW

$C_6H_4N_2$ <u>3-Cyanopyridine</u>   L

Metal ion	Equilibrium	Log K 25°, 0
Ag^+	$ML_2/M.L^2$	2.90

Bibliography: 55MB

CN

$C_6H_4N_2$ <u>4-Cyanopyridine</u>   L

Metal ion	Equilibrium	Log K 25°, 0
Ag^+	$ML_2/M.L^2$	3.08

Bibliography: 55MB

$C_{11}H_9O_2N_3$ <u>1,3-Dihydroxy-4-(2-pyridylazo)benzene</u> H_2L
                                     <u>(2,4'-pyridylazoresorcinol</u>, <u>PAR</u>)

Metal ion	Equilibrium	Log K 25°, 0.1	Log K 20°, 0.1
H^+	HL/H.L	(12.31)	12.31
	H_2L/HL.H	5.50	5.71
	H_3L/H_2L.H	2.69	2.75
Sc^{3+}	MHL/M.HL		6.25
La^{3+}	MHL/M.HL		2.69
Pr^{3+}	MHL/M.HL		3.35
Nd^{3+}	MHL/M.HL		3.45
Sm^{3+}	MHL/M.HL		3.49
Eu^{3+}	MHL/M.HL		3.50
Gd^{3+}	MHL/M.HL		3.52
Tb^{3+}	MHL/M.HL		3.43
Dy^{3+}	MHL/M.HL		3.48
Ho^{3+}	MHL/M.HL		3.60
Er^{3+}	MHL/M.HL		3.66
Tm^{3+}	MHL/M.HL		3.77
Yb^{3+}	MHL/M.HL		3.77
Lu^{3+}	MHL/M.HL		3.81
UO_2^{2+}	MHL/M.HL	12.5	
	$M(HL)_2$/M.$(HL)^2$	20.9	
Co^{2+}	MHL/M.HL	10.0	
	$M(HL)_2$/M.$(HL)^2$	17.1	
Cu^{2+}	MHL/M.HL	14.8	
	$M(HL)_2$/M.$(HL)^2$	23.9	

PAR   (continued)

Metal ion	Equilibrium	Log K 25°, 0.1	Log K 20°, 0.1
Zn^{2+}	MHL/M.HL	10.5	
	$M(HL)_2/M.(HL)^2$	17.1	
Pb^{2+}	MHL/M.HL	8.6	
	$M(HL)_2/M.(HL)^2$	15.7	

Bibliography:

H^+ 62GN,71EK

Sc^{3+}-Lu^{3+} 71EK

UO_2^{2+}-Pb^{2+} 62GN

Other references: 59KL,61HS,61I,62SH,63HS, 66BV,66DM,66HS,67AD,67SIN,67SN,68TF

$C_5H_5O_2N$ 2,3-Dihydroxypyridine   HL

Metal ion	Equilibrium	Log K 25°, 1.0
H^+	HL/H.L	8.69
	$H_2L/HL.H$	0.11
Fe^{3+}	ML/M.L	12.13
	MHL/M.HL	2.13

Bibliography: 72CA   Other reference: 70GD

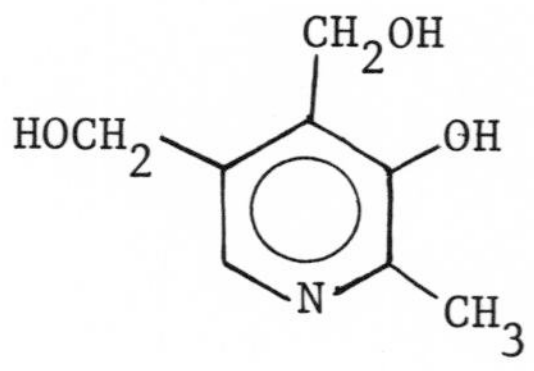

$C_8H_{11}O_3N$ 3-Hydroxy-4,5-bis(hydroxymethyl)-2-methylpyridine   HL

(pyridoxine)

Metal ion	Equilibrium	Log K 25°, 0.15	Log K 30°, 0.5
H^+	HL/H.L	8.84	(8.92)
	$H_2L/HL.H$	4.88	4.90
UO_2^{2+}	ML/M.L		11.49
	$ML_2/M.L^2$		19.97
	$ML_3/M.L^3$		23.73
Fe^{3+}	ML/M.L		13.7[a]

———————————

[a] 20°, 0.25

Bibliography:
H^+ 54WN,71SE Fe^{3+} 60C
UO_2^{2+} 71SE

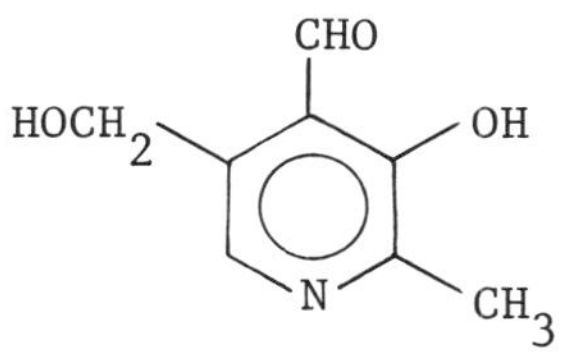

$$\begin{array}{c} \text{CHO} \\ \text{HOCH}_2 \diagdown \diagup \text{OH} \\ \diagup \diagdown \\ \text{N} \quad \text{CH}_3 \end{array}$$

$C_8H_9O_3N$		3-Hydroxy-5-hydroxymethyl-2-methylpyridine-4-carboxaldehyde	HL
		(pyridoxal)	

Metal ion	Equilibrium	Log K 25°, 0.1	Log K 25°, 0.5
H^+	HL/H.L	8.57 ±0.02	8.39
	H_2L/HL.H	4.10 ±0.01	4.10
Ni^{2+}	ML/M.L		1.85
Cu^{2+}	ML/M.L		3.51
	$ML_2/M.L^2$		7.0
Zn^{2+}	ML/M.L		2.32

Bibliography:

H^+ 54WN,55MS,66LH Ni^{2+}-Zn^{2+} 66LH

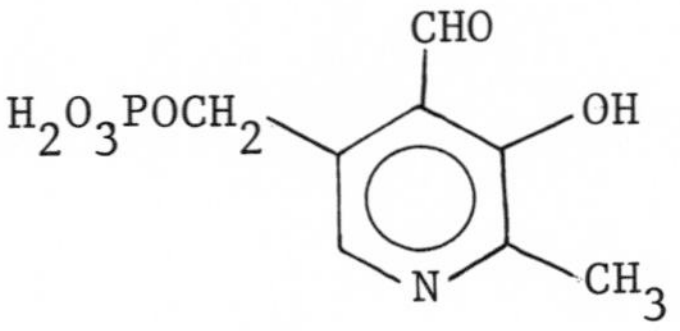

$C_8H_{10}O_6NP$ Pyridoxal-5-(dihydrogenphosphate) H_3L

Metal ion	Equilibrium	Log K 25°, 0.1	Log K 25°, 0.5	Log K 25°, 2.0
H^+	HL/H.L	8.45	(7.99)	8.17
	H_2L/HL.H	6.01	5.83	5.75
	H_3L/H_2L.H	3.44	3.48	3.58
	H_4L/H_3L.H	1.4		1.64
Zn^{2+}	ML/M.L		3.6	
	MHL/ML.H		6.3	
	MH_2L/MHL.H		5.6	

Bibliography:

H^+ 64AM,70FE Zn^{2+} 70FE

$C_8H_{12}O_2N_2$ 4-Aminomethyl-3-hydroxy-5-hydroxymethyl-2-methylpyridine   HL

(pyridoxamine)

Metal ion	Equilibrium	Log K 25°, 0.1
H^+	HL/H.L	10.13 +0.4
	$H_2L/HL.H$	8.01 +0.08
	$H_3L/H_2L.H$	3.37 +0.05
Mn^{2+}	ML/M.L	3.56
Co^{2+}	ML/M.L	5.09
	$ML_2/M.L^2$	9.60
Ni^{2+}	ML/M.L	6.00
	$ML_2/M.L^2$	10.92
Cu^{2+}	ML/M.L	10.20
	$ML_2/M.L^2$	15.97
Zn^{2+}	ML/M.L	5.68
Cd^{2+}	ML/M.L	4.59

Bibliography:
H^+ 54WN,57GM Mn^{2+}-Zn^{2+} 57GM

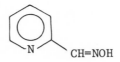

C$_6$H$_6$ON$_2$ Pyridine-2-carboxaldehyde oxime   HL

Metal ion	Equilibrium	Log K 25°, 0.1	Log K 20°, 0.5	Log K 25°, 0	ΔH 25°, 0	ΔS 25°, 0
H$^+$	HL/H.L	9.90	9.92	10.18 −0.01	−2.1	39
	H$_2$L/HL.H	3.5	3.69	3.55 ±0.05	−4.8	0
Mn^{2+}	ML/M.L		5.2[a]			
	ML$_2$/M.L^2		9.1[a]			
Fe^{2+}	MHL$_3$/ML$_3$.H			7.13	(−1)[b]	(29)
	MH$_2$L$_3$/MHL$_3$.H			3.36	(−1)[c]	(12)
Co^{2+}	ML/M.L		8.7[a]			
	ML$_2$/M.L^2		17.4[a]			
Ni^{2+}	ML/M.L		9.4[a]			
	ML$_2$/M.L^2		16.5[a]			
	ML$_3$/M.L^3		22.0[a]			
Cu^{2+}	M(OH)$_2$L/M.(OH)^2.L		19.85			
	ML$_2$/M.L^2		18.6			
	MHL$_2$/ML$_2$.H		7.17			
	MH$_2$L$_2$/MHL$_2$.H		2.37			
	MOHL$_2$/ML$_2$.OH		3.3			
Cu$^+$	ML$_2$/M.L^2		14.4			
	MH$_2$L$_2$/M.(HL)^2		11.05			
Zn^{2+}	ML/M.L		5.8[a]			
	ML$_2$/M.L^2		11.1[a]			

---

[a]25°, 0.3;   [b] 18-34°, 0;   [c] 25-33°, 0

Bibliography:

H$^+$ 58C,61GF,62BE,73P   Cu^{2+},Cu$^+$ 73P

Mn^{2+},Co^{2+},Ni^{2+},Zn^{2+} 66BE   Other references: 61LL,62IJ,62K

Fe^{2+} 62HI

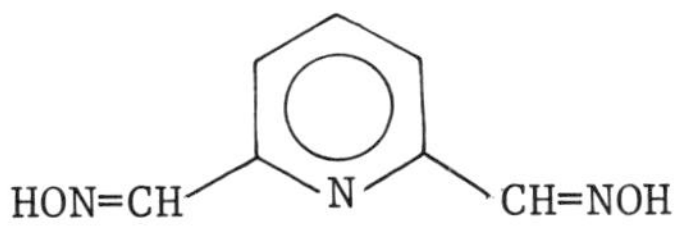

$$HON=CH \quad \boxed{N} \quad CH=NOH$$

$C_7H_7O_2N_3$ Pyridine-2,6-dicarboxaldehyde dioxime H_2L

Metal ion	Equilibrium	Log K 25°, 0.05	Log K 25°, 0	ΔH 25°, 0.05	ΔS 25°, 0.05
H^+	HL/H.L	10.54	10.88	(-7)[a]	(25)
	H_2L/HL.H	9.91	10.08	(-5)[a]	(29)
Fe^{2+}	$MHL_2/ML_2.H$	7.04	7.40	(0)[a]	(32)

[a] 20-35°, 0.05

Bibliography: 65HI   Other reference: 63BFa

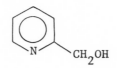

C_6H_7ON $\qquad\qquad$ 2-(Hydroxymethyl)pyridine $\qquad\qquad$ L

Metal ion	Equilibrium	Log K 25°, 0.1	Log K 25°, 0.6	Log K 20°, 1.0
H^+	$L/H_{-1}L.H$			(13.9)
	$HL/H.L$	4.89	4.95	5.15
				4.86[a]
Mn^{2+}	$ML/M.L$	1		
Co^{2+}	$ML/M.L$	2.1		
Ni^{2+}	$ML/M.L$	2.79	2.90	
	$ML_2/M.L^2$	5.39	5.26	
	$ML_3/M.L^3$		7.04	
Cu^{2+}	$ML/M.L$	3.41	3.56	3.75
	$ML_2/M.L^2$	6.22	6.23	6.68
	$ML_3/M.L^3$		8.00	8.40
	$ML_4/M.L^4$		8.3	
	$M(H_{-1}L)_2/M.(H_{-1}L)^2$			(23.0)
Cu^+	$ML_2/M.L^2$			9.7
Ag^+	$ML/M.L$		2.14	
	$ML_2/M.L^2$		4.37	
Zn^{2+}	$ML/M.L$	1.9		

[a] 25°, 0

Bibliography:
H^+,Cu^{2+} 65MT,67SB,67TT,73P Cu^+ 73P
Mn^{2+},Co^{2+},Zn^{2+} 65MT Ag^+ 67SB
Ni^{2+} 65MT,67SB   Other references: 55LFa,68TM

C_6H_7ON 3-(Hydroxymethyl)pyridine   L

Metal ion	Equilibrium	Log K 25°, 0.6	Log K 20°, 0.5
H^+	HL/H.L	5.04	5.10
			4.90[a]
Ni^{2+}	ML/M.L	1.85	
	$ML_2/M.L^2$	2.99	
	$ML_3/M.L^3$	5.0	
Cu^{2+}	ML/M.L	2.43	2.49
	$ML_2/M.L^2$	4.27	4.37
	$ML_3/M.L^3$	5.0	
Cu^+	$ML_2/M.L^2$		7.15
Ag^+	ML/M.L	2.01	
	$ML_2/M.L^2$	4.09	

[a] 25°, 0

Bibliography:
H^+,Cu^{2+} 67SB,67TT,73P Cu^+ 73P
Ni^{2+},Ag^+ 67SB

C$_6$H$_7$ON   4-(Hydroxymethyl)pyridine   L

Metal ion	Equilibrium	Log K 25°, 0.6	Log K 25°, 0
H$^+$	HL/H.L	5.51	5.33
Ni^{2+}	ML/M.L	1.97	
	ML$_2$/M.L^2	3.02	
Cu^{2+}	ML/M.L	2.65	
	ML$_2$/M.L^2	4.53	
	ML$_3$/M.L^3	5.7	
Ag$^+$	ML/M.L	2.15	
	ML$_2$/M.L^2	4.23	

Bibliography:

H$^+$ 67SB,67TT   Ni^{2+}-Ag$^+$ 67SB

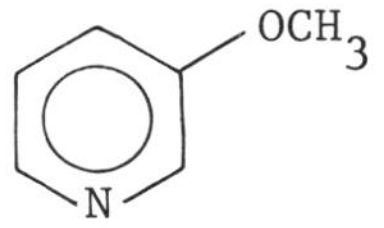

C_6H_7ON 3-Methoxypyridine   L

Metal ion	Equilibrium	Log K 30°, 0.1	Log K 25°, 0
H^+	HL/H.L		4.91
Ag^+	ML/M.L		1.58
	$ML_2/M.L^2$		3.67
Zn^{2+}	ML/M.L	0.90	
	$ML_2/M.L^2$	1.40	
	$ML_3/M.L^3$	1.53	
Cd^{2+}	ML/M.L	1.11	

Bibliography:

H^+, Ag^+ 55MB Zn^{2+}, Cd^{2+} 66DK

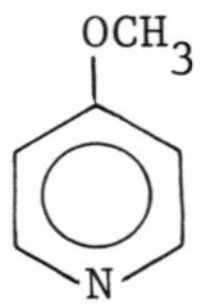

$$\text{OCH}_3$$

C$_6$H$_7$ON   4-Methoxypyridine   L

Metal ion	Equilibrium	Log K 30°, 0.1	Log K 25°, 0
H$^+$	HL/H.L		6.47
Ag$^+$	ML/M.L		2.28
	ML$_2$/M.L^2		4.44
Zn^{2+}	ML/M.L	1.53	
	ML$_2$/M.L^2	2.31	
	ML$_3$/M.L^3	3.08	
Cd^{2+}	ML/M.L	1.65	

Bibliography:

H$^+$,Ag$^+$ 55MB   Zn^{2+},Cd^{2+} 66DK

C$_6$H$_5$ON   Pyridine-2-carboxaldehyde   L

Metal ion	Equilibrium	Log K 25°, 0	ΔH 25°, 0	ΔS 25°, 0
H$^+$	HL/H.L	3.84	-6.5	-4
Cu^{2+}	ML/M.L	2.65		
	ML$_2$/M.L^2	4.34		
	ML/MOHL.H	4.29		
	ML$_2$/MOHL$_2$.H	3.89		
	MOHL$_2$/M(OH)$_2$L$_2$.H	5.16		

Bibliography:

H$^+$ 61GF

Cu^{2+} 71GR

Other reference: 62KI

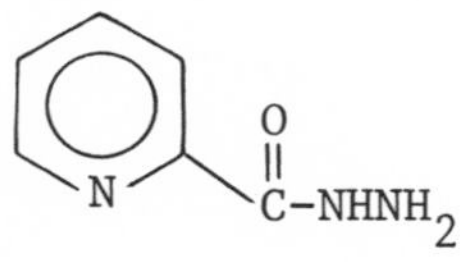

C$_6$H$_7$ON$_3$ Pyridine-2-carboxylic acid hydrazide   (picolinoylhydrazine)   HL

Metal ion	Equilibrium	Log K 25°, 1.0
H$^+$	HL/H.L	12.11
	H$_2$L/HL.H	2.93
	H$_3$L/H$_2$L.H	1.12
Cu^{2+}	MHL/M.HL	3.72
	MHL/ML.H	1.20
	ML$_2$.H^2/MHL.HL	-0.57

Bibliography:

H$^+$ 63NT

Cu^{2+} 63NK

Other reference: 56A

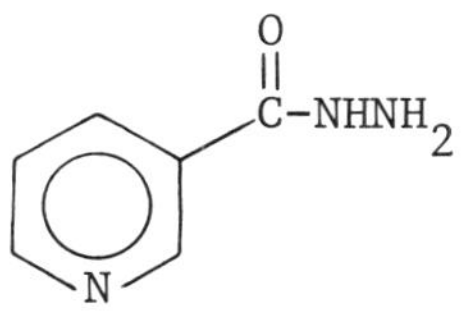

$$\overset{\displaystyle O}{\underset{\displaystyle}{\parallel}}$$

C-NHNH$_2$ (attached to pyridine ring)

$C_6H_7ON_3$ <u>Pyridine-3-carboxylic acid hydrazide</u>   (nicotinoylhydrazine)   HL

Metal ion	Equilibrium	Log K 25°, 1.0
H^+	HL/H.L	11.35
	H_2L/HL.H	3.49
	H_3L/HL.H	2.12
Cu^{2+}	MHL/M.HL	3.09
	$M(HL)_2$/M.$(HL)^2$	6.30
	MHL/M.HL	4.92
	$M(HL)_2$/MHL_2.H	1.56
	MHL_2/ML_2.H	4.26

Bibliography:

H^+ 63NT

Cu^{2+} 63NK

Other reference: 56A

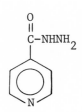

$C_6H_7ON_3$ Pyridine-4-carboxylic acid hydrazide   (isonicotinoylhydrazine)   HL

Metal ion	Equilibrium	Log K 25°, 1.0
H^+	HL/H.L	10.89
	$H_2L/HL.H$	3.67
	$H_3L/H_2L.H$	1.99
Mn^{2+}	MHL/M.HL	1.04
	MHL/ML.H	7.79
Co^{2+}	MHL/M.HL	1.64
	MHL/ML.H	7.97
Ni^{2+}	MHL/M.HL	2.59
	MHL/ML.H	7.86
Cu^{2+}	MHL/M.HL	3.15
	$M(HL)_2/M.(HL)^2$	5.08
	MHL/ML.H	4.02
	$M(HL)_2/MHL_2.H$	2.11
	$MHL_2/ML_2.H$	3.81
Zn^{2+}	MHL/M.HL	1.86
	MHL/ML.H	(5.57)
Cd^{2+}	MHL/M.HL	1.09
	MHL/ML.H	8.55

Bibliography:
H^+ 63NT   Other references: 53Aa,63TN,65KSD
$Mn^{2+}-Cd^{2+}$ 63NK

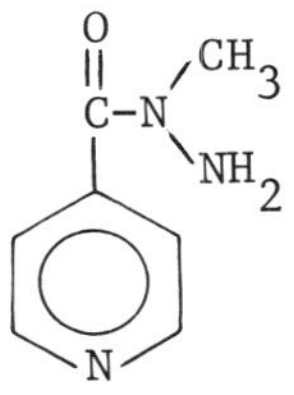

$$\begin{array}{c}
O \quad CH_3 \\
\| \quad / \\
C-N \\
\quad \backslash NH_2
\end{array}$$

$C_7H_9ON_3$ <u>Pyridine-4-carboxylic acid 1-methylhydrazide</u>   L

<u>(1-isonicotinoyl-1-methylhydrazine)</u>

Metal ion	Equilibrium	Log K 25°, 1.0
H^+	HL/H.L	4.03
	$H_2L/HL.H$	0.89
Cu^{2+}	ML/M.L	3.02
	$ML_2/M.L^2$	5.89
	$ML_3/M.L^3$	7.96

Bibliography:

H^+ 63NT Cu^{2+} 63NK

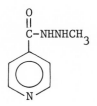

C$_7$H$_9$ON$_3$ <u>Pyridine-4-carboxylic acid 2-methylhydrazide</u>   HL

<u>(1-isonicotinoyl-2-methylhydrazine)</u>

Metal ion	Equilibrium	Log K 25°, 1.0
H$^+$	HL/H.L	10.82
	H$_2$L/HL.H	3.90
	H$_3$L/H$_2$L.H	2.32
Cu^{2+}	MHL/M.HL	3.54
	M(HL)$_2$/M.(HL)^2	7.08
	MHL/ML.H	4.22
	M(HL)$_2$/MHL$_2$.H	2.35

Bibliography:

H$^+$ 63NT   Cu^{2+} 63NK

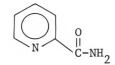

C$_6$H$_6$ON$_2$ <u>Pyridine-2-carboxylic acid amide</u>   (picolinamide)

Metal ion	Equilibrium	Log K 25°, 0.16
H$^+$	HL/H.L	1.8
Ni^{2+}	ML$_2$/M(H$_{-1}$L)L.H	7.88
	M(H$_{-1}$L)/M(H$_{-1}$L)$_2$.H	9.33
Cu^{2+}	ML$_2$/M(H$_{-1}$L)L.H	4.98
	M(H$_{-1}$L)L/M(H$_{-1}$L)$_2$.H	6.40

Bibliography: 65CM   Other references: 54JU,72CM,73MW

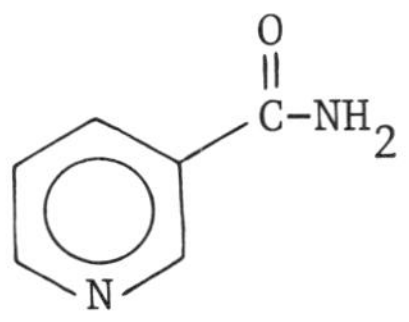

$C_6H_6ON_2$ Pyridine-3-carboxylic acid amide (nicotinamide) L

Metal ion	Equilibrium	Log K 25°, 0.5	Log K 25°, 0
H^+	HL/H.L	3.47	3.35[a]
Co^{2+}	ML/M.L	1.00	
	$ML_2/M.L^2$	1.60	
Ni^{2+}	ML/M.L	1.49	
	$ML_2/M.L^2$	2.43	
	$ML_3/M.L^3$	3.00	
Cu^{2+}	ML/M.L	1.79	
	$ML_2/M.L^2$	2.83	
	$ML_3/M.L^3$	3.30	
Ag^+	$ML_2/M.L^2$		3.22
Zn^{2+}	ML/M.L	0.78	
	$ML_2/M.L^2$	1.18	

[a] 20°, 0

Bibliography:
H^+ 51JW,71WL
Co^{2+}-Cu^{2+},Zn^{2+} 71WL
Ag^+ 55MB

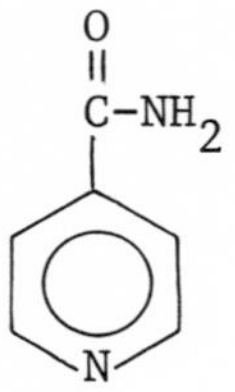

$$\text{O} \atop \text{C-NH}_2$$

$C_6H_6ON_2$ Pyridine-4-carboxylic acid amide   (isonicotinamide)   L

Metal ion	Equilibrium	Log K 25°, 1.0	Log K 25°, 0
H^+	HL/H.L	3.68	3.61^a
Cu^{2+}	ML/M.L	2.33	
	$ML_2/M.L^2$	3.38	
Ag^+	$ML_2/M.L^2$		3.01

_____

[a] 20°, 0

Bibliography:

H^+ 54JU,63NK Ag^+ 55MB
Cu^{2+} 63NK

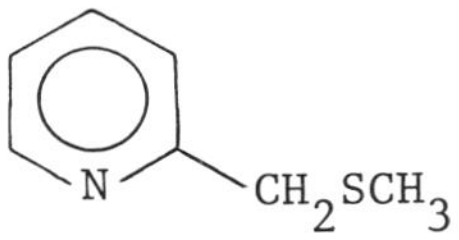

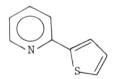

C₇H₉NS   2-(Methylthiomethyl)pyridine   L

Metal ion	Equilibrium	Log K 25°, 0.1
H⁺	HL/H.L	4.53
Ni²⁺	ML/M.L	2.06
Cu²⁺	ML/M.L	3.27

Bibliography: 64KS

C₉H₇NS   2-(2-Thienyl)pyridine   L

Metal ion	Equilibrium	Log K 25°, 0.1
H⁺	HL/H.L	3.69
Cu²⁺	ML/M.L	0.2

Bibliography: 72SW

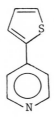

C_9H_7NS 3-(2-Thienyl)pyridine   L

Metal ion	Equilibrium	Log K 25°, 0.1
H^+	HL/H.L	4.41
Cu^{2+}	ML/M.L	2.15

Bibliography: 72SW

C_9H_7NS 4-(2-Thienyl)pyridine   L

Metal ion	Equilibrium	Log K 25°, 0.1
H^+	HL/H.L	5.48 ±0.00
Ni^{2+}	ML/M.L	1.91
Cu^{2+}	ML/M.L	2.57 +0.01
Zn^{2+}	ML/M.L	1.10

Bibliography:
H^+, Cu^{2+} 64KS,72SW Ni^{2+}, Zn^{2+} 64KS

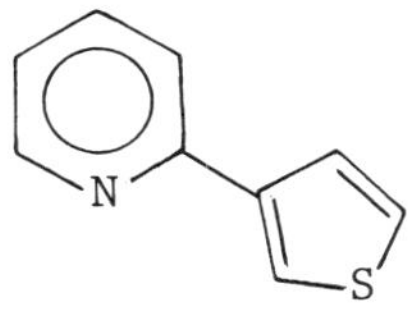

C$_9$H$_7$NS   2-(3-Thienyl)pyridine   L

Metal ion	Equilibrium	Log K 25°, 0.1
H$^+$	HL/H.L	4.67
Cu^{2+}	ML/M.L	0.7

Bibliography: 72SW

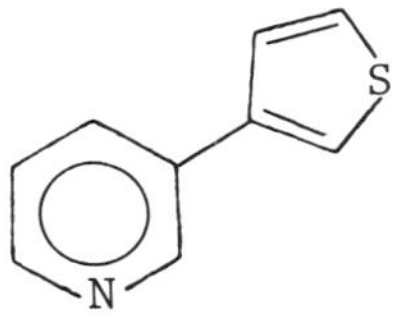

C$_9$H$_7$NS   3-(3-Thienyl)pyridine   L

Metal ion	Equilibrium	Log K 25°, 0.1
H$^+$	HL/H.L	4.81
Cu^{2+}	ML/M.L	2.30

Bibliography: 72SW

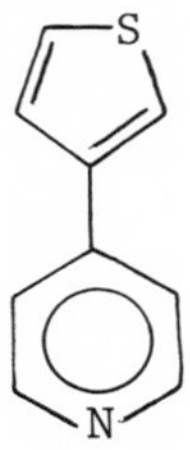

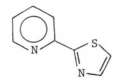

C₉H₇NS   4-(3-Thienyl)pyridine   L

Metal ion	Equilibrium	Log K 25°, 0.1
H⁺	HL/H.L	5.60
Cu²⁺	ML/M.L	2.67

Bibliography: 72SW

2-(2-Pyridyl)-1,3-thiazole   L

C₈H₆N₂S

Metal ion	Equilibrium	Log K 25°, 0.1	Log K 25°, 0.5
H⁺	HL/H.L	2.17	2.15
Fe²⁺	ML/M.L	2.7	
	ML₂/M.L²	6.0	
	ML₃/M.L³	8.8	
Ni²⁺	ML/M.L	5.10	
	ML₂/M.L²	9.97	
	ML₃/M.L³	14.09	
Cu²⁺	ML/M.L	5.72	5.65
	ML₂/M.L²	9.59	

Bibliography:
H⁺,Cu²⁺ 65KSE,68EH   Fe²⁺,Ni²⁺ 68EH

$C_8H_6N_2S$ 4-(2-Pyridyl)-1,3-thiazole   L

Metal ion	Equilibrium	Log K 25°, 0.1
H^+	HL/H.L	4.05 ±0.06
Fe^{2+}	ML/M.L	4.06
	$ML_2/M.L^2$	7.10
	$ML_3/M.L^3$	12.61
Co^{2+}	ML/M.L	5.00
	$ML_2/M.L^2$	9.35
	$ML_3/M.L^3$	13.10
Ni^{2+}	ML/M.L	5.93
	$ML_2/M.L^2$	12.19
	$ML_3/M.L^3$	17.52
Cu^{2+}	ML/M.L	7.13 ±0.08
	$ML_2/M.L^2$	11.48
	$ML_3/M.L^3$	14.75
Zn^{2+}	ML/M.L	4.17
	$ML_2/M.L^2$	7.77
	$ML_3/M.L^3$	10.63
Hg^{2+}	ML/M.L	8.73
	$ML_2/M.L^2$	14.97

Bibliography:
H^+,Cu^{2+} 65KSE,68EH $Fe^{2+}-Ni^{2+},Zn^{2+},Hg^{2+}$ 68EH

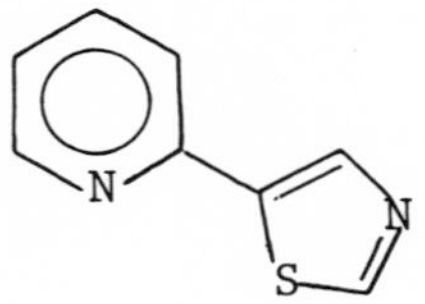

$C_8H_6N_2S$ 5-(2-Pyridyl)-1,3-thiazole   L

Metal ion	Equilibrium	Log K 25°, 0.1
H^+	HL/H.L	2.68
Ni^{2+}	ML/M.L	1.35
Cu^{2+}	ML/M.L	1.70
Zn^{2+}	ML/M.L	0.3

Bibliography: 65KSE

A. PYRIDINES

C$_5$H$_6$N$_2$ <u>2-Aminopyridine</u> L

Metal ion	Equilibrium	Log K 25°, 0.2	Log K 25°, 0.5	Log K 25°, 1.0	ΔH 25°, 0	ΔS 25°, 0.2
H$^+$	HL/H.L	6.70	6.83 ±0.05	6.93	-8.40	2.5
					-8.39[b]	3.1[b]
Cu^{2+}	ML/M.L		1.71[a]			
	ML$_2$/M.L^2		3.25[a]			
Cu$^+$	ML/M.L			5.28[c]		
	ML$_2$/M.L^2			8.00[c]		
Ag$^+$	ML/M.L		2.38[a]			
	ML$_2$/M.L^2	4.85	4.79[a]			

[a] 25°, 0.6; [b] 25°, 0.5; [c] 20°, 1.0

Bibliography:

H$^+$ 67SB,72CP,72CS,73BEM,73BN

Cu^{2+} 67SB

Cu$^+$ 72CP

Ag$^+$ 67SB,73BN

Other references: 67RB

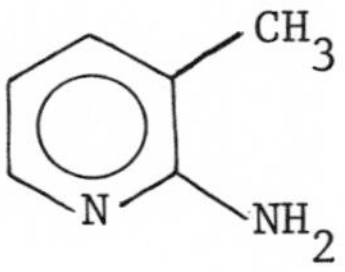

$C_6H_8N_2$ 2-Amino-3-methylpyridine   L

Metal ion	Equilibrium	Log K 25°, 0.6
H^+	HL/H.L	7.08
Cu^{2+}	ML/M.L	1.91
Ag^+	ML/M.L	2.42
	$ML_2/M.L^2$	4.85

Bibliography: 67SB

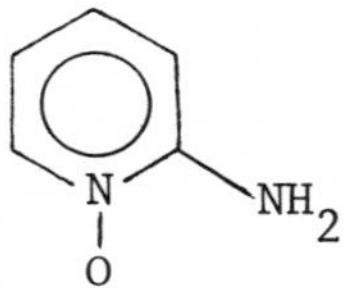

$C_5H_6ON_2$ 2-Aminopyridine 1-oxide   HL

Metal ion	Equilibrium	Log K 25°, 0.1	Log K 25°, 0.5
H^+	HL/H.L	16.63	
	$H_2L/HL.H$	2.46[a]	2.48
Cu^{2+}	ML/M.L	13.11	
	$ML_2/M.L^2$	24.79	
	MHL/ML.H	4.37	

———————

[a] 25°, 0.2

Bibliography: 63SB,63SBE

C$_5$H$_6$N$_2$ 3-Aminopyridine   L

Metal ion	Equilibrium	Log K 25°, 0.1	Log K 25°, 0.5	Log K 25°, 0	ΔH 25°, 0	ΔS 25°, 0
H$^+$	HL/H.L	6.06	6.18	6.03	−6.43	6.0
				6.33[b]	−6.66[c]	5.9[c]
Ni^{2+}	ML/M.L		1.97[a]			
	ML$_2$/M.L^2		3.23[a]			
	ML$_3$/M.L^3		4.1[a]			
Cu^{2+}	ML/M.L		2.80[a]	2.91[d]		
	ML$_2$/M.L^2		4.84[a]	5.18[d]		
	ML$_3$/M.L^3		6.48[a]	7.06[d]		
	ML$_4$/M.L^4		7.5[a]			
Cu$^+$	ML/M.L			5.47[d]		
	ML$_2$/M.L^2			7.97[d]		
Ag$^+$	ML/M.L		2.21[a]			
	ML$_2$/M.L^2		4.41[a]			
Zn^{2+}	ML/M.L	1.34[e]				
	ML$_2$/M.L^2	2.16[e]				
	ML$_3$/M.L^3	2.78[e]				
Cd^{2+}	ML/M.L	1.52[e]				
	ML$_2$/M.L^2	2.19[e]				
	ML$_3$/M.L^3	2.88[e]				

[a] 25°, 0.6;   [b] 25°, 1.0;   [c] 25°, 0.5;   [d] 20°, 1.0;   [e] 30°, 0.1

Bibliography:

H$^+$ 63DK,67SB,72CS,73BEM,73CPa   Cu$^+$ 73CPa
Ni^{2+},Ag$^+$ 67SB   Zn^{2+} 66DK
Cu^{2+} 67SB,73CPa   Cd^{2+} 63DK

C$_5$H$_6$N$_2$ 4-Aminopyridine   L

Metal ion	Equilibrium	Log K 25°, 0.2	Log K 25°, 0.5	Log K 25°, 0	ΔH 25°, 0	ΔS 25°, 0
H$^+$	HL/H.L	9.14	9.25	9.114	-11.28 ±0.03	4.0
				9.39[a]	-11.34[b]	4.3[b]
Cu$^+$	ML/M.L			7.03[c]		
	ML$_2$/M.L^2			11.53[c]		
Ag$^+$	ML$_2$/M.L^2	6.04				

---

[a] 25°, 1.0;   [b] 25°, 0.5;   [c] 20°, 1.0

Bibliography:

H$^+$ 60BH,72CS,72CS,73BEM,73BN   Ag$^+$ 73BN
Cu$^+$ 72CP

$C_6H_8N_2$ 2-(Aminomethyl)pyridine   (2-picolylamine)   L

Metal ion	Equilibrium	Log K 25°, 0.1	Log K 25°, 0.5	Log K 25°, 0	ΔH 25°, 0.5	ΔS 25°, 0.5
H^+	HL/H.L	8.61 ±0.04	8.74	8.65	-10.98	3.2
		8.75[a]	8.91[b]			
	$H_2L/HL.H$	2.00 ±0.1	2.25		-2.95	0.4
		2.04[a]	2.46[b]			
Mn^{2+}	ML/M.L	2.66[a]				
Fe^{2+}	ML/M.L			3.82[c]		
	$ML_2/M.L^2$			7.16[c]		
Co^{2+}	ML/M.L	5.5 ±0.2	5.54	5.49	-7.4	0
	$ML_2/M.L^2$	10.19	10.33	10.03	-14.9	-3
	$ML_3/M.L^3$	13.70	13.83	13.43	-21.7	-10
Ni^{2+}	ML/M.L	7.11 ±0.0	7.11	7.18	-9.9	-1
	$ML_2/M.L^2$	13.34	13.52	13.38	-19.3	-3
	$ML_3/M.L^3$	18.54	18.66	18.39	-29.0	-12
Cu^{2+}	ML/M.L	9.5 ±0.2	9.89[b]	9.55	-9.9[d]	10[e]
	$ML_2/M.L^2$	17.2 ±0.3	17.90[b]	17.46	-19.9[d]	12[e]
Cu^+	$M(OH)_2L/M.(OH)^2.L$		18.65[f]			
	$MOHL_2/M.OH.L^2$		19.6[f]			
	$M_2(OH)_2L_2/M.(OH)^2.L^2$		36.1[f]			
Ag^+	ML/M.L	(4.11)[a]	3.1			
	$ML_2/M.L^2$		7.17			
	MHL/ML.H		6.92			
	$M_2L/ML.H$		2.58			
	$M_2L_2/ML_2.M$		4.05			

[a] 20°, 0.1;   [b] 25°, 1.0;   [c] 30°, 0;   [d] 25°, 0.3;   [e] 25°, 0.1;   [f] 20°, 1.0

2-(Aminomethyl)pyridine (continued)

Metal ion	Equilibrium	Log K 25°, 0.1	Log K 25°, 0.5	Log K 25°, 0	ΔH 25°, 0.5	ΔS 25°, 0.5
Zn^{2+}	ML/M.L	5.28 −0.1	5.37	5.29	−6.5	3
	$ML_2/M.L^2$	9.44	9.84	9.62	−12.9	2
	$ML_3/M.L^3$	12.34	12.64	12.67	−19.8	−9
Cd^{2+}	ML/M.L	4.4 ±0.1	4.76	4.67	−5.8	2
	$ML_2/M.L^2$	8.15	8.70	8.54	−11.5	1
	$ML_3/M.L^3$	11.11	11.29	11.1	−17.7	−8
Hg^{2+}	$ML_2/M.L^2$	20.08[a]				

————————

[a] 20°, 0.1

Bibliography:

H^+ 59GF,64LM,67AW,67HW,71GE,71HG,74CP, 74GE

Mn^{2+},Hg^{2+} 71A

Fe^{2+} 59GF

Co^{2+},Zn^{2+},Cd^{2+} 59GF,64LM,71A,71GE

Ni^{2+} 59GF,64LM,67HWa,71A,71GE

Cu^{2+} 59GF,64LM,67HWa,71A,71HG,74CP

Cu^+ 74CPa

Ag^+ 71A,74GE

Other reference: 60HJ

$C_7H_{10}N_2$ 2-Aminomethyl-6-methylpyridine   L

H_3C—[pyridine N]—CH_2NH_2

Metal ion	Equilibrium	Log K 20°, 0.1
H^+	HL/H.L	8.90
	H_2L/HL.H	3.08
Mn^{2+}	ML/M.L	1.95
Co^{2+}	ML/M.L	3.82
Ni^{2+}	ML/M.L	5.15
	$ML_2/M.L^2$	8.80
	$ML_3/M.L^3$	11
Cu^{2+}	ML/M.L	7.35
	$ML_2/M.L^2$	13.80
Ag^+	ML/M.L	4.4
Zn^{2+}	ML/M.L	4
Cd^{2+}	ML/M.L	4.35
	$ML_2/M.L^2$	7.4

Bibliography: 71A

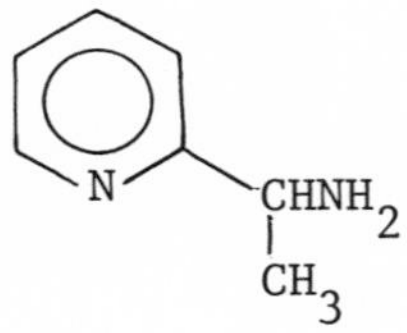

$C_7H_{10}N_2$ DL-2-(1-Aminoethyl)pyridine   L

Metal ion	Equilibrium	Log K 25°, 0.1
H^+	HL/H.L	9.64
	$H_2L/HL.H$	3.86
Cu^{2+}	ML/M.L	7.59
	$ML_2/M.L^2$	$(13.29)^a$

_____

[a] Optical isomerism not stated.

Bibliography: 71SH

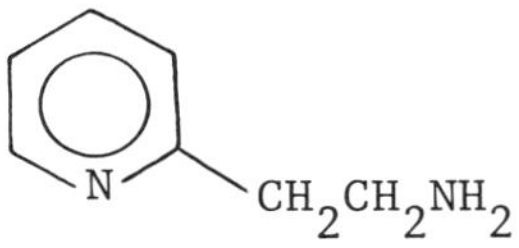

$C_7H_{10}N_2$ 2-(2-Aminoethyl)pyridine   L

Metal ion	Equilibrium	Log K 25°, 0.1	Log K 25°, 0.5	Log K 25°, 0	ΔH 25°, 0.5	ΔS 25°, 0.5
H^+	HL/H.L	9.59	9.70 9.84[a]	9.60	-12.12	3.7
	$H_2L/HL.H$	3.92	4.16 4.37[a]	3.75	-4.60	3.6
Ni^{2+}	ML/M.L	5.2	5.37	5.19	-7.1	1
	$ML_2/M.L^2$		8.74	8.43	-13.2	-4
	$ML_3/M.L^3$		11.2[b]		-21[b]	-18
Cu^{2+}	ML/M.L	7.3	7.64 7.71[a]	7.48	-9.4	4
	$ML_2/M.L^2$	12.9	13.23 13.34[a]		-17.7	1
Cu^+	ML/M.L		10.90[c]			

[a] 25°, 1.0;   [b] 25°, 0.3;   [c] 20°, 1.0

Bibliography:

H^+ 59GF,64LM,67HW,71GE,74CP Cu^+ 74CP
Ni^{2+} 59GF,64LM,67HWa,71GE   Other reference: 60HJ
Cu^{2+} 59GF,64LM,67HWa,71GE,74CP

$C_7H_{10}N_2$ 2-(Methylaminomethyl)pyridine   L

Metal ion	Equilibrium	Log K 25°, 0.5	Log K 25°, 0	ΔH 25°, 0.5	ΔS 25°, 0.5
H^+	HL/H.L	9.01	8.96	−9.88	8.1
	H_2L/HL.H	1.91		−2.34	0.9
Fe^{2+}	ML/M.L		3.53^a		
	ML_2/M.L^2		6.26^a		
Co^{2+}	ML/M.L	5.22	5.14	−6.8	1
	ML_2/M.L^2	9.20	8.98	−12.7	−1
	ML_3/M.L^3		11.5	$(−15)^b$	$(2)^c$
Ni^{2+}	ML/M.L	6.91	6.82	−8.5	3
	ML_2/M.L^2	12.44	12.23	−17.0	0
	ML_3/M.L^3		15.1	$(−21)^b$	$(−2)^c$
Cu^{2+}	ML/M.L	9.07	9.09	−10.7	6
	ML_2/M.L^2	15.78	15.76	−18.8	9
	ML_3/M.L^3		18.5		
Zn^{2+}	ML/M.L	4.96	4.88	−5.8	3
	ML_2/M.L^2	8.58		−10.2	5
Cd^{2+}	ML/M.L	4.49	4.55	−5.2	3
	ML_2/M.L^2	7.84	8.02	−10.0	3
	ML_3/M.L^3		10.7		

---

[a] 30°, 0;   [b] 10-40°, 0;   [c] 25°, 0

Bibliography:
H^+,Co^{2+}-Cd^{2+} 59GF,71GE Fe^{2+} 59GF

$C_8H_{12}N_2$ 2-(Methylaminomethyl)-6-methylpyridine   L

Metal ion	Equilibrium	Log K 25°, 0.5	Log K 25°, 0	ΔH 25°, 0.5	ΔS 25°, 0.5
H^+	HL/H.L	9.05	8.82	−9.70	8.9
	$H_2L/HL.H$	2.92		−3.93	0.2
Co^{2+}	ML/M.L	3.57	3.53	−4.4	2
	$ML_2/M.L^2$	4.98		−6.7	0
Ni^{2+}	ML/M.L	4.74	4.61	−6.0	2
	$ML_2/M.L^2$	7.26		−10.6	−2
Cu^{2+}	ML/M.L	6.88	6.55	−7.0	8
	$ML_2/M.L^2$	12.32	11.65	−13.5	11
Ag^+	ML/M.L	3.59			
	$ML_2/M.L^2$	7.26			
	MHL/ML.H	6.85			
	$M_2L/ML.M$	1.48			
	$M_2L_2/ML_2.M$	3.40			

Bibliography:

H^+ 61RF,71GE,74GE Ag^+ 74GE

Co^{2+}-Cu^{2+} 61RF,71GE

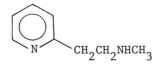

C$_8$H$_{12}$N$_2$ 2-(2-Methylaminoethyl)pyridine   L

Metal ion	Equilibrium	Log K 25°, 0.5	ΔH 25°, 0.5	ΔS 25°, 0.5
H$^+$	HL/H.L	9.96	−10.84	9.2
	H$_2$L/HL.H	4.02	−4.41	3.6
Ni^{2+}	ML/M.L	4.65	−6.3	0
Cu^{2+}	ML/M.L	6.87		

Bibliography: 71GE   Other reference: 61RF

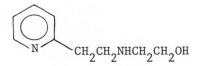

C$_9$H$_{14}$ON$_2$ 2-[2-(2-Hydroxyethylamino)ethyl]pyridine   L

Metal ion	Equilibrium	Log K 25°, 0.1
H$^+$	HL/H.L	7.94
	H$_2$L/HL.H	3.15
Co^{2+}	ML/M.L	5.3
Ni^{2+}	ML/M.L	7.1
Cu^{2+}	ML/M.L	9.2
Zn^{2+}	ML/M.L	5.2
Cd^{2+}	ML/M.L	4.4

Bibliography: 64LM

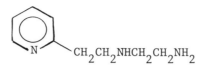

$C_9H_{15}N_3$ 2-[2-(2-Aminoethylamino)ethyl]pyridine   L

                        (N-(2,2'-pyridylethyl)ethylenediamine)

Metal ion	Equilibrium	Log K 25°, 0.1
H^+	HL/H.L	9.51
	$H_2L/HL.H$	6.59
	$H_3L/H_2L.H$	3.50
Co^{2+}	ML/M.L	7.0
Ni^{2+}	ML/M.L	9.4
Cu^{2+}	ML/M.L	13.4
Zn^{2+}	ML/M.L	6.7
Cd^{2+}	ML/M.L	6.2

Bibliography: 64LM

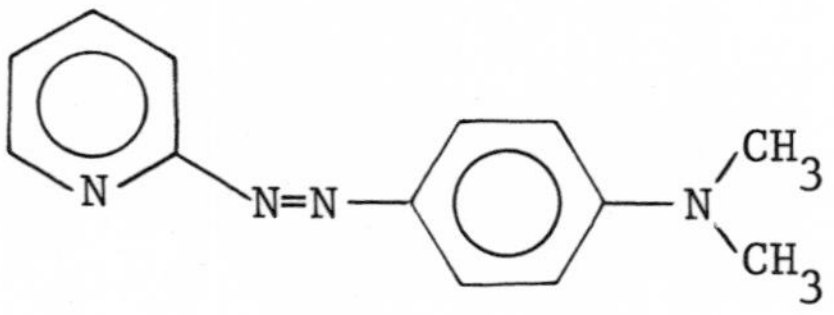

$C_{13}H_{14}N_4$ 2-[4-(Dimethylamino)phenylazo]pyridine   L

Metal ion	Equilibrium	Log K 25°, 0.15	Log K 25°, 0.3	ΔH 25°, 0.15	ΔS 25°, 0.15
H^+	HL/H.L	4.5			
	H_2L/HL.H	2.0			
Mn^{2+}	ML/M.L	0.7			
Co^{2+}	ML/M.L	3.33	3.34		
Ni^{2+}	ML/M.L	4.24	4.11		
Cu^{2+}	ML/M.L	5.21			
Zn^{2++}	ML/M.L	2.36	2.36	(-2)[a]	(5)
Hg^{2+}	ML/M.L	5.06			

_____

[a] 25-35°, 0.15

Bibliography:
H^+,Mn^{2+},Hg^{2+} 53KL Cu^{2+} 53KL
Co^{2+} 53KL,71CHa Zn^{2+} 53KL,71CH
Ni^{2+} 53KL,72CH   Other reference: 64W

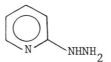

| $C_5H_7N_3$ | | 2-Hydrazinopyridine (2-pyridylhydrazine) | L |

Metal ion	Equilibrium	Log K 20°, 0.1
H^+	HL/H.L	7.24
Mn^{2+}	ML/M.L	2.64
Co^{2+}	ML/M.L	5.89
	$ML_2/M.L^2$	10.86
	$ML_3/M.L^3$	14.88
Ni^{2+}	ML/M.L	7.06
	$ML_2/M.L^2$	13.87
	$ML_3/M.L^3$	19.66
Zn^{2+}	ML/M.L	5.40
	$ML_2/M.L^2$	9.95
	$ML_3/M.L^3$	13.30
Cd^{2+}	ML/M.L	4.36
	$ML_2/M.L^2$	8.18

Bibliography: 71A

$$\text{N}^+\text{CH}_2\text{CH}_2\text{NH}_2$$

$\text{C}_7\text{H}_{11}\text{N}_2{}^+$		N-2-Aminoethylpyridinium (nitrate)		L^+

Metal ion	Equilibrium	Log K 25°, 0.5	ΔH 25°, 0.5	ΔS 25°, 0.5
H^+	HL/H.L	6.98		
Ag^+	ML/M.L	2.08	−5.56	−9.1
	$\text{ML}_2/\text{M.L}^2$	4.46	−11.1	−17

Bibliography:

H$^+$ 69VD   Ag$^+$ 69VD,72VT

$$\text{N}^+\text{CH}_2\text{CH}_2\text{CH}_2\text{NH}_2$$

$\text{C}_8\text{H}_{13}\text{N}_2{}^+$		N-3-Aminopropylpyridinium (nitrate)		L^+

Metal ion	Equilibrium	Log K 25°, 0.5	ΔH 25°, 0.5	ΔS 25°, 0.5
H^+	HL/H.L	9.08		
Ag^+	ML/M.L	2.91	−5.19	−4.0
	$\text{ML}_2/\text{M.L}^2$	6.01	−12.9	−16

Bibliography:

H$^+$ 69VD   Ag$^+$ 69VD,72VT

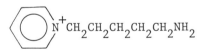

$C_9H_{15}N_2^+$		N-4-Aminobutylpyridinium (nitrate)		L^+

Metal ion	Equilibrium	Log K 25°, 0.5	ΔH 25°, 0.5	ΔS 25°, 0.5
H^+	HL/H.L	9.91		
Ag^+	ML/M.L	3.24	−5.43	−3.4
	$ML_2/M.L^2$	6.61	−13.5	−15

Bibliography:

H^+ 69VD Ag^+ 69VD,72VT

$C_{10}H_{17}N_2^+$		N-5-Aminopentylpyridinium (nitrate)		L^+

Metal ion	Equilibrium	Log K 25°, 0.5	ΔH 25°, 0.5	ΔS 25°, 0.5
H^+	HL/H.L	10.43		
Ag^+	ML/M.L	3.49	−5.81	−3.5
	$ML_2/M.L^2$	7.22	−13.5	−12

Bibliography:

H^+ 69VD Ag^+ 69VD,72VT

C$_9$H$_7$N   Benzo[b]pyridine   (quinoline)   L

Metal ion	Equilibrium	Log K 25°, 0.1	Log K 25°, 0	ΔH 25°, ∿0	ΔS 25°, 0
H$^+$	HL/H.L	4.97	4.81	−5.4	4
Cu^{2+}	ML/M.L	2.65			
CH$_3$Hg$^+$	ML/M.L	4.05			

Bibliography:
H$^+$ 57BI,60SP,74A   Other reference: 67N
Cu^{2+},CH$_3$Hg$^+$ 74A

C$_9$H$_7$N   Benzo[c]pyridine   (isoquinoline)   L

Metal ion	Equilibrium	Log K 30°, 0.1	ΔH 25°, ∿0
H$^+$	HL/H.L		−5.9
Zn^{2+}	ML/M.L	1.08	
	ML$_2$/M.L^2	1.65	
	ML$_3$/M.L^3	2.00	
Cd^{2+}	ML/M.L	1.23	

Bibliography:
H$^+$ 60SP   Zn^{2+},Cd^{2+} 66DK

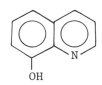

C_9H_7ON		8-Hydroxyquinoline	(oxine)			HL
Metal ion	Equilibrium	Log K 25°, 0.1	Log K 25°, 0.5	Log K 25°, 0	ΔH 25°, 0.1	ΔS 25°, 0.1
H^+	HL/H.L	9.66 ±0.03	9.62 9.58[a]	9.81	(−7)[b]	(21)
	H_2L/HL.H	4.99 ±0.04	5.09 5.17[a]	4.91	(−5)[c]	(6)
Mg^{2+}	ML/M.L	4.31[d]	3.91[e]	4.74		
Ca^{2+}	ML/M.L	2.82[d]	2.44[e]	3.27		
Sr^{2+}	ML/M.L	2.11[d]	1.55[e]	2.56		
Ba^{2+}	ML/M.L	1.62[d]	1.26[e]	2.07		
La^{3+}	ML/M.L	5.9				
	ML_2/M.L^2	11.5				
	ML_3/M.L^3	17.0				
Sm^{3+}	ML/M.L	6.8				
	ML_2/M.L^2	13.3				
	ML_3/M.L^3	19.5				
Th^{4+}	ML/M.L	10.5				
	ML_2/M.L^2	20.4				
	ML_3/M.L^3	29.9				
	ML_4/M.L^4	38.8				
UO_2^{2+}	ML_3/M.L^3	23.8				
Mn^{2+}	ML/M.L	6.24				
Co^{2+}	ML/M.L			8.65		
Ni^{2+}	ML/M.L			9.27		
Cu^{2+}	ML/M.L	12.1		12.56[f]		
	ML_2/M.L^2	23.0				

[a] 25°, 1.0;   [b] 20-25°, 0;   [c] 25-45°, 0.1;   [d] 20°, 0.1;   [e] 20°, 1.0;   [f] 20°, 0

8-Hydroxyquinoline (continued)

Metal ion	Equilibrium	Log K 25°, 0.1	Log K 25°, 0.5	Log K 25°, 0	ΔH 25°, 0.1	ΔS 25°, 0.1
Fe^{3+}	$ML/M \cdot L$	13.69	13.0	14.52		
	$ML_2/M \cdot L^2$	26.3	25.3			
	$ML_3/M \cdot L^3$		36.9			
Ag^+	$ML/M \cdot L$	5.20[d]				
	$ML_2/M \cdot L^2$	9.56[d]				
CH_3Hg^+	$ML/M \cdot L$	8.8				
Zn^{2+}	$ML/M \cdot L$	8.52		8.56[f]		
	$ML_2/M \cdot L^2$	15.8				
Cd^{2+}	$ML/M \cdot L$			7.78[f]		
Pb^{2+}	$ML/M \cdot L$			9.02		
Ga^{3+}	$ML/M \cdot L$	14.5[d]				
	$ML_2/M \cdot L^2$	28.0[d]				
	$ML_3/M \cdot L^3$	40.5[d]				
In^{3+}	$ML/M \cdot L$	12.0 ±0.0				
	$ML_2/M \cdot L^2$	23.9 +0.1				
	$ML_3/M \cdot L^3$	35.4 ±0.1				
Ge(IV)	$M(OH)_2L_2/M(OH)_4 \cdot (HL)^2$		6.61		-18.1[g]	-31[g]

---

[a] 20°, 0.1; [f] 20°, 0; [g] 25°, 0.5

Bibliography:

H^+ 51N,51NL,52D,71ML,74A	Fe^{3+} 49SS,65ZK,68TS
Mg^{2+},Sr^{2+},Ba^{2+} 52N	Ag^+ 65H
Ca^{2+} 51N	CH_3Hg^+ 74A
La^{3+},Sm^{3+} 54D,60R	Cd^{2+} 52NP
Th^{4+} 53D,60R	Ga^{3+} 65Sb
UO_2^{2+} 53DD	In^{3+} 65ZL,68SA
Mn^{2+} 70HZ	Ge(IV) 67TMC
Co^{2+},Ni^{2+},Pb^{2+} 53N	Other references: 53A,55D,55LFa,57TBa,66HE,
Cu^{2+},Zn^{2+} 52NP,64FF	66KF,68KD,68RS,68TK,69SR,70FK,70O,
	72HMZ,73DN,65CF

$C_{10}H_9ON$ 2-Methyl-8-hydroxyquinoline   HL

Metal ion	Equilibrium	Log K 25°, 0.1	Log K 25°, 0.5	Log K 25°, 0
H^+	HL/H.L	10.04	9.96	10.16
	$H_2L/HL.H$	5.63	5.70	5.608
Ge(IV)	$M(OH)_2L_2/M(OH)_4 \cdot (HL)^2$		3.4	

Bibliography:

H^+ 55Na   Other references: 65CF,68RS,69SR

Ge(IV)   67TMC

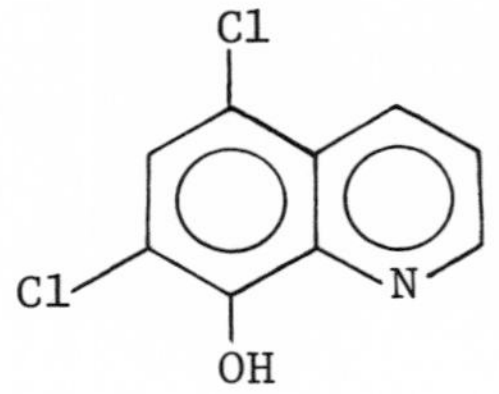

C$_9$H$_5$ONCl   5,7-Dichloro-8-hydroxyquinoline   HL

Metal ion	Equilibrium	Log K 25°, 0.1	Log K 25°, 0.5	Log K 25°, 0
H$^+$	HL/H.L	7.4		7.617
	H$_2$L/HL.H	2.9		2.887
Am^{3+}	ML$_3$/M.L^3	21.9		
Cf^{3+}	ML/M.L	6.5		
	ML$_2$/M.L^2	15.1		
	ML$_3$/M.L^3	22.6		
Th^{4+}	ML/M.L	11.4		
	ML$_2$/M.L^2	21.8		
	ML$_3$/M.L^3	31.2		
	ML$_4$/M.L^4	39.6		
Ge(IV)	M(OH)$_2$L$_2$/M(OH)$_4$.(HL)^2		6.7	

Bibliography:
H$^+$ 53Na,56DD
Am^{3+},Cf^{3+} 69FK
Th^{4+} 56DD,60R

Ge(IV)   67TMC
Other references: 66RG,68RS

| $C_9H_7O_4NS$ | 8-Hydroxyquinoline-5-sulfonic acid (sulfoxine) | | | | H_2L |

Metal ion	Equilibrium	Log K 25°, 0.1	Log K 25°, 0.5	Log K 25°, 0	ΔH 25°, 0.1	ΔS 25°, 0.1
H^+	HL/H.L	8.42 ±0.07	8.23 ±0.00	8.757	-4.0	25
					-4.8[a]	21[a]
	$H_2L/HL.H$	3.93 ±0.05	3.86 ±0.03	4.112	-4.4	3
					-4.4[a]	3[a]
Mg^{2+}	ML/M.L	4.02 ±0.04		4.79		
	$ML_2/M.L^2$	7.63		8.2		
Ca^{2+}	ML/M.L	2.66		3.52		
Sr^{2+}	ML/M.L	1.98		2.75		
Ba^{2+}	ML/M.L	1.56		2.31		
La^{3+}	ML/M.L	5.42[b]	5.25[c]	5.63	(-2)[d]	(18)
	$ML_2/M.L^2$	9.89[b]	9.74[c]	10.1	(-7)[d]	(22)
	$ML_3/M.L^3$	13.41[b]	13.46[c]	13.8	(-11)[d]	(24)
Ce^{3+}	ML/M.L	5.90[b]	5.47[c]	6.05	(-4)[d]	(14)
	$ML_2/M.L^2$	10.80[b]	10.16[c]	11.1	(-8)[d]	(23)
	$ML_3/M.L^3$	14.70[b]	14.08[c]	15.0	(-12)[d]	(27)
Pr^{3+}	ML/M.L	6.02[b]	5.65[c]	6.17	(-3)[d]	(18)
	$ML_2/M.L^2$	11.14[b]	10.63[c]	11.37	(-6)[d]	(31)
	$ML_3/M.L^3$	15.34[b]	14.92[c]	15.7	(-11)[d]	(33)
Nd^{3+}	ML/M.L	6.07[b]	5.78[c]	6.3	(-3)[d]	(18)
	$ML_2/M.L^2$	11.26[b]	10.89[c]	11.6	(-6)[d]	(31)
	$ML_3/M.L^3$	15.58[b]	15.33[c]	16.0	(-9)[d]	(41)
Sm^{3+}	ML/M.L	6.41[b]	6.11[c]	6.58	(-3)[d]	(19)
	$ML_2/M.L^2$	11.99[b]	11.53[c]	12.28	(-6)[d]	(35)
	$ML_3/M.L^3$	16.78[b]	16.26[c]	17.04	(-9)[d]	(47)

[a] 25°, 0.5; [b] 25°, 0.14; [c] 25°, 0.41; [d] 25-40°, 0

8-Hydroxyquinoline-5-sulfonic acid (continued)

Metal ion	Equilibrium	Log K 25°, 0.1	Log K 25°, 0.5	Log K 25°, 0	ΔH 25°, 0.1	ΔS 25°, 0.1
Gd^{3+}	$ML/M.L$	6.50^b	6.25^c	6.64	$(-4)^d$	(16)
	$ML_2/M.L^2$	12.19^b	11.84^c	12.37	$(-8)^d$	(29)
	$ML_3/M.L^3$	17.08^b	16.76^c	17.3	$(-12)^d$	(38)
Er^{3+}	$ML/M.L$	7.05^b	6.63^c	7.16	$(-5)^d$	(16)
	$ML_2/M.L^2$	13.20^b	12.53^c	13.34	$(-11)^d$	(24)
	$ML_3/M.L^3$	18.45^b	17.70^c	18.56	$(-17)^d$	(27)
Th^{4+}	$ML/M.L$	9.56				
	$ML_2/M.L^2$	18.29				
	$ML_3/M.L^3$	25.91				
	$ML_4/M.L^4$	32.02				
	$ML_3/MOHL_3.H$	6.2				
	$(ML_3)^2/(MOHL_2)_2.H^2$	8.9				
UO_2^{2+}	$ML/M.L$	8.52				
	$ML_2/M.L^2$	15.68				
	$ML_2/MOHL_2.H$	6.68				
	$(ML_2)^2/(MOHL_2)_2.H$	11.7				
Mn^{2+}	$ML/M.L$	5.67		6.94		
	$ML_2/M.L^2$	10.72				
Co^{2+}	$ML/M.L$	8.11		8.82		
	$ML_2/M.L^2$	15.06		15.9		
Ni^{2+}	$ML/M.L$	9.02		9.75	-6.4	20
	$ML_2/M.L^2$	16.77		18.5	-14.8	27
Cu^{2+}	$ML/M.L$	11.92	11.57	(11.53)	-8.1	27
	$ML_2/M.L^2$	21.87	21.63	(21.6)	-17.6	41

---

^{b}25°, 0.14; ^{c} 25°, 0.41; ^{d} 25-40°, 0

8-Hydroxyquinoline-5-sulfonic acid   (continued)

Metal ion	Equilibrium	Log K 25°, 0.1	Log K 25°, 0.5	Log K 25°, 0	ΔH 25°, 0.1	ΔS 25°, 0.1
Fe^{3+}	$ML/M.L$	11.6				
	$ML_2/M.L^2$	22.8				
	$ML/MOHL.H$	3.02				
	$MOHL/M(OH)_2L.H$	3.94				
	$ML_2/MOHL_2.H$	5.02				
	$(ML_2)^2/(MOHL_2)_2.H^2$	5.45				
VO^{2+}	$ML/M.L$	11.79			$(-2)^e$	(45)
	$ML/MOHL.H$	6.45				
	$(MOHL)_2/(MOHL)^2$	4.84				
CH_3Hg^+	$ML/M.L$	8.3				
$C_6H_5Hg^+$	$ML/M.L$	8.9				
Zn^{2+}	$ML/M.L$	7.54		8.65	-5.1	17
			7.45 -0.05		-5.5^a	16^a
	$ML_2/M.L^2$	14.32		16.2	-9.6	33
			14.0 +0.5		-8.0^a	37^a
Cd^{2+}	$ML/M.L$			7.70		
	$ML_2/M.L^2$			14.2		
Pb^{2+}	$ML/M.L$			8.53		
	$ML_2/M.L^2$			16.1		
Ge(IV)	$M(OH)_2L_2/M(OH)_4.(HL)^2$	6.55			-18.5^a	-32^a

---

[a] 25°, 0.5; [e] 25-35°, 0.1

Bibliography:

H^+ 52NE,54NU,66LM,66MM,67SPa,67TMC,68GF, 69CM,74A

Mg^{2+} 54NU,59RG

Ca^{2+}-Ba^{2+} 54NU

La^{3+}-Er^{3+} 58FO

Th^{4+},UO_2^{2+} 59RG

Mn^{2+},Co^{2+} 54NU,59RG

Ni^{2+} 54NU,59RG,68GF

Cu^{2+} 54NU,59RG,66LM,67SPa,68GF

Fe^{3+} 59RG

VO^{2+} 66MM

CH_3Hg^+,$C_6H_5Hg^+$ 74A

Zn^{2+} 54NU,59RG,67TMC,68GF,69SV

Cd^{2+},Pb^{2+} 54NU

Ge(IV)   67TMC

Other references: 49MM,53A,63FF,65BY,65CF, 66LA

C$_{10}$H$_9$O$_4$NS   2-Methyl-8-hydroxyquinoline-5-sulfonic acid   H$_2$L

Metal ion	Equilibrium	Log K 25°, 0.5	ΔH 25°, 0.5	ΔS 25°, 0.5
H$^+$	HL/H.L	8.72	−5.2	22
	H$_2$L/HL.H	4.63	−4.1	7
Zn^{2+}	ML/M.L	7.32	(−8.7)	(4)
	ML$_2$/M.L^2	14.0		
Ge(IV)	M(OH)$_2$L$_2$/M(OH)$_4$.(HL)^2	2.2		

Bibliography: 67TMC   Other reference: 63FF

$$SO_3H$$

$C_9H_6O_6N_2S$ 7-Nitro-8-hydroxyquinoline-5-sulfonic acid H_2L

Metal ion	Equilibrium	Log K 25°, 0.1	Log K 25°, 0.5	Log K 25°, 0
H^+	HL/H.L	5.39	5.23	5.75
	$H_2L/HL.H$	1.80	1.85	1.95
Mg^{2+}	ML/M.L	2.46	2.09	3.28
	$ML_2/M.L^2$			4.7
Ca^{2+}	ML/M.L	1.42	1.03	2.26
	$ML_2/M.L^2$			4.5
Sr^{2+}	ML/M.L	1.21	0.78	2.07
	$ML_2/M.L^2$			4.5
Ba^{2+}	ML/M.L	0.94	0.54	1.78
	$ML_2/M.L^2$			3.1
Mn^{2+}	ML/M.L			4.76
	$ML_2/M.L^2$			7.8
Co^{2+}	ML/M.L			6.06
Ni^{2+}	ML/M.L			7.05
	$ML_2/M.L^2$			13.4
Cu^{2+}	ML/M.L			8.86
	$ML_2/M.L^2$			15.2
Zn^{2+}	ML/M.L			5.96
	$ML_2/M.L^2$			12.2
Cd^{2+}	ML/M.L			5.17
	$ML_2/M.L^2$			9.3
Pb^{2+}	ML/M.L			5.92

Bibliography: 55NU

C$_9$H$_6$O$_4$NIS   7-Iodo-8-hydroxyquinoline-5-sulfonic acid   (ferron)   H$_2$L

Metal ion	Equilibrium	Log K 25°, 0.1	Log K 25°, 0.5	Log K 25°, 0	ΔH 25°, 0.5	ΔS 25°, 0.5
H$^+$	HL/H.L	7.08 ±0.03	6.93 ±0.03 6.90[a]	7.42	-3.7	19
	H$_2$L/HL.H	2.43 ±0.08	2.22 +0.01 2.43[a]	2.51	-2.2	3
Ca^{2+}	ML/M.L	2.36	2.01[a]	3.07		
	ML$_2$/M.L^2			4		
Co^{2+}	ML/M.L	7.3				
	ML$_2$/M.L^2	13.6				
	ML$_3$/M.L^3	18.6				
Ni^{2+}	ML/M.L	8.2				
	ML$_2$/M.L^2	15.2				
	ML$_3$/M.L^3	20.8				
Fe^{3+}	ML/M.L	8.9				
	ML$_2$/M.L^2	17.3				
	ML$_3$/M.L^3	25.2				
CH$_3$Hg$^+$	ML/M.L	8.1				
C$_6$H$_5$Hg$^+$	ML/M.L	8.7				
Zn^{2+}	ML/M.L	7.1	6.85		-5.0	15
	ML$_2$/M.L^2	13.2	13.0		-6.0	39
Hg^{2+}	ML$_2$/M.L^2	21.2				
Al^{3+}	ML/M.L	7.6				
	ML$_2$/M.L^2	14.7				
	ML$_3$/M.L^3	20.3				
	ML$_2$/MOHL$_2$.H	5.0				

---

[a] 25°, 1.0

Ferron (continued)

Metal ion	Equilibrium	Log K 25°, 0.1	Log K 25°, 0.5	Log K 25°, 0	ΔH 25°, 0.5	ΔS 25°, 0.5
Ga^{3+}	ML/M.L	14.7				
	$ML_2/M.L^2$	23.9				
	$ML_3/M.L^3$	29.6				
	MHL/ML.H	3.7				
	$M(OH)_2L/ML.(OH)^2$	17.6				
	$MOHL_2/ML_2.OH$	7.1				
In^{3+}	ML/M.L	15.1				
	$ML_2/M.L^2$	24.2				
	$ML_3/M.L^3$	30.7				
	$M(OH)_2L/ML.(OH)^2$	19.2				
	$MOHL_2/ML_2.OH$	8.5				
Tl^{3+}	ML/M.L	18.9				
	$ML_2/M.L^2$	27.7				
	$ML_3/M.L^3$	35.4				
	$M(OH)_2L/ML.(OH)^2$	14.1				
Ge(IV)	$M(OH)_2L_2/M(OH)_4.(HL)^2$		6.78		-17.5	-28

Bibliography:

H^+ 52NE,53EN,61LS,67TMC,69BN,73LS,74MO,74A Al^{3+} 61LS

Ca^{2+} 53EN Ga^{3+} 74MO

Co^{2+},Ni^{2+} 63S In^{3+},Ti^{3+} 74MOa

Fe^{3+} 61SL   Ge(IV)   67TMC

CH_3Hg^+,$C_6H_5Hg^+$,Hg^{2+} 74A   Other references: 60S,67LR,67MB,71LS,71Ma,

Zn^{2+} 63S,67TMC   72PB,73DN

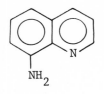

$C_9H_8N_2$ 8-Aminoquinoline   L

Metal ion	Equilibrium	Log K 20°, 0.1
H^+	HL/H.L	4.04
Mg^{2+}	ML/M.L	1.43
Ca^{2+}	ML/M.L	1.49
Sr^{2+}	ML/M.L	1.27
Co^{2+}	ML/M.L	2.66
Ni^{2+}	ML/M.L	4.90
	$ML_2/M.L^2$	8.54
	$ML_3/M.L^3$	11.83
Cu^{2+}	ML/M.L	6.06
	$ML_2/M.L^2$	10.79
	$ML_3/M.L^3$	14.48
Zn^{2+}	ML/M.L	2.42
Cd^{2+}	ML/M.L	2.37

Bibliography: 57WS   Other references: 64SM,69RW

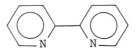

$C_{10}H_8N_2$ 2,2'-Bipyridyl   L

Metal ion	Equilibrium	Log K 25°, 0.1	Log K 25°, 1.0	Log K 25°, 0	ΔH 20°, 0.1	ΔS 25°, 0.1
H^+	HL/H.L	4.42 ±0.03	4.67 −0.01	4.35	−3.66	8.3
			4.51[a]		−4.0[b]	8[d]
					−3.55[c]	
	$M_2L/HL.H$	1.5				
Mn^{2+}	ML/M.L	2.62 ±0.0			−3.5	0
			2.61		(−5.7)[b]	(−7)[d]
	$ML_2/M.L^2$	4.62	4.47		−6.1[b]	1[d]
	$ML_3/M.L^3$	5.6	6.0		−6.2[b]	5[d]
Fe^{2+}	ML/M.L	4.20	4.65[b]	4.36		
	$ML_2/M.L^2$	7.90				
	$ML_3/M.L^3$	17.2 ±0.2			−31.4	−27
			17.49		−28.0[b]	−14[d]
Co^{2+}	ML/M.L	5.8 ±0.1			−8.2	−1
			5.81		−7.2[b]	2[d]
	$ML_2/M.L^2$	11.24 ±0.01			−15.2	0
			11.31		−14.4[b]	3[d]
	$ML_3/M.L^3$	15.9 ±0.1			−21.3	1
			16.18		−19.7[b]	8[d]
Ni^{2+}	ML/M.L	7.04 ±0.03			−9.6	0
			7.06		−8.9[b]	2[d]
	$ML_2/M.L^2$	13.85 ±0.08			−19.0	0
			14.01		−17.8[b]	4[d]
	$ML_3/M.L^3$	20.16 ±0.03			−28.2	−2
			20.47		−26.7[b]	4[d]

[a] 25°, 0.5;   [b] 30°, 1.0;   [c] 25°, 0.1;   [d] 25°, 1.0

2,2'-Bipyridyl (continued)

Metal ion	Equilibrium	Log K 25°, 0.1	Log K 25°, 1.0	Log K 25°, 0	ΔH 20°, 0.1	ΔS 25°, 0.1
Cu^{2+}	$ML/M.L$	6.33				
	$ML/MOHL.H$	7.9				
	$ML/M(OH)_2L.H^2$	17.67				
	$(MOHL)_2.H^2/(ML)^2$	10.81				
Fe^{3+}	$M_2(OH)_2L_4.H^2/M^2.L^4$					
		16.29[e]				
Cu^+	$ML_2/M.L^2$	12.95[e]	13.18[f]			
Ag^+	$ML/M.L$	3.03[g]	3.0			
	$ML_2/M.L^2$	6.67[g]	7.11			
CH_3Hg^+	$ML/M.L$	5.86				
Zn^{2+}	$ML/M.L$	5.13 ±0.09			−7.1	0
			5.34		−6.3[b]	3[d]
	$ML_2/M.L^2$	9.5 ±0.1			−12.5	2
			9.96		−11.8[b]	6[d]
	$ML_3/M.L^3$	13.2 ±0.2			−17.5	2
			13.97		−15.9[b]	11[d]
Cd^{2+}	$ML/M.L$	4.18 ±0.06			−5.1	2
	$ML_2/M.L^2$	7.7 ±0.1			−9.4	4
	$ML_3/M.L^3$	10.3 ±0.1			−14.0	0
Hg^{2+}	$ML/M.L$	9.64[e]				
	$ML_2/M.L^2$	16.7[e] −0.1				
	$ML_3/M.L^3$	19.5[e]				
Pb^{2+}	$ML/M.L$	2.9				
Ga^{3+}	$ML/M.L$		4.52			
	$ML_2/M.L^2$		7.70			
In^{3+}	$ML/M.L$		4.75			
	$ML_2/M.L^2$		8.00			
Tl^{3+}	$ML/M.L$		9.40			
	$ML_2/M.L^2$		16.10			
	$ML_3/M.L^3$		20.05			

---

[b] 30°, 1.0;   [d] 25°, 1.0;   [e] 20°, 0.1;   [f] 25°, 0.3;   [g] 35°, 0.1

D. DIPYRIDINES

2,2'-Bipyridyl (continued)

Bibliography:

H^+ 55N,56YY,59GM,61JW,62CM,63A,63Aa,
         65DD,66PS,69CM,70EHP,71U,74A

Mn^{2+} 62IM,63A,63Aa,65DD,74HM

Fe^{2+} 49K,50K,62IM,62Aa,63Aa,65DD

Co^{2+},Ni^{2+} 62IM,63A,63Aa,65DD

Cu^{2+} 59GM

Cu^+ 61JW,63A

Fe^{3+} 62Aa

Ag^+ 67L,72KM

CH_3Hg^+ 74A

Zn^{2+} 62CL,62IM,63A,63Aa,65DD

Cd^{2+} 62IM,63A,63Aa

Hg^{2+},Pb^{2+} 63A

Ga^{3+} 72KS

In^{3+} 71KMF,72KMF

Tl^{3+} 61KM,62KM

Other references: 47DM,50BG,50OL,54SS,55LF,
        55MBa,55SK,55YY,57MC,58CS,62AB,69P,
        70D

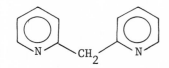

$C_{11}H_{10}N_2$ Methylenedi-2-pyridine   (di-2-pyridylmethane)   L

Metal ion	Equilibrium	Log K 20°, 0.1
H^+	HL/H.L	5.18
	$H_2L/HL.H$	2.69
Co^{2+}	ML/M.L	3.46
	$ML_2/M.L^2$	6.28
	MHL/ML.H	3.6
	$M_2L/ML.M$	1.9
Ni^{2+}	ML/M.L	5.02
	$ML_2/M.L^2$	9.17
Cu^{2+}	ML/M.L	6.7
	$ML_2/M.L^2$	11.8
Ag^+	ML/M.L	3.33
	$ML_2/M.L^2$	6.41
	MHL/ML.H	2.9
	$M_2L/ML.M$	1.4
Zn^{2+}	ML/M.L	2.81
	$ML_2/M.L^2$	5.20
	MHL/ML.H	4.4
	$M_2L/ML.M$	2.1
Cd^{2+}	ML/M.L	3.04
	$ML_2/M.L^2$	5.58
	MHL/ML.H	4.3
	$M_2L/ML.M$	2.1
Hg^{2+}	ML/M.L	7.8
	$ML_2/M.L^2$	14.3

Bibliography: 70BA

D. DIPYRIDINES

$$\text{N} \quad \text{CH}_2\text{CH}_2 \quad \text{N}$$

$C_{12}H_{12}N_2$ Ethylenedi-2-pyridine   (1,2-di-2-pyridylethane)   L

Metal ion	Equilibrium	Log K 20°, 0.1
H^+	HL/H.L	5.80
	H_2L/HL.H	3.99
Co^{2+}	ML/M.L	1.3
	MHL/ML.H	5.5
Ni^{2+}	ML/M.L	1.4
	MHL/ML.H	5.5
Cu^{2+}	ML/M.L	3.41
	ML_2/M.L^2	5.58
	MHL/ML.H	3.7
	M_2L/ML.M	1.4
Ag^+	ML/M.L	3.20
	ML_2/M.L^2	5.93
	MHL/ML.H	3.9
	M_2L/ML.M	1.6
Zn^{2+}	ML/M.L	1.4
	MHL/ML.H	5.4
Cd^{2+}	ML/M.L	1.3
	MHL/ML.H	5.5
Hg^{2+}	ML/M.L	7.0
	ML_2/M.L^2	11.4

Bibliography: 70BA

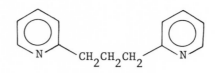

$C_{13}H_{14}N_2$ <u>Trimethylenedi-2-pyridine</u>   <u>(1,3-di-2-pyridylpropane)</u>   L

Metal ion	Equilibrium	Log K 20°, 0.1
H^+	HL/H.L	6.15
	H_2L/HL.H	4.80
Co^{2+}	ML/M.L	1.3
	MHL/ML.H	5.9
Ni^{2+}	ML/M.L	1.4
	MHL/ML.H	5.8
Cu^{2+}	ML/M.L	2.8
Ag^+	ML/M.L	3.44
	$ML_2/M.L^2$	6.43
Zn^{2+}	ML/M.L	1.3
	MHL/ML.H	5.9
Cd^{2+}	ML/M.L	1.3
	MHL/ML.H	5.9
Hg^{2+}	ML/M.L	7.8
	$ML_2/M.L^2$	11.8

Bibliography: 70BA

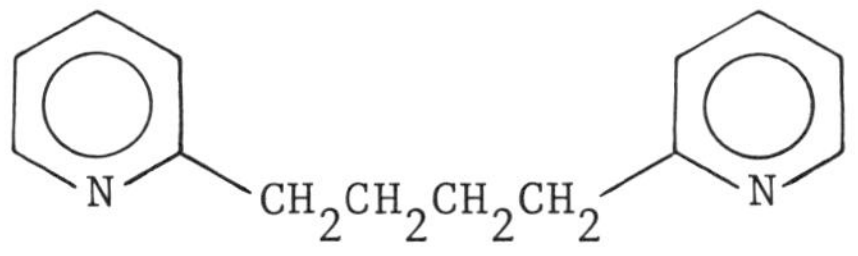

$C_{14}H_{16}N_2$ <u>Tetramethylenedi-2-pyridine</u>   <u>(1,4-di-2-pyridylbutane)</u>   L

Metal ion	Equilibrium	Log K 20°, 0.1
H^+	HL/H.L	6.25
	H_2L/HL.H	5.27
Co^{2+}	ML/M.L	1.2
Ni^{2+}	ML/M.L	1.2
Cu^{2+}	ML/M.L	2.6
Ag^+	ML/M.L	3.72
	$ML_2/M.L^2$	6.44
Zn^{2+}	ML/M.L	1.0
Cd^{2+}	ML/M.L	1.1
Hg^{2+}	ML/M.L	8.3
	$ML_2/M.L^2$	11.5

Bibliography: 70BA

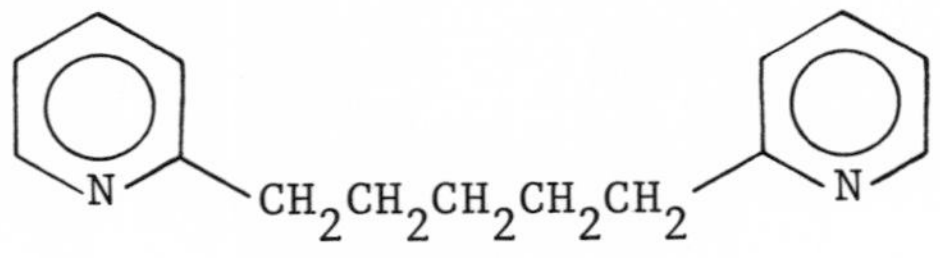

$C_{15}H_{18}N_2$ Pentamethylenedi-2-pyridine   (1,5-di-2-pyridylpentane)   L

Metal ion	Equilibrium	Log K 20°, 0.1
H^+	HL/H.L	6.33
	H_2L/HL.H	5.45
Co^{2+}	ML/M.L	1.0
Ni^{2+}	ML/M.L	1.0
Cu^{2+}	ML/M.L	2.7
Ag^+	ML/M.L	4.52
	MHL/ML.H	3.5
Zn^{2+}	ML/M.L	1.0
Cd^{2+}	ML/M.L	1.0
Hg^{2+}	ML/M.L	9.6

Bibliography: 70BA

$C_{16}H_{20}N_2$ <u>Hexamethylenedi-2-pyridine</u>   (1,6-di-2-pyridylhexane)   L

Metal ion	Equilibrium	Log K 20°, 0.1
H^+	HL/H.L	6.38
	H_2L/HL.H	5.61
Co^{2+}	ML/M.L	1.6
	MHL/ML.H	6.1
Ni^{2+}	ML/M.L	1.6
	MHL/ML.H	6.1
Cu^{2+}	ML/M.L	2.9
Ag^+	ML/M.L	4.1
	MHL/ML.H	4.2
Zn^{2+}	ML/M.L	1.5
	MHL/ML.H	6.0
Cd^{2+}	ML/M.L	1.5
	MHL/ML.H	6.1
Hg^{2+}	ML/M.L	9.2

Bibliography: 70BA

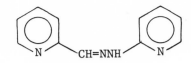

C$_{11}$H$_{11}$N$_4$ 2-(2-Pyridylmethylenehydrazino)pyridine   L
                        (pyridine-2-aldehyde 2'-pyridylhydrazone)

Metal ion	Equilibrium	Log K 20°, 0.1	Log K 25°, 0	ΔH 25°, 0	ΔS 25°, 0
H$^+$	HL/H.L	5.83	5.62	(-7)[a]	(2)
	H$_2$L/HL.H	3.21	2.91	(-5)[a]	(-3)
Mn^{2+}	ML/M.L	3.68			
	ML$_2$/M.L^2	6			
Fe^{2+}	ML$_2$/M.L^2	17.2	16.57	(-25)[a]	(-8)
	ML$_2$/M(H$_{-1}$L)L.H		6.08	(-6)[a]	(8)
	M(H$_{-1}$L)L/M(H$_{-1}$L)$_2$.H		7.39	(-6)[a]	(14)
Cu^{2+}	ML/M.L	11.0			
	ML$_2$/M.L^2	15.3			
	MHL/ML.H	6.12			
Zn^{2+}	ML/M.L	6.21	5.82	(-11)[a]	(-10)
	ML$_2$/M.L^2	11.79	11.08	(-14)[a]	(4)
Cd^{2+}	ML/M.L	5.43			
	ML$_2$/M.L^2	10.45			

[a] 5-60°, 0

Bibliography:
H$^+$,Fe^{2+},Zn^{2+} 68GG,71A   Other reference: 64GHL
Mn^{2+},Cu^{2+},Cd^{2+} 71A

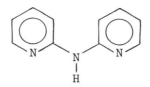

$C_{10}H_9N_3$ Iminodi-2-pyridine   (di-2-pyridylamine)   L

Metal ion	Equilibrium	Log K 20°, 0.1
H^+	HL/H.L	7.14
Mn^{2+}	ML/M.L	2
Co^{2+}	ML/M.L	4.72
	$ML_2/M.L^2$	8.92
Ni^{2+}	ML/M.L	6.25
	$ML_2/M.L^2$	12.03
Cu^{2+}	ML/M.L	8.05
	$ML_2/M.L^2$	13.56
Zn^{2+}	ML/M.L	3.75
	$ML_2/M.L^2$	6.95
Cd^{2+}	ML/M.L	2.6

Bibliography: 71A

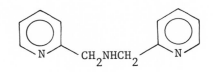

$C_{12}H_{13}N_3$ <u>Iminobis(methylene-2-pyridine)</u>   <u>(di-2-picolylamine</u>, <u>DPA</u>)   L

Metal ion	Equilibrium	Log K 25°, 0.1
H^+	HL/H.L	7.29 ±0.01
	$H_2L/HL.H$	2.60 ±0.00
	$H_3L/H_2L.H$	1.12 +0.01
Mn^{2+}	ML/M.L	4.16
	$ML_2/M.L^2$	7.07
Co^{2+}	ML/M.L	7.74
	$ML_2/M.L^2$	13.05
Ni^{2+}	ML/M.L	8.70 +0.1
	$ML_2/M.L^2$	16.60
Cu^{2+}	ML/M.L	14.4[a] −5
	$ML_2/M.L^2$	19.0[a] −5
	MOHL/ML.OH	5.06 ±0.05
Ag^+	ML/M.L	5.5
	$ML_2/M.L^2$	8.6
Zn^{2+}	ML/M.L	7.57
	$ML_2/M.L^2$	11.93
Cd^{2+}	ML/M.L	6.44
	$ML_2/M.L^2$	11.74

[a] 25°, 0.05

Bibliography:
H^+ 67RBB,74NR Cu^{2+} 68RB,73YB,74NR
Mn^{2+},Co^{2+},Ag^+-Cd^{2+} 68RB   Other reference: 68G
Ni^{2+} 68RB,74NR

$C_{18}H_{18}N_4$ Nitrilotris(methylene-2-pyridine)   (tri-2-picolylamine,   TPA)   L

Metal ion	Equilibrium	Log K 20°, 0.1	ΔH 20°, 0.1	ΔS 20°, 0.1
H^+	HL/H.L	6.17		
	$H_2L/HL.H$	4.35		
	$H_3L/H_2L.H$	2.55		
Mn^{2+}	ML/M.L	5.6	−6.2	5
Fe^{2+}	ML/M.L	8.7	−7.8	13
Co^{2+}	ML/M.L	11.4	−11.2	14
Ni^{2+}	ML/M.L	14.5	−16.2	11
Cu^{2+}	ML/M.L	16.2	−17.5	14
Zn^{2+}	ML/M.L	11.0	−9.8	17

Bibliography:

H^+ 67AW Mn^{2+}-Zn^{2+} 70WA

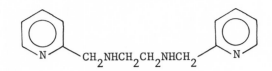

$C_{14}H_{18}N_4$ <u>Ethylenebis(iminomethylene-2-pyridine)</u>   L

                                     <u>(N,N'-di-2-picolylethylenediamine)</u>

Metal ion	Equilibrium	Log K 25°, 0.1	
H$^+$	HL/H.L	8.23	+0.05
	H$_2$L/HL.H	5.45	+0.02
	H$_3$L/H$_2$L.H	1.81	+0.2
	H$_4$L/H$_3$L.H	1.62	+0.2
Mn^{2+}	ML/M.L	5.9	
Co^{2+}	ML/M.L	12.8	−0.8
Ni^{2+}	ML/M.L	14.4	−2
Cu^{2+}	ML/M.L	16.3	
Zn^{2+}	ML/M.L	11.5	−0.1
Cd^{2+}	ML/M.L	9.9	

Bibliography:

H$^+$,Co^{2+},Ni^{2+},Zn^{2+} 64LM,68G   Mn^{2+},Cu^{2+},Cd^{2+} 64LM

$C_{26}H_{28}N_6$ Ethylenedinitrilotetrakis(methylene-2-pyridine)   L

(N,N,N',N'-tetra-2-picolylethylenediamine,   TPEN)

Metal ion	Equilibrium	Log K 20°, 0.1	ΔH 20°, 0.1	ΔS 20°, 0.1
H^+	HL/H.L	7.19		
	$H_2L/HL.H$	4.85		
	$H_3L/H_2L.H$	3.32		
	$H_4L/H_3L.H$	2.85		
Mn^{2+}	ML/M.L	10.3	−11.4	8
Fe^{2+}	ML/M.L	14.6	−16.7	10
Co^{2+}	ML/M.L		−17.2	
Ni^{2+}	ML/M.L	(18.0)	−23.6	(2)
Cu^{2+}	ML/M.L	20.6	−23.1	16
Zn^{2+}	ML/M.L	(18.0)	−14.8	(32)

Bibliography:

H^+ 67AW Mn^{2+}-Zn^{2+} 70WA

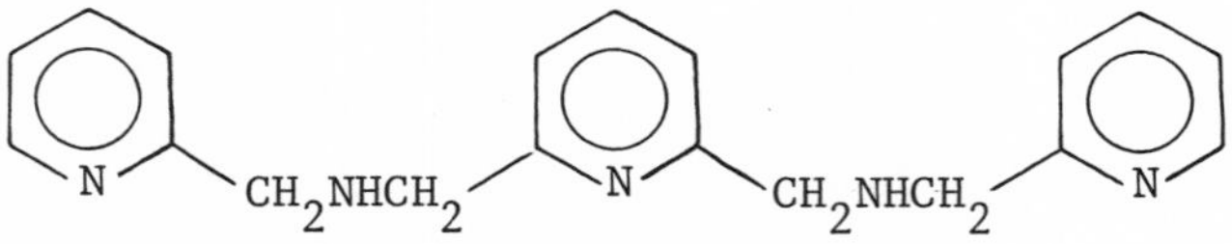

$C_{19}H_{21}N_5$ <u>Pyridine-2,6-bis(methyleneiminomethylene-2-pyridine)</u>   L

Metal ion	Equilibrium	Log K 25°, 0.1
H^+	HL/H.L	7.55
	H_2L/HL.H	6.86
Co^{2+}	ML/M.L	14.8
Cu^{2+}	ML/M.L	18.4
Zn^{2+}	ML/M.L	12.0

Bibliography: 68G

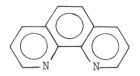

$C_{12}H_8N_2$ 1,10-Phenanthroline   L

Metal ion	Equilibrium	Log K 25°, 0.1	Log K 25°, 0.5	Log K 25°, 0	ΔH 20°, 0.1	ΔS 25°, 0.1
H^+	HL/H.L	4.93 ±0.04	5.02 ±0.01	4.86	−3.95	9.3
			5.12[a]		−3.60[b] +0.1	10.2[b]
	$H_2L/HL.H$	1.9				
Mg^{2+}	ML/M.L	1.2[c]				
Ca^{2+}	ML/M.L	0.7[c]				
Mn^{2+}	ML/M.L	4.0 ±0.2			−3.5	7
	$ML_2/M.L^2$	7.3 ±0.2			−7.0	10
	$ML_3/M.L^3$	10.3 ±0.2			−9.0	17
Fe^{2+}	ML/M.L	5.85 ±0.01				
	$ML_2/M.L^2$	11.15 ±0.05				
	$ML_3/M.L^3$	21.0 ±0.1		20.7 ±0.2	−33.0	−15
Co^{2+}	ML/M.L	7.08 ±0.1			−9.1	2
	$ML_2/M.L^2$	13.72 ±0.03			−15.8	10
	$ML_3/M.L^3$	19.8 ±0.3			−23.8	11
Ni^{2+}	ML/M.L	8.6	8.65		−11.2	2
	$ML_2/M.L^2$	16.7	17.08		−20.5	8
	$ML_3/M.L^3$	24.3	24.91		−30.0	11
Cu^{2+}	ML/M.L	7.4				
	$ML/M(OH)_2L.H^2$	17.3				
	$(MOHL)_2.H^2/(ML)^2$	10.69				

---

[a] 25°, 1.0; [b] 25°, 0; [c] 20°, 0.1

1,10-Phenanthroline (continued)

Metal ion	Equilibrium	Log K 25°, 0.1	Log K 25°, 0.5	Log K 25°, 0	ΔH 20°, 0.1	ΔS 25°, 0.1
Fe^{3+}	$ML/M.L$	6.5^c				
	$ML_2/M.L^2$	11.4^c				
	$ML_3/M.L^3$	14.1^d	14.6^e	13.8	$(-10)^f$	$(30)^b$
	$M_2L_4/M_2OHL_4.H$	6.5^c				
	$M_2OHL/M_2(OH)_2L_4.H$	4.4^c				
Cu^+	$ML_2/M.L^2$		15.82^g			
Ag^+	$ML/M.L$	5.02	5.0^a			
	$ML_2/M.L^2$	12.06 ± 0.01	12.11^a			
CH_3Hg^+	$ML/M.L$	7.15				
Zn^{2+}	$ML/M.L$	6.4 ± 0.1	6.73		-7.5	4
				6.2	-7.5^b	3^b
	$ML_2/M.L^2$	12.2 ± 0.1			-15.0	5
				(12.1)	-12.3^b	$(14)^b$
	$ML_3/M.L^3$	17.1 ± 0.1			-19.3	14
				(17.3)	-15.2^b	$(28)^b$
Cd^{2+}	$ML/M.L$	5.8 ± 0.1			-6.3	5
	$ML_2/M.L^2$	10.6 ± 0.1			-13.1	5
	$ML_3/M.L^3$	14.6 ± 0.3			-16.1	13
Hg^{2+}	$ML_2/M.L^2$	19.65^c				
	$ML_3/M.L^3$	23.35^c				
Pb^{2+}	$ML/M.L$	4.65^c				
$(CH_3)_2Sn^{2+}$	$ML/M.L$	3.88				
Ga^{3+}	$ML/M.L$		5.58^a			
	$ML_2/M.L^2$		9.21^a			
In^{3+}	$ML/M.L$		5.70^a			
	$ML_2/M.L^2$		10.04^a			
	$ML_3/M.L^3$		14.0^a			

---

[a] 25°, 1.0; [b] 25°, 0; [c] 20°, 0.1; [d] 25°, 0.15; [e] 25°, 1.5; [f] 25-45°, 0; [g] 25°, 0.3

1,10-Phenanthroline   (continued)

Metal ion	Equilibrium	Log K 25°, 0.1	Log K 25°, 0.5	Log K 25°, 0	ΔH 20°, 0.1	ΔS 25°, 0.1
Tl^{3+}	$ML/M.L$		11.57^{a}			
	$ML_2/M.L^2$		18.30^{a}			
	$ML_3/M.L^3$		24.3^{a}			

---

[a] 25°, 1.0

Bibliography:

H^+ 48LK,55IM,56NU,56YY,59GM,61JW,63A,
           63Aa,63BB,63YT,66PS,67SPa,69CM,70E,
           71PD,74A

Mg^{2+},Ca^{2+},Hg^{2+},Pb^{2+} 63A

Mn^{2+} 62MB,63A,63Aa,63DB,72BB

Fe^{2+} 59BB,62A,62IM,62MB,63Aa,64LA,72Ba

Co^{2+} 62IM,62MB,63A,63Aa

Ni^{2+} 59A,63Aa,67SPa

Cu^{2+} 59GM

Fe^{3+} 48LK,62Aa,64LA

Cu^+ 61JW

Ag^+ 63DB,72BB,72KM

CH_3Hg^+ 74A

$(CH_3)_2Sn^{2+}$ 63TY,63YT

Zn^{2+} 51KL,55IM,59BB,62IM,63A,63Aa,63DB,
           66FB,67SPa,70E,72BB

Cd^{2+} 59A,63A,63Aa,63DB,66FB,72BB

Ga^{3+} 72KS

In^{3+} 72KMF

Tl^{3+} 61KM,62KM

Other references: 46DN,50KL,51CL,52BG,54PB,
           55M,55MBa,55SK,56MB,57MC,57TB,61HD,
           67LAb,71KMF,72BBB,73DN

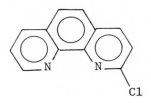

$C_{12}H_7N_2Cl$ 2-Chloro-1,10-phenanthroline   L

Metal ion	Equilibrium	Log K 25°, 0.1	Log K 25°, 0.3
H^+	HL/H.L	4.18	4.20
	$H_2L/HL.H$	0.22	
Fe^{2+}	$ML_3/M.L^3$	11.6	
Ni^{2+}	ML/M.L	4.58	
	$ML_2/M.L^2$	9.26	
	$ML_3/M.L^3$	13.6	
Cu^{2+}	ML/M.L	5.07	(5.60)
	$ML_2/M.L^2$	10.07	(10.45)
	$ML_3/M.L^3$	13.9	
Cu^+	$ML_2/M.L^2$		14.6
Zn^{2+}	ML/M.L	3.3	
	$ML_2/M.L^2$	6.6	

Bibliography:
H^+,Cu^{2+} 61JW,71IG Cu^+ 61JW
Fe^{2+},Ni^{2+},Zn^{2+} 71IG

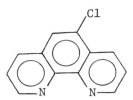

$C_{12}H_7N_2Cl$ <u>5-Chloro-1,10-phenanthroline</u>   L

Metal ion	Equilibrium	Log K 25°, 0.1
H^+	HL/H.L	4.07
Fe^{2+}	ML/M.L	5.71 ±0.01
	$ML_2/M.L^2$	10.72
	$ML_3/M.L^3$	19.7 −4
Ag^+	ML/M.L	4.70
	$ML_2/M.L^2$	11.04
Zn^{2+}	ML/M.L	5.85

Bibliography:

H^+,Zn^{2+} 59BB Ag^+ 74B

Fe^{2+} 59BB,72Ba   Other reference: 52BG

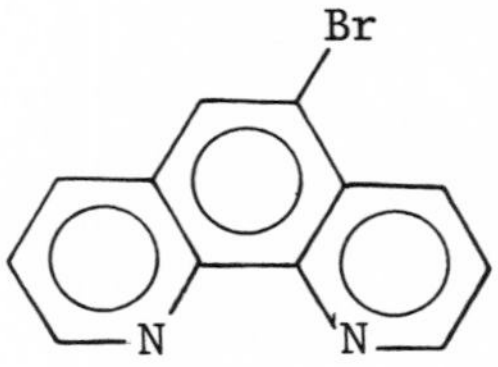

$C_{12}H_7N_2Br$ <u>5-Bromo-1,10-phenanthroline</u>   L

Metal ion	Equilibrium	Log K 25°, 0.1
H^+	HL/H.L	4.09
Fe^{2+}	ML/M.L	5.65
	$ML_2/M.L^2$	10.78
	$ML_3/M.L^3$	(16.0)
Ag^+	ML/M.L	5.30
	$ML_2/M.L^2$	11.77

Bibliography:

H^+	59BB	Fe^{2+}	72Ba
Ag^+	74B		

$C_{12}H_7O_2N_3$ 5-Nitro-1,10-phenanthroline   L

Metal ion	Equilibrium	Log K 25°, 0.1	Log K 25°, 0.3	Log K 25°, 0	ΔH 25°, 0	ΔS 25°, 0
H^+	HL/H.L	3.22	3.25	(3.23)	(−2)[a]	(8)
Fe^{2+}	ML/M.L	5.06		4.57[b]		
	$ML_3/M.L^3$			15.6	(−25)[a]	(−12)
Co^{2+}	ML/M.L	6.3				
	$ML_2/M.L^2$	11.8				
	$ML_3/M.L^3$	16.5				
Ni^{2+}	ML/M.L	7.0				
	$ML_2/M.L^2$	13.4				
	$ML_3/M.L^3$	20.4				
Cu^{2+}	ML/M.L	8.0				
	$ML_2/M.L^2$	13.5				
	$ML_3/M.L^3$	17.7				
Fe^{3+}	ML/M.L			(7.46)		
Zn^{2+}	ML/M.L	5.4				

[a] 25-45°, 0;   [b] 35°, 0

Bibliography:

H^+ 59BB,61JW,64LAa Fe^{3+} 64LAa

Fe^{2+} 59BB,64LAa   Other references: 52BG,67LAa

Co^{2+}-Cu^{2+},Zn^{2+} 59BB

$C_{13}H_{10}N_2$ 2-Methyl-1,10-phenanthroline   L

Metal ion	Equilibrium	Log K 25°, 0.1	Log K 25°, 0.3
H^+	HL/H.L	5.31	5.30
Mn^{2+}	ML/M.L	3.0	
	$ML_2/M.L^2$	5.5	
	$ML_3/M.L^3$	7.9	
Fe^{2+}	ML/M.L	4.2	
	$ML_2/M.L^2$	7.9	
	$ML_3/M.L^3$	10.8	
Co^{2+}	ML/M.L	5.1	
	$ML_2/M.L^2$	10.0	
	$ML_3/M.L^3$	13.9	
Ni^{2+}	ML/M.L	5.95	
	$ML_2/M.L^2$	11.8	
	$ML_3/M.L^3$	16.7	
Cu^{2+}	ML/M.L	7.40	7.4
	$ML_2/M.L^2$	13.85	13.8
Cu^+	$ML_2/M.L^2$		16.95
Zn^{2+}	ML/M.L	4.96	
	$ML_2/M.L^2$	9.36	
	$ML_3/M.L^3$	12.7	
Cd^{2+}	ML/M.L	5.15	
	$ML_2/M.L^2$	9.65	
	$ML_3/M.L^3$	13.3	

Bibliography:

H^+ 53IC,61JW Fe^{2+} 53IC

Mn^{2+},Co^{2+},Ni^{2+},Zn^{2+},Cd^{2+} 62IM Cu^{2+} 61JW,62IM

                                                     Cu^+ 61JW

$C_{13}H_{10}N_2$ 5-Methyl-1,10-phenanthroline   L

Metal ion	Equilibrium	Log K 25°, 0.1	Log K 25°, 0	ΔH 25°, 0	ΔS 25°, 0
H^+	HL/H.L	5.27 ±0.01	5.17	(-4)[a]	(10)
Mn^{2+}	ML/M.L	4.28			
	$ML_2/M.L^2$	7.6			
	$ML_3/M.L^3$	15.3			
Fe^{2+}	ML/M.L	6.00 ±0.05	(6.11)[b]		
	$ML_2/M.L^2$	11.45			
	$ML_3/M.L^3$	22.1 ±0.2	(21.3)[b]	(-34)[c]	(-17)[b]
Co^{2+}	ML/M.L	7.14			
	$ML_2/M.L^2$	14.0			
	$ML_3/M.L^3$	20.6			
Ni^{2+}	ML/M.L	8.30			
	$ML_2/M.L^2$	17.0			
	$ML_3/M.L^3$	24.7			
Cu^{2+}	ML/M.L	8.55			
	$ML_2/M.L^2$	15.0			
	$ML_3/M.L^3$	20.1			
Ag^+	ML/M.L	7.30			
	$ML_2/M.L^2$	13.39			
Zn^{2+}	ML/M.L	6.62			
	$ML_2/M.L^2$	12.6			
	$ML_3/M.L^3$	18.3			
Cd^{2+}	ML/M.L	6.13			
	$ML_2/M.L^2$	11.0			

---

[a] 15-35°, 0;   [b] 22°, 0;   [c] 15-45°, 0

Bibliography:

H^+ 56YS,62MB,67LA

Mn^{2+},Co^{2+}-Cu^{2+},Zn^{2+},Cd^{2+} 62MB

Fe^{2+} 59BB,62MB,67LA,72Ba

Ag^+ 74B

Other reference: 52BG

$C_{14}H_{12}N_2$ 2,9-Dimethyl-1,10-phenanthroline   L

Metal ion	Equilibrium	Log K 25°, 0.1	Log K 25°, 0.3
H^+	HL/H.L	5.85	5.88
Co^{2+}	ML/M.L	4.2	
	$ML_2/M.L^2$	7.0	
Ni^{2+}	ML/M.L	5.0	
	$ML_2/M.L^2$	8.5	
Cu^{2+}	ML/M.L	5.2	6.1
	$ML_2/M.L^2$	11.0	11.7
Cu^+	$ML_2/M.L^2$		19.1
Zn^{2+}	ML/M.L	4.1	
	$ML_2/M.L^2$	7.7	
Cd^{2+}	ML/M.L	4.1	
	$ML_2/M.L^2$	7.4	
	$ML_3/M.L^3$	10.4	

Bibliography:
H^+,Cu^{2+} 61JW,62IMa Cu^+ 61JW
$Co^{2+},Ni^{2+},Zn^{2+},Cd^{2+}$ 62IMa   Other reference: 56YS

$C_{14}H_{12}N_2$ 4,7-Dimethyl-1,10-phenanthroline   L

Metal ion	Equilibrium	Log K 25°, 0.1
H^+	HL/H.L	5.95 −0.01
Fe^{2+}	ML/M.L	5.60
Co^{2+}	ML/M.L	8.08
	$ML_2/M.L^2$	16.1
	$ML_3/M.L^3$	24.5
Ni^{2+}	ML/M.L	8.44
	$ML_2/M.L^2$	16.6
	$ML_3/M.L^3$	25.0
Cu^{2+}	ML/M.L	8.76
	$ML_2/M.L^2$	16.0
	$ML_3/M.L^3$	22.0
Zn^{2+}	ML/M.L	6.90
	$ML_2/M.L^2$	13.1
	$ML_3/M.L^3$	19.1

Bibliography:
H^+ 56YS,63BM Fe^{2+}-Zn^{2+} 63BM

$C_{14}H_{12}N_2$ 5,6-Dimethyl-1,10-phenanthroline   L

Metal ion	Equilibrium	Log K 25°, 0.1
H^+	HL/H.L	5.60
Fe^{2+}	ML/M.L	6.37
Co^{2+}	ML/M.L	7.47
	$ML_2/M.L^2$	15.5
	$ML_3/M.L^3$	16.6
Ni^{2+}	ML/M.L	8.25
	$ML_2/M.L^2$	16.6
	$ML_3/M.L^3$	24.8
Cu^{2+}	ML/M.L	8.71
	$ML_2/M.L^2$	15.7
	$ML_3/M.L^3$	21.1
Zn^{2+}	ML/M.L	6.87
	$ML_2/M.L^2$	12.9
	$ML_3/M.L^3$	18.6

Bibliography: 63BM

| C$_4$H$_4$N$_2$ | 1,2-Diazine (pyridazine) | | | L |

Metal ion	Equilibrium	Log K 25°, 0.1	ΔH 25°, 0.1	ΔS 25°, 0.1
Ag$^+$	ML/M.L	1.48	-7.39	18.0
	ML$_2$/M.L^2	2.82	-8.06	14.1

Bibliography: 73BE

| C$_4$H$_4$N$_2$ | 1,4-Diazine (pyrazine) | | | | L |

Metal ion	Equilibrium	Log K 25°, 0.1	Log K 25°, 1.0	ΔH 25°, 0.1	ΔS 25°, 0.1
Ni^{2+}	ML/M.L		1.01	(-3)[a]	(-6)[b]
Ag$^+$	ML/M.L	1.38		-4.07	7.3
	ML$_2$/M.L^2	2.41		-8.10	16.1

[a] 10-35°, 1.0; [b] 25°, 1.0

Bibliography:
Ni^{2+} 72MS   Other reference: 62ST
Ag^{2+} 73BE

$C_4H_4N_2$ <u>1,3-Diazine</u>   <u>(pyrimidine)</u>   L

Metal ion	Equilibrium	Log K 25°, 0.1	ΔH 25°, 0.1	ΔS 25°, 0.1
Ag^+	ML/M.L	1.61	−4.21	6.7
	$ML_2/M.L^2$	2.98	−8.35	14.4

Bibliography: 73BE

NH₂ structure (cytosine)

$C_4H_5ON_2$ <u>4-Amino-2-pyrimidone</u>   <u>(cytosine)</u>   L

Metal ion	Equilibrium	Log K 25°, 0.1	Log K 25°, 0.05	Log K 25°, 0	ΔH 25°, 0	ΔS 25°, 0
H^+	$L/H_{-1}L.H$	11.71		12.15	−11.5	17
	HL/H.L	4.4 ±0.0	4.49	4.58	−5.1 −4.5[a]	4
Cu^{2+}	ML/M.L	2.0[b]				
Hg^{2+}	$ML_2/M.L^2$	10.90[c]				

[a] 25°, 0.1; [b] 25°, 0.16; [c] 27°, 0.1

Bibliography:
H^+ 60RS,61FB,64SPB,67CR,69RW Hg^{2+} 61FB
Cu^{2+} 61M

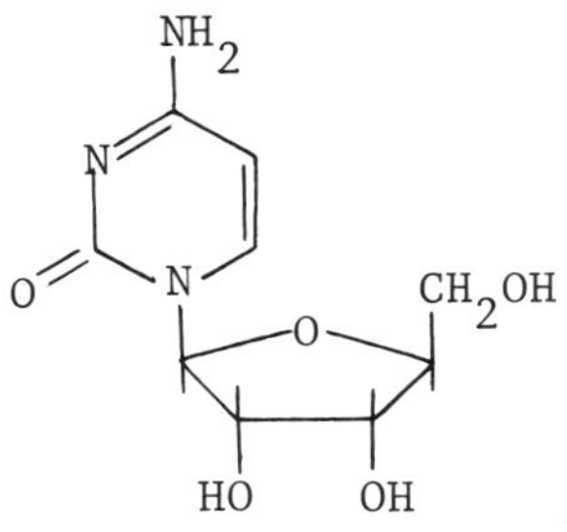

$C_9H_{13}O_5N_3$ 1-(β-D-Ribofuranosyl)cytosine   (cytidine)   L

Metal ion	Equilibrium	Log K 25°, 0.1	Log K 20°, 1.0	Log K 25°, 0	ΔH 25°, 0	ΔS 25°, 0
H^+	L/H$_{-1}$L.H			12.5	−10.3	23
	HL/H.L	4.11 ±0.02	4.09 ±0.00	4.08	−5.11	−1.5
Cu^{2+}	ML/M.L	1.4[a]				

———————————

[a] 25, 0.16

Bibliography:

H^+ 61M,64SPB,65FB,66LHa,67CR,70CR Cu^{2+} 61M

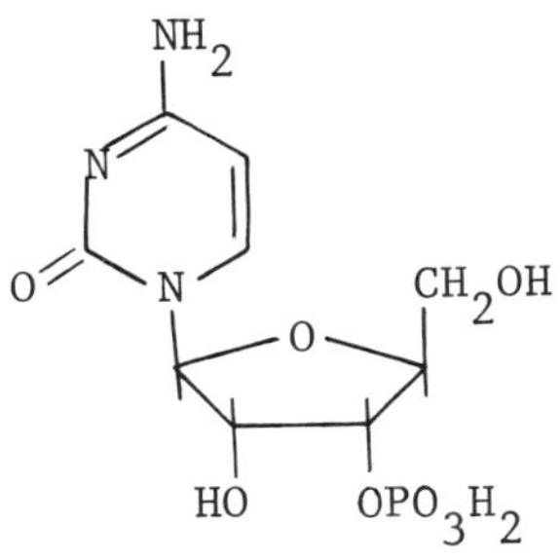

———————————————————————————————————————

$C_9H_{14}O_8N_3P$ Cytidine-3'-(dihydrogenphosphate)   (CMP-3) H_2L

Metal ion	Equilibrium	Log K 25°, 0.16
Sr^{2+}	ML/M.L	1.6

Bibliography: 54S

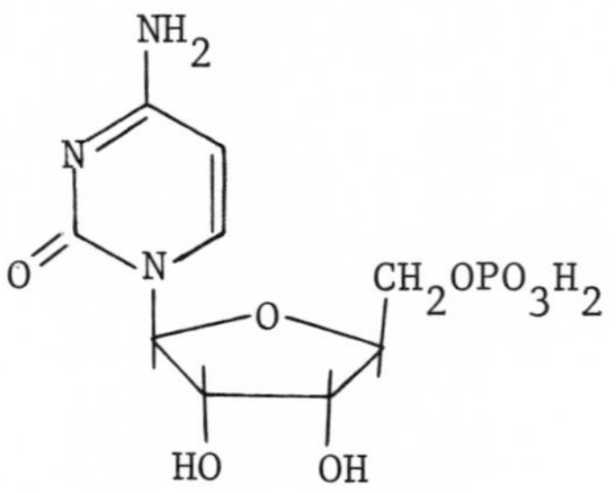

$C_9H_{14}O_8N_3P$ Cytidine-5'-(dihydrogenphosphate)   (CMP-5) H_2L

Metal ion	Equilibrium	Log K 25°, 0.1	Log K 15°, 0.1	Log K 25°, 0	ΔH 25°, 0	ΔS 25°, 0
H^+	HL/H.L	6.3	6.27	6.62	$(1)^a$	(33)
	H_2L/HL.H	4.5	4.52			
Mg^{2+}	ML/M.L		1.75			
Ni^{2+}	ML/M.L		(1.90)			
Zn^{2+}	ML/M.L	2.54				

———————

[a] 10-37°, 0

Bibliography:
H^+ 56BL,65PE,72FS Zn^{2+} 58WS
Mg^{2+},Ni^{2+} 72FS

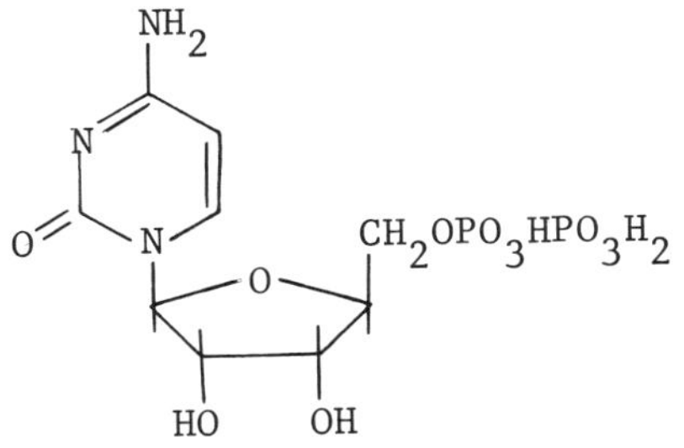

$C_9H_{15}O_{11}N_3P_2$ Cytidine-5'-(trihydrogendiphosphate)   (CDP) H_3L

Metal ion	Equilibrium	Log K 25°, 0.1	Log K 15°, 0.1	Log K 25°, 0	ΔH 25°, 0	ΔS 25°, 0
H^+	HL/H.L	6.4	6.38	7.18	(1)[a]	(36)
	H_2L/HL.H	4.6	4.56			
Mg^{2+}	ML/M.L		3.22			
	MHL/ML.H		4.76			
Ni^{2+}	ML/M.L		(3.48)			
	$ML_2/M.L^2$		(5.47)			
	MHL/ML.H		(4.77)			

[a] 10-37°, 0

Bibliography:
H^+ 56BL,65PE,72FS Mg^{2+},Ni^{2+} 72FS

$C_9H_{16}O_{14}N_3P_3$ Cytidine-5'-(tetrahydrogentriphosphate)   (CTP) H_4L

Metal ion	Equilibrium	Log K 25°, 0.1	Log K 15°, 0.1	Log K 25°, 0	ΔH 25°, 0	ΔS 25°, 0
H^+	HL/H.L	6.6	6.63	7.65	(2)[a]	(42)
	H_2L/HL.H	4.8	4.85			
Mg^{2+}	ML/M.L	4.01[b]	4.03			
	MHL/ML.H		4.75			
Ca^{2+}	ML/M.L	3.81[b]				
Mn^{2+}	ML/M.L	4.78[b]				
Co^{2+}	ML/M.L	4.48[b]				
Ni^{2+}	ML/M.L		(4.41)			
	MHL/ML.H		(4.90)			
Cu^{2+}	ML/MOHL.H	7.6				

[a] 10-37°, 0;   [b] 23°, 0.1

Bibliography:
H^+ 56BL,65PE,72FS Ni^{2+} 72FS
Mg^{2+} 58W,72FS Cu^{2+} 68S
Ca^{2+}-Co^{2+} 58W

$C_9H_{13}O_9N_2P$		Uridine-5'-(dihydrogenphosphate) (UMP)			H_2L

Metal ion	Equilibrium	Log K 25°, 0.1	Log K 25°, 0	ΔH 25°, 0	ΔS 25°, 0
H^+	$L/H_{-1}L.H$	9.5			
	$HL/H.L$	6.4	6.63	$(1)^a$	(34)
Mg^{2+}	$ML/M.L$	2.25^b		1.8^c	16^b

^{a} 10-37°, 0; ^{b} 23°, 0.1; ^{c} 30°, 0.2

Bibliography:

H^+ 56BL,65PE	Mg^{2+} 58W,73SBa

$C_9H_{14}O_{12}N_2P_2$		Uridine-5'-(trihydrogendiphosphate) (UDP)			H_3L

Metal ion	Equilibrium	Log K 25°, 0.1	Log K 25°, 0	ΔH 25°, 0	ΔS 25°, 0
H^+	$L/H_{-1}L.H$	9.4			
	$HL/H.L$	6.5	7.16	$(1)^a$	(36)
Mg^{2+}	$ML/M.L$	3.17^b		3.2^c	25^b

^{a} 10-37°, 0; ^{b} 23°, 0.1; ^{c} 30°, 0.2

Bibliography:

H^+ 56BL,65PE	Mg^{2+} 58W,73SBa

$C_9H_{15}O_{15}N_2P_3$ <u>Uridine-5'-(tetrahydrogentriphosphate)</u>   <u>(UTP)</u> H_4L

Metal ion	Equilibrium	Log K 25°, 0.1	Log K 25°, 0	ΔH 25°, 0	ΔS 25°, 0
H^+	$L/H_{-1}L.H$	9.5 -0.1			
	$HL/H.L$	6.6	7.58	(2)[a]	(41)
Mg^{2+}	$ML/M.L$	4.02[b]		4.4[c]	33[b]
Ca^{2+}	$ML/M.L$	3.71[b]			
Mn^{2+}	$ML/M.L$	4.78[b]			
Co^{2+}	$ML/M.L$	4.55[b]			
Cu^{2+}	$ML/MOHL.H$	7.8			
	$MOHL/M(OH)_2L.H$	8.4			

[a] 10-37°, 0;   [b] 23°, 0.1;   [c] 30°, 0.2

Bibliography:

H^+ 56BL,65PE,68S $Ca^{2+}-Co^{2+}$ 58W

Mg^{2+} 58W,73SBa Cu^{2+} 68S

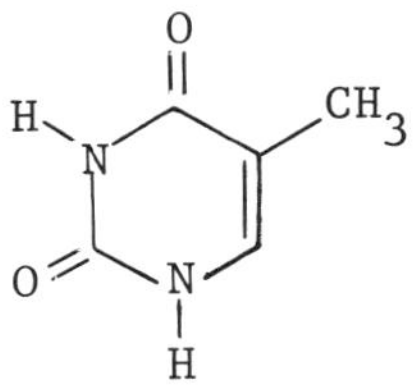

$C_5H_6O_2N_2$ 5-Methyl-2,4-dioxopyrimidine   (thymine)   HL

Metal ion	Equilibrium	Log K 25°, 0.1	Log K 25°, 0	ΔH 25°, 0	ΔS 25°, 0
H^+	HL/H.L	9.65 ±0.0	9.90	-8.2	18
Hg^{2+}	$ML_2/M.L^2$	21.2[a]			

---

[a] 27°, 0.1

Bibliography:

H^+ 61FB,66LB,67CR,70CR Hg^{2+} 61FB

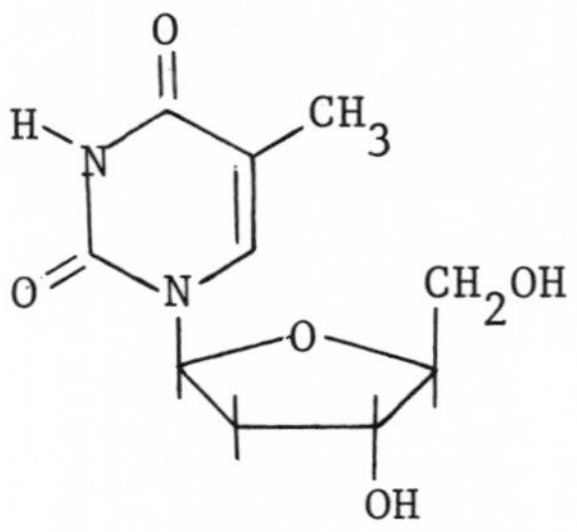

$C_{10}H_{14}O_5N_2$ 1-(2-Deoxy-β-D-ribofuranosyl)thymine   (thymidine)   HL

Metal ion	Equilibrium	Log K 25°, 0.1	Log K 20°, 1.0	Log K 25°, 0	ΔH 25°, 0	ΔS 25°, 0
H^+	$L/H_{-1}L.H$			12.85	−12.4	17
	$HL/H.L$	9.44	9.42 ±0.00	9.79	−7.32	20.3
CH_3Hg^+	$ML/MOH.HL$	4.35				
Hg^{2+}	$ML_2/M.L^2$	21.2[a]				

----

[a] 27°, 0.1

Bibliography:
H^+ 61FB,65FB,66GD,67CR Hg^{2+} 61FB
CH_3Hg^+ 66GD

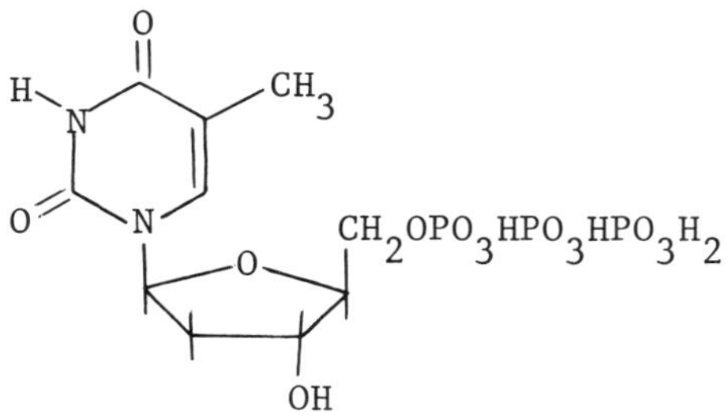

$C_{10}H_{17}O_{13}N_2P_3$ Thymidine-5'-(tetrahydrogentriphosphate)   (TTP) H_4L

Metal ion	Equilibrium	Log K 25°, 0.1
H^+	$L/H_{-1}L.H$	9.7
Cu^{2+}	$ML/MOHL.H$	7.6
	$MOHL/M(OH)_2L.H$	8.2

Bibliography: 68S

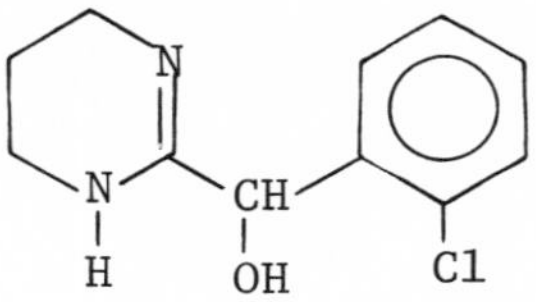

$C_{11}H_{13}ON_2Cl$ DL-2-(2-Chlorophenylhydroxymethyl)-1,4,5,6-tetrahydropyrimidine   HL

Metal ion	Equilibrium	Log K 25°, 0.1
H^+	HL/H.L	12.04
	$H_2L/HL.H$	11.08
Cu^{2+}	ML/M.L	12.62
	$ML_2/M.L^2$	(23.06)

Bibliography: 68JK

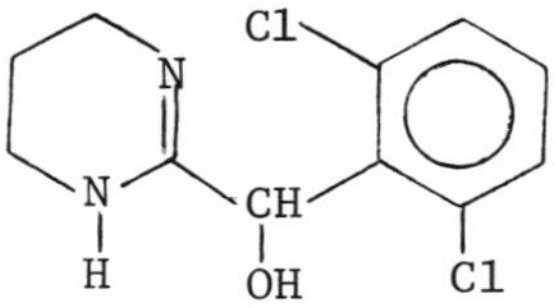

$C_{11}H_{12}ON_2Cl_2$ DL-2-(2,6-Dichlorophenylhydroxymethyl)-1,4,5,6-tetrahydropyrimidine   HL

Metal ion	Equilibrium	Log K 25°, 0.1
H^+	HL/H.L	11.87
	$H_2L/HL.H$	10.81
Cu^{2+}	ML/M.L	12.16
	$ML_2/M.L^2$	(21.97)

Bibliography: 68JK

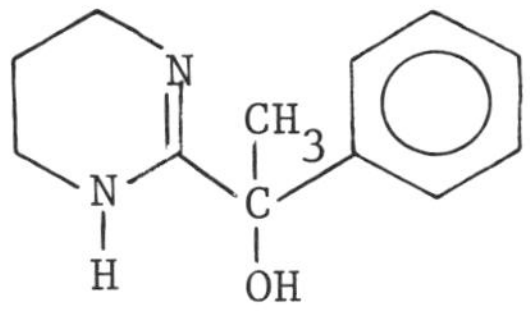

$C_{12}H_{16}ON_2$ DL-2-(1-Phenyl-1-hydroxyethyl)-1,4,5,6-tetrahydropyrimidine   HL

Metal ion	Equilibrium	Log K 25°, 0.1
H^+	HL/H.L	12.31
	H_2L/HL.H	11.49
Cu^{2+}	ML/M.L	13.18
	ML_2/M.L^2	24.31

Bibliography: 68JK

| $C_5H_5N_5$ | | | 6-Aminopurine   (adenine) | | | L |

<table>
<tr><td>Metal
ion</td><td>Equilibrium</td><td>Log K
25°, 0.1</td><td>Log K
37°, 0.15</td><td>Log K
25°, 0</td><td>ΔH
25°, 0</td><td>ΔS
25°, 0</td></tr>
<tr><td>H^+</td><td>$L/H_{-1}L.H$</td><td>9.67 ±0.02</td><td>9.26</td><td>9.87</td><td>-9.4 ±0.3</td><td>14</td></tr>
<tr><td></td><td>HL/H.L</td><td>4.07 ±0.03</td><td>3.84</td><td>4.20</td><td>-4.9 ±0.1</td><td>3</td></tr>
<tr><td></td><td></td><td></td><td></td><td></td><td>-4.1[a] ±0.1</td><td></td></tr>
<tr><td>Co^{2+}</td><td>ML/M.L</td><td></td><td>1.38</td><td></td><td></td><td></td></tr>
<tr><td>Ni^{2+}</td><td>ML/M.L</td><td></td><td>1.47</td><td></td><td></td><td></td></tr>
<tr><td>Cu^{2+}</td><td>ML/M.L</td><td></td><td>2.68</td><td></td><td></td><td></td></tr>
<tr><td>Zn^{2+}</td><td>ML/M.L</td><td></td><td>1.62</td><td></td><td></td><td></td></tr>
<tr><td>Hg^{2+}</td><td>$ML_2/M.L^2$</td><td>11.5[b]</td><td></td><td></td><td></td><td></td></tr>
</table>

_____

[a] 25°, 0.1;   [b] 27°, 0.1

Bibliography:

H^+ 51AS,60RS,61FB,62IC,64SPB,66LB,69RW, Hg^{2+} 61FB
         70CR,71TK,72ZB,74MWa
Co^{2+}-Zn^{2+} 74MWa   Other references: 58HF,60AS,71KK

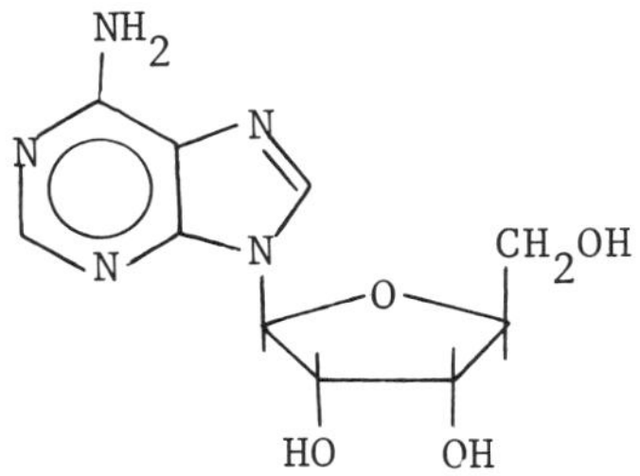

$C_{10}H_{13}O_4N_5$ 9-(β-D-Ribofuranosyl)adenine   (adenosine)   L

Metal ion	Equilibrium	Log K 25°, 0.1	Log K 20°, 1.0	Log K 25°, 0	ΔH 25°, 0	ΔS 25°, 0
H^+	$L/H_{-1}L.H$			12.35	-9.7	24
	HL/H.L	3.44 ±0.04	3.56	3.50	-3.9	3
					-3.8^a	
Ag^+	ML/M.L			2.02	$(-5)^b$	(-9)
	$ML_2/M.L^2$			3.86	$(-9)^b$	(-13)
Zn^{2+}	ML/M.L	1.51^c				
Hg^{2+}	$ML_2/M.L^2$	8.50^d				

---

[a] 25°, 0.1;   [b] 10-40°, 0;   [c] 20°, 0.1;   [d] 27°, 0.1

Bibliography:

H^+ 57WS,60RS,61FB,64SPB,65FB,66IR,69RWa, Zn^{2+} 57WS
     70CR Hg^{2+} 61FB

Ag^+ 68PG

                                    Other reference: 64SB

$$NH_2$$

(chemical structure of Adenosine-2'-(dihydrogenphosphate) with purine ring, ribose, CH_2OH, H_2O_3PO, and OH groups)

| $C_{10}H_{14}O_7N_5P$ | Adenosine-2'-(dihydrogenphosphate) | (AMP-2) | H_2L |

Metal ion	Equilibrium	Log K 25°, 0.1	ΔH 25°, 0.1	ΔS 25°, 0.1
H^+	HL/H.L	6.02 ±0.01		
	H_2L/HL.H	3.72 ±0.02		
Mg^{2+}	ML/M.L	1.93 −0.1	(3)[a]	(21)
Ca^{2+}	ML/M.L	1.83	(−1)[a]	(6)
Sr^{2+}	ML/M.L	1.74	(−1)[a]	(4)
Ba^{2+}	ML/M.L	1.71	(−2)[a]	(1)
Mn^{2+}	ML/M.L	2.38	(−1)[a]	(8)
Co^{2+}	ML/M.L	2.24	(−1)[a]	(8)
Ni^{2+}	ML/M.L	2.81 −0.8	(−1)[a]	(9)
Cu^{2+}	ML/M.L	3.16	(−2)[a]	(8)
Zn^{2+}	ML/M.L	2.64	(−1)[a]	(8)

---

[a] 0-40°, 0.1

Bibliography:
H^+,Mg^{2+},Ni^{2+} 67TM,72FS   Other reference: 51AS
$Ca^{2+}-Co^{2+},Cu^{2+},Zn^{2+}$ 67TM

$C_{10}H_{14}O_7N_5P$ Adenosine-3'-(dihydrogenphosphate)   (AMP-3) H_2L

Metal ion	Equilibrium	Log K 25°, 0.1	ΔH 25°, 0.1	ΔS 25°, 0.1
H^+	HL/H.L	5.83		
	H_2L/HL.H	3.65		
Mg^{2+}	ML/M.L	1.89 −0.2	(3)[a]	(20)
Ca^{2+}	ML/M.L	1.80	(−1)[a]	(6)
Sr^{2+}	ML/M.L	1.71 −0.3	(−1)[a]	(4)
Ba^{2+}	ML/M.L	1.69	(−2)[a]	(1)
Mn^{2+}	ML/M.L	2.28 −0.4	(−1)[a]	(8)
Co^{2+}	ML/M.L	2.20 −0.1	(−1)[a]	(8)
Ni^{2+}	ML/M.L	2.79	(−1)[a]	(10)
Cu^{2+}	ML/M.L	2.96	(−2)[a]	(8)
Zn^{2+}	ML/M.L	2.60 ±0.1	(−1)[a]	(8)

[a] 0-40°, 0.1

Bibliography:
H^+ 62TMa,67TM
Mg^{2+}-Ba^{2+} 54S,58WS,62TMa,67TM
Mn^{2+}-Zn^{2+} 58WS,62TMa,66DT,67TM
Other reference: 51AS

$C_{10}H_{14}O_7N_5P$ Adenosine-5'-(dihydrogenphosphate)   (AMP-5) H_2L

Metal ion	Equilibrium	Log K 25°, 0.1	Log K 25°, 0	ΔH 25°, 0	ΔS 25°, 0.1
H^+	$L/(H_{-1}L).H$		13.06	-10.9	23[d]
	HL/H.L	6.19 ±0.05	6.67	1.8	34
	$H_2L/HL.H$	3.80 ±0.04		-4.2	4
Li^+	ML/M.L	0.6[a]			
Na^+	ML/M.L	0.5[a] ±0.0			
K^+	ML/M.L	0.2[a]			
Mg^{2+}	ML/M.L	1.97 ±0.2		1.8[c]	15
Ca^{2+}	ML/M.L	1.80 ±0.05		(-1)[b]	(6)
Sr^{2+}	ML/M.L	1.79 -0.3		(-1)[b]	(4)
Ba^{2+}	ML/M.L	1.73 -0.6		(-2)[b]	(1)
Mn^{2+}	ML/M.L	2.35 ±0.05		(-1)[b]	(8)
Co^{2+}	ML/M.L	2.56 ±0.03		(-1)[b]	(8)
Ni^{2+}	ML/M.L	2.84 -0.3		(-1)[b]	(10)
Cu^{2+}	ML/M.L	3.18 -0.1		(-2)[b]	(8)
Zn^{2+}	ML/M.L	2.72 -0.5		(-1)[b]	(8)

[a] 25°, 0.2 $(C_3H_7)_4NCl$;   [b] 0-40°, 0.1;   [c] 30°, 0.2;   [d] 25°, 0

Bibliography:
H^+ 56MS,61TD,62CI,62TMa,64SB,65PE,66IR, Mg^{2+}-Ba^{2+} 56MS,57N,58W,58WS,60Oa,62TMa,
   67TM,72FS   67TM,69BS,72FS
Li^+-K^+ 56SA,61TD Mn^{2+}-Zn^{2+} 58W,62TMa,64SB,66DT,67TM,72FS
                                                   Other references: 51AS,56SAa

$C_{10}H_{15}O_{10}N_5P_2$ Adenosine-5'-(trihydrogendiphosphate)   (ADP) H_3L

Metal ion	Equilibrium	Log K 25°, 0.1	Log K 25°, 0	ΔH 25°, 0	ΔS 25°, 0.1
H^+	HL/H.L	6.40 ±0.05	7.20	1.3	34
	H_2L.HL.H	3.96 ±0.03		-4.1	4
Li^+	ML/M.L	1.2[a]			
Na^{2+}	ML/M.L	0.8[a] -0.1			
K^+	ML/M.L	0.7[a] ±0.0			
Mg^{2+}	ML/M.L	3.17 ±0.1	4.10	3.2[c]	25
	MHL/ML.H	4.91 ±0.2	4.38	(-1)[b]-1	(19)
Ca^{2+}	ML/M.L	2.81 ±0.05		(-1)[b]	(10)
	MHL/ML.H	5.16		(-1)[b]	(20)
Sr^{2+}	ML/M.L	2.54		(-2)[b]	(5)
	MHL/ML.H	5.43		(0)[b]	(25)
Ba^{2+}	ML/M.L	2.36		(-3)[b]	(1)
	MHL/ML.H	5.52		(0)[b]	(25)
Mn^{2+}	ML/M.L	4.05 ±0.1		(-2)[b]	(12)
	MHL/ML.H	4.17		(-1)[b]	(16)
Co^{2+}	ML/M.L	4.20 -0.5		(-2)[b]	(13)
	MHL/ML.H	4.25		(-1)[b]	(16)
Ni^{2+}	ML/M.L	4.50 -0.4		(-2)[b]	(13)
	MHL/ML.H	4.24 +0.3		(-1)[b]	(16)
Cu^{2+}	ML/M.L	5.90		(-4)[b]	(13)
	MHL/ML.H	3.16		(0)[b]	(14)

[a] 25°, 0.2, $(C_3H_7)_4NCl$;   [b] 0-40°, 0.1;   [c] 30°, 0.2

ADP   (continued)

Metal ion	Equilibrium	Log K 25°, 0.1		Log K 25°, 0	ΔH 25°, 0	ΔS 25°, 0.1
	ML/MOHL.H	7.08			$(-9)^b$	(2)
	$(ML)_2/(MOHL)_2.H^2$	10.73			$(-11)^b$	(11)
	$(MOHL)_2/(MOHL)^2$	3.42			$(-6)^b$	(-5)
Zn^{2+}	ML/M.L	4.28	-0.1		$(-2)^b$	(13)
	MHL/ML.H	4.20	+0.6		$(-1)^b$	(16)
	ML/MOHL.H	8.51			$(-10)^b$	(6)
	$(ML)_2/(MOHL)_2.H^2$	13.68			$(-13)^b$	(17)
	$(MOHL)_2/(MOHL)^2$	3.34			$(-5)^b$	(-2)

^{b} 0-40°, 0.1

Bibliography:

H^+ 56MS,62CI,62TMa,63GP,65PE,72FS

Li^+-K^+ 54M,56SA

$Mg^{2+}-Ba^{2+}$ 56MS,57N,58W,58WS,59B,62TMa,
    63GPa,66PG,67TM,69BS,72FS

$Mn^{2+}-Zn^{2+}$ 62TMa,67TM,72FS

Other references: 51AS,53DN,56SAa,61HB,62AM,
    71MMW

$C_{10}H_{16}O_{13}N_5P_3$	Adenosine-5'-(tetrahydrogentriphosphate)		(ATP)		H_4L
Metal ion	Equilibrium	Log K 25°, 0.1	Log K 25°, 0	ΔH 25°, 0	ΔS 25°, 0.1
H^+	HL/H.L	6.51 ±0.03	7.68	1.2	34
	H_2L/HL.H	4.06 ±0.02		-3.7	6
Li^+	ML/M.L	1.6[a] ±0.1			
Na^+	ML/M.L	1.1[a] ±0.1	2.36		
K^+	ML/M.L	1.0[a] ±0.1	2.34		
Rb^+	ML/M.L	0.9[b]			
Cs^+	ML/M.L	0.8[b]			
Mg^{2+}	ML/M.L	4.06 ±0.2	5.83	4.5[d]	34
	MHL/ML.H	4.55 +0.3	5.44	(0)[c]-1	(21)
Ca^{2+}	ML/M.L	3.77 ±0.2	(6.37)	(-1)[c]	(12)
	MHL/ML.H	4.69		(0)[c]	(21)
	M_2L/ML.M		(3.04)		
Sr^{2+}	ML/M.L	3.54 -0.4		(-3)[c]	(6)
	MHL/ML.H	5.04		(1)[c]	(26)
Ba^{2+}	ML/M.L	3.29		(-4)[c]	(12)
	MHL/ML.H	5.09		(1)[c]	(27)
Mn^{2+}	ML/M.L	4.76 ±0.03		(-3)[c]	(12)
	MHL/ML.H	4.14 +0.4		(0)[c]	(19)
Co^{2+}	ML/M.L	4.63 ±0.03		(-2)[c]	(14)
	MHL/ML.H	4.19		(0)[c]	(19)
Ni^{2+}	ML/M.L	5.02 -0.5		(-3)[c]	(15)
	MHL/ML.H	4.23 +0.3		(0)[c]	(19)

[a] 25°, 0.2 $(C_4H_7)_4NCl$;   [b] 25°, 0.32 CsCl;   [c] 0-40°, 0.1;   [d] 30°, 0.2

ATP   (continued)

Metal ion	Equilibrium	Log K 25°, 0.1	Log K 25°, 0	ΔH 25°, 0	ΔS 25°, 0.1
Cu^{2+}	ML/M.L	6.13 ±0.3		(-4)^c	(14)
	MHL/ML.H	3.52 +0.4		(1)^c	(19)
	ML/MOHL.H	6.47 +1		(-8)^c	(3)
	MOHL/M(OH)$_2$L.H	7.45		(-4)^c	(19)
	(ML)^2/(MOHL)$_2$.H^2	10.35		(-10)^c	(14)
	(MOHL)$_2$/(MOHL)^2	2.59		(-6)^c	(-10)
Zn^{2+}	ML/M.L	4.85 ±0.4		(-3)^c	(13)
	MHL/ML.H	4.35 +0.4		(0)^c	(20)

---

^{c} 0-40°, 0.1

Bibliography:
H$^+$ 56MS,62CI,62TM,63GP,65PE,66TM
Li$^+$-Cs$^+$ 54M,56SA,64OP,65BC,70MR
Mg^{2+}-Ba^{2+} 53DN,56MS,58W,58WS,59B,600a,
    62HBa,62TM,63GPa,66PG,66TM,69BS,72FS,
    72MR

Mn^{2+}-Zn^{2+} 58W,58WS,61B,62HBa,62TM,64SBE,
    66TM,67SBM,72FS

Other references: 51AS,56SAa,57N,61HB,61Na,
    61OP,62AM,64OP,66PSa,68S,71MMW,72TRa,
    73LJ,73SR

$C_{10}H_{17}O_{16}N_5P_4$ Adenosine-5'-(pentahydrogentetraphosphate) H_5L

Metal ion	Equilibrium	Log K 25°, 0.2	Log K 20°, 0.1
H^+	HL/H.L	(6.58)	6.79
	H_2L/HL.H		4.09
Li^+	ML/M.L	1.9[a]	
Na^+	ML/M.L	1.4[a]	
K^+	ML/M.L	1.3[a]	
Mg^{2+}	ML/M.L		4.22
	MHL/ML.H		5.3

[a] $(C_3H_7)_4NCl$ used as background electrolyte.

Bibliography:
H^+ 56SA,57SA Mg^{2+} 57SA
Li^+-K^+ 56SA

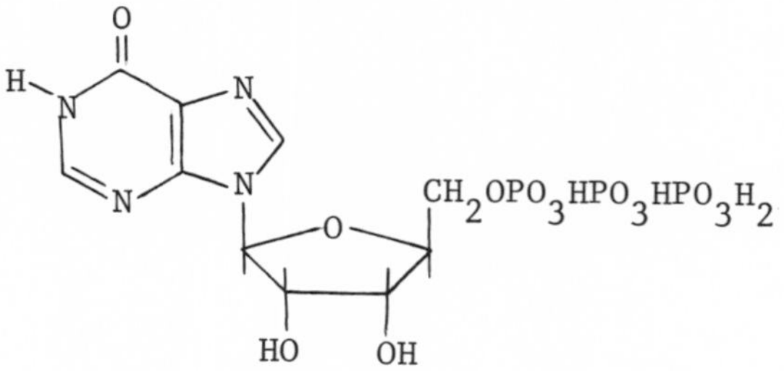

$$CH_2OPO_3HPO_3HPO_3H_2$$

$C_{10}H_{15}O_{14}N_4P_3$ <u>Inosine-5'-(tetrahydrogentriphosphate)</u>   <u>(ITP)</u> H_4L

Metal ion	Equilibrium	Log K 25°, 0.1	Log K 25°, 0	ΔH 25°, 0	ΔS 25°, 0
H^+	$L/H_{-1}L.H$	9.1			
	$HL/H.L$		7.68	(2)[a]	(42)
Mg^{2+}	$ML/M.L$	4.04[b]		4.5[c]	34[b]
Ca^{2+}	$ML/M.L$	3.76[b]			
Mn^{2+}	$ML/M.L$	4.57[b]			
Co^{2+}	$ML/M.L$	4.74[b]			
Cu^{2+}	$ML/MOHL.H$	7.4			
	$MOHL/M(OH)_2L.H$	9.2			

[a] 10-37°, 0;   [b] 23°, 0.1;   [c] 30°, 0.2

Bibliography:

H^+ 65PE,68S

Mg^{2+} 58W,73SBa

Ca^{2+}-Co^{2+} 58W

Cu^{2+} 68S

Other reference: 73TR

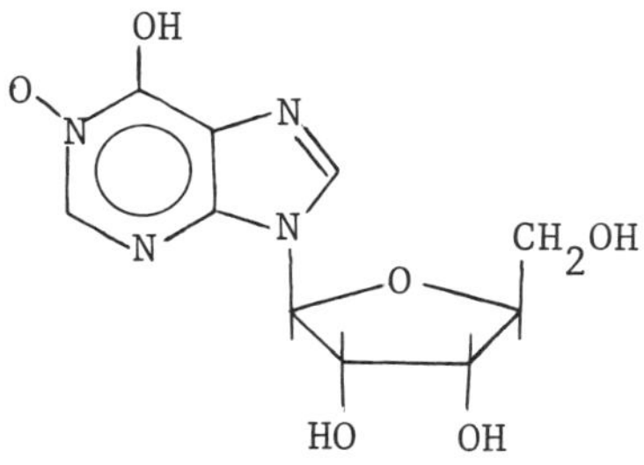

$C_{10}H_{12}O_6N_4$ Inosine N(1)-oxide   L

Metal ion	Equilibrium	Log K 25°, 0.1
H^+	HL/H.L	5.29
Mg^{2+}	ML/M.L	1.7
Ca^{2+}	ML/M.L	1.5
Ba^{2+}	ML/M.L	1.2
Mn^{2+}	ML/M.L	2.5
Co^{2+}	ML/M.L	3.46
Ni^{2+}	ML/M.L	3.50
Cu^{2+}	ML/M.L	5.27
Zn^{2+}	ML/M.L	3.60

Bibliography:

H^+ 65S Mg^{2+}–Zn^{2+} 65Sa

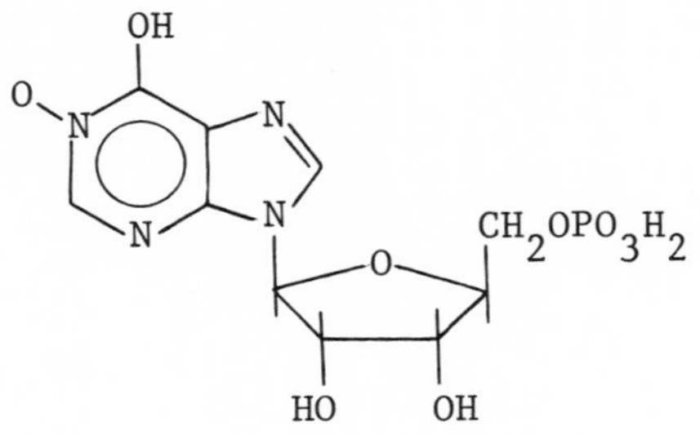

$C_{10}H_{13}O_9N_4$ Inosine-5'-(dihydrogenphosphate) N(1)-oxide H_2L

Metal ion	Equilibrium	Log K 25°, 0.1
H^+	HL/H.L	6.16
	H_2L/HL.H	5.28
Mg^{2+}	ML/M.L	2.1
Ca^{2+}	ML/M.L	2.0
Ba^{2+}	ML/M.L	1.6
Mn^{2+}	ML/M.L	2.85
Co^{2+}	ML/M.L	3.73
Ni^{2+}	ML/M.L	3.90
Cu^{2+}	ML/M.L	5.46
Zn^{2+}	ML/M.L	3.83

Bibliography:

H^+ 65S Mg^{2+}-Zn^{2+} 65Sa

$C_{10}H_{13}O_8N_5P$ Guanosine-5'-(dihydrogenphosphate)   (GMP) H_2L

Metal ion	Equilibrium	Log K 25°, 0.1	Log K 25°, 0	ΔH 25°, 0	ΔS 25°, 0
H^+	HL/H.L	6.1	6.66	(2)[a]	(37)
	H_2L/HL.H	2.4			
Mg^{2+}	ML/M.L			1.7[b]	

[a] 10-37°, 0;   [b] 30°, 0.2

Bibliography: H^+ 56BL,65PE Mg^{2+} 73SBa

$C_{10}H_{14}O_{11}N_5P_2$ Guanosine-5'-(trihydrogendiphosphate)   (GDP) H_3L

Metal ion	Equilibrium	Log K 25°, 0.1	Log K 25°, 0	ΔH 25°, 0	ΔS 25°, 0
H^+	$L/H_{-1}L$.H	9.6			
	HL/H.L	6.3	7.19	(2)[a]	(40)
	H_2L/HL.H	2.9			
Mg^{2+}	ML/M.L			3.4[b]	

[a] 10-37°, 0;   [b] 30°, 0.2

Bibliography:
H^+ 56BL,65PE Mg^{2+} 73SBa

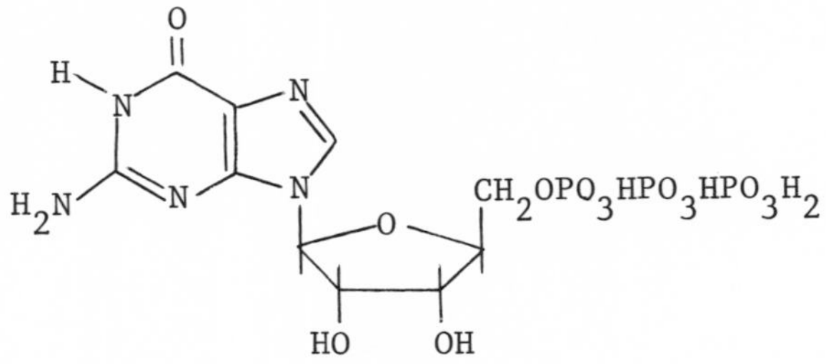

$C_{10}H_{15}O_{14}N_5P_3$ <u>Guanosine-5'-(tetrahydrogentriphosphate)</u>   <u>(GTP)</u> H_4L

Metal ion	Equilibrium	Log K 25°, 0.1	Log K 25°, 0	ΔH 25°, 0	ΔS 25°, 0.1
H^+	$L/H_{-1}L.H$	9.4 ±0.1			
	$HL/H.L$	6.5	7.65	(2)[a]	(37)
	$H_2L/HL.H$	3.3			
Mg^{2+}	$ML/M.L$	4.02[b]		4.3[c]	33[b]
Ca^{2+}	$ML/M.L$	3.58[b]			
Mn^{2+}	$ML/M.L$	4.73[b]			
Co^{2+}	$ML/M.L$	4.63[b]			
Cu^{2+}	$ML/MOHL.H$	7.6			
	$MOHL/M(OH)_2L.H$	9.3			

---

[a] 10–37°, 0;   [b] 23°, 0.1;   [c] 30°, 0.2

Bibliography:

H^+ 56BL,65PE,68S

Mg^{2+} 58W,73SBa

$Ca^{2+}-Co^{2+}$ 58W

Cu^{2+} 68S

Other reference: 73TR

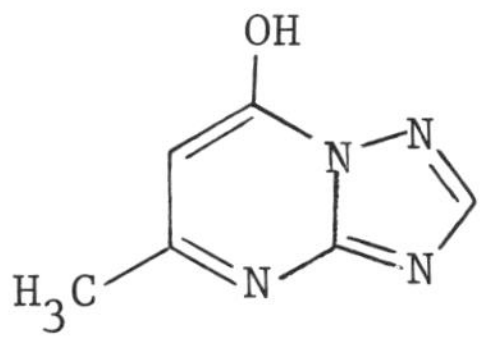

$C_6H_6ON_4$ 5-Methyl-7-hydroxy-1',2',4'-triazolo[1,5-a]pyrimidine   HL

Metal ion	Equilibrium	Log K 20°, 0.1
H^+	HL/H.L	6.22
Co^{2+}	ML/M.L	2.15
Ni^{2+}	ML/M.L	2.53
	$ML_2/M.L^2$	5.11
Cu^{2+}	ML/M.L	3.19
	$ML_2/M.L^2$	5.90

Bibliography: 66OC

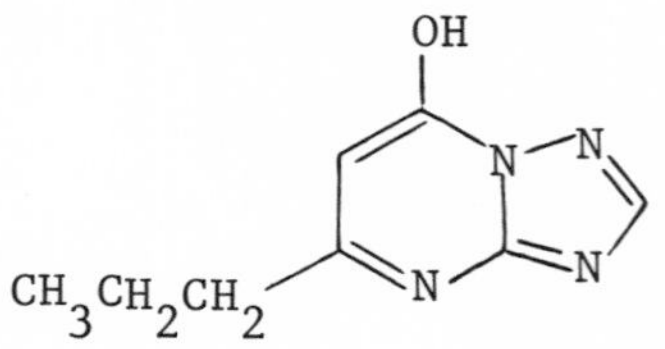

$C_8H_{10}ON_4$ 5-Propyl-7-hydroxy-1',2',4'-triazolo[1,5-a]pyrimidine   HL

Metal ion	Equilibrium	Log K 20°, 0.1
H^+	HL/H.L	6.31
Co^{2+}	ML/M.L	2.39
Ni^{2+}	ML/M.L	2.78
	$ML_2/M.L^2$	5.65
Cu^{2+}	ML/M.L	3.41
	$ML_2/M.L^2$	6.45

Bibliography: 660C

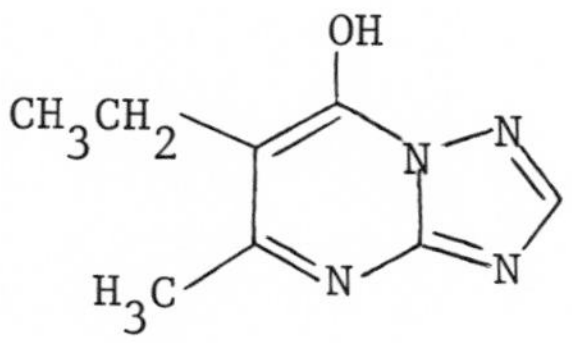

$C_8H_{10}ON_4$ 5-Methyl-6-ethyl-7-hydroxy-1',2',4'-triazolo[1,5-a]pyrimidine   HL

Metal ion	Equilibrium	Log K 20°, 0.1
H^+	HL/H.L	6.84
Co^{2+}	ML/M.L	2.40
Ni^{2+}	ML/M.L	2.73
	$ML_2/M.L^2$	5.60
Cu^{2+}	ML/M.L	3.46
	$ML_2/M.L^2$	6.84

Bibliography: 660C

$$H_2O_2PCH_2NHCH_2CH_2NHCH_2PO_2H_2$$

$C_4H_{14}O_4N_2P_2$ Ethylenebis(iminomethylenephosphonous acid) H_2L

Metal ion	Equilibrium	Log K 25°, 0.1
H^+	HL/H.L	8.08
	H_2L/HL.H	4.98
La^{3+}	ML/M.L	4.16
Co^{2+}	ML/M.L	5.95
Ni^{2+}	ML/M.L	7.52
Cu^{2+}	ML/M.L	10.72
Fe^{3+}	ML/M.L	10.29
Zn^{2+}	ML/M.L	6.16

Bibliography: 71MMa

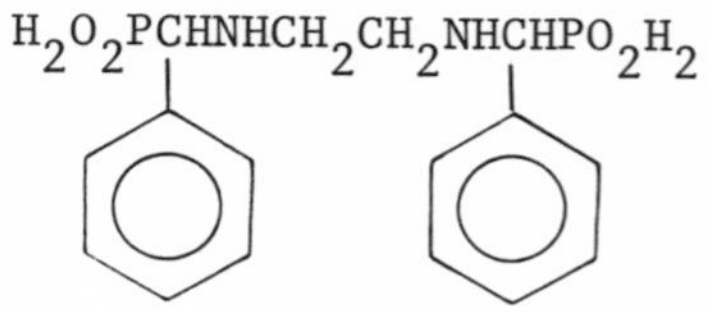

$$H_2O_2PCHNHCH_2CH_2NHCHPO_2H_2$$

$C_{16}H_{22}O_4N_2P_2$ Ethylenebis[imino(phenyl)methylenephosphonous acid] H_2L

Metal ion	Equilibrium	Log K 25°, 0.1
H^+	HL/H.L	7.58
	H_2L/HL.H	4.32
Ni^{2+}	ML/M.L	6.17
Cu^{2+}	ML/M.L	10.10
Zn^{2+}	ML/M.L	5.52

Bibliography: 68MRD

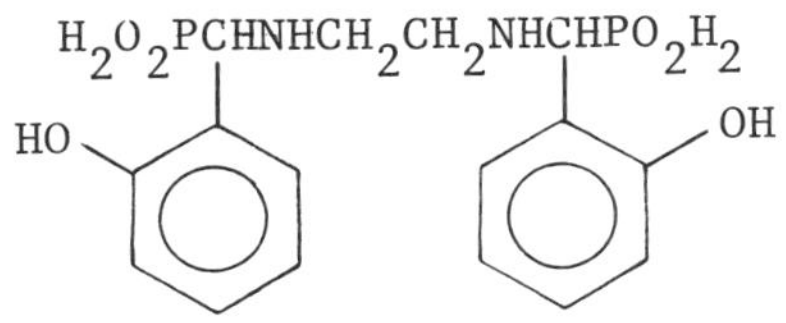

$$H_2O_2PCHNHCH_2CH_2NHCHPO_2H_2$$

HO   OH

$C_{16}H_{22}O_6N_2P_2$ Ethylenebis[imino(2-hydroxyphenyl)methylenephosphonous acid] H_4L

Metal ion	Equilibrium	Log K 25°, 0.1
H^+	HL/H.L	11.25
	$H_2L/HL.H$	10.84
	$H_3L/H_2L.H$	7.57
	$H_4L/H_3L.H$	4.61
Ni^{2+}	ML/M.L	15.04
	MHL/ML.H	7.80
	$MH_2L/MHL.H$	6.00
Cu^{2+}	ML/M.L	20.13
	MHL/ML.H	7.70
	$MH_2L/MHL.H$	5.25
Fe^{3+}	ML/M.L	31
Zn^{2+}	ML/M.L	13.37
	$MH_2L/ML.H^2$	14.38

Bibliography: 68MRD

$$H_2O_2PCH_2 \diagdown NCH_2CH_2N \diagup CH_2PO_2H_2$$
$$H_2O_2PCH_2 \diagup \diagdown CH_2PO_2H_2$$

$C_6H_{20}O_8N_2P_4$ <u>Ethylenedinitrilotetrakis(methylenephosphonous acid)</u> H_4L

Metal ion	Equilibrium	Log K 25°, 0.1
H^+	HL/H.L	6.87
	H_2L/HL.H	2.43
Mg^{2+}	ML/M.L	1.94
La^{3+}	ML/M.L	6.44
Co^{2+}	ML/M.L	7.29
Ni^{2+}	ML/M.L	8.44
Cu^{2+}	ML/M.L	9.75
Zn^{2+}	ML/M.L	7.60

Bibliography: 71MMa

$$\begin{array}{l} NH_2 \\ | \\ CH_2PO_3H_2 \end{array}$$

| CH_6O_3NP | | Aminomethylphosphonic acid | | H_2L |

Metal ion	Equilibrium	Log K 25°, 0.1	Log K 25°, 0.5
H^+	HL/H.L	10.00	9.97
	H_2L/HL.H	5.38	5.32
Mg^{2+}	ML/M.L	2.04	
	MHL/ML.H	9.31	
Ca^{2+}	ML/M.L	1.84	
	MHL/ML.H	9.17	
Co^{2+}	ML/M.L	4.18	
	ML_2/M.L^2	8.1	
	MHL/ML.H	7.53	
	MHL_2/ML_2.H	8.6	
	MH_2L_2/MHL_2.H	6.6	
Ni^{2+}	ML/M.L	5.18	4.94
	ML_2/M.L^2	9.0	8.5
	MHL/ML.H	6.46	
	MHL_2/ML_2.H	7.8	
	MH_2L_2/MHL_2.H	6.8	
Cu^{2+}	ML/M.L	7.95	7.77
	ML_2/M.L^2	14.6	14.1
	MHL/ML.H	4.63	
	MHL_2/ML_2.H	5.7	
	MH_2L_2/MHL_2.H	5.0	
Zn^{2+}	ML/M.L	5.64	
	MHL/ML.H	6.04	

Bibliography:
H^+,Ni^{2+},Cu^{2+} 71GD,71WN $Mg^{2+}-Co^{2+},Zn^{2+}$ 71WN

$$\overset{\overset{\displaystyle NH_2}{\displaystyle |}}{CH_3CHPO_3H_2}$$

$C_2H_8O_3NP$ DL-1-Aminoethylphosphonic acid H_2L

Metal ion	Equilibrium	Log K 25°, 0.1
H^+	HL/H.L	10.15
	$H_2L/HL.H$	5.57
Cu^{2+}	ML/M.L	8.24
	$ML_2/M.L^2$	$(15.3)^a$
	MHL/ML.H	4.87
	$MHL_2/ML_2.H$	5.7
	$MH_2L_2/MHL_2.H$	5.1

---

[a] Optical isomerism not stated

Bibliography: 72WN

$$\overset{\overset{\displaystyle NH_2}{\displaystyle |}}{CH_3CH_2CHPO_3H_2}$$

$C_3H_{10}O_3NP$ DL-1-Aminopropylphosphonic acid H_2L

Metal ion	Equilibrium	Log K 25°, 0.1
H^+	HL/H.L	10.22
	$H_2L/HL.H$	5.64
Cu^{2+}	ML/M.L	8.80
	$ML_2/M.L^2$	$(16.1)^a$
	MHL/ML.H	4.54
	$MHL_2/ML_2.H$	5.6
	$MH_2L_2/MHL_2.H$	5.0

---

[a] Optical isomerism not stated.

Bibliography: 72WN

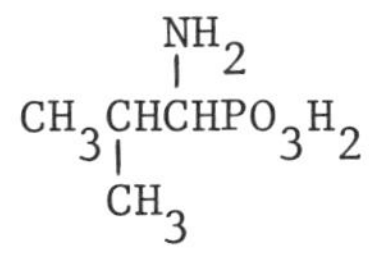

$$\begin{array}{c} NH_2 \\ | \\ CH_3CHCHPO_3H_2 \\ | \\ CH_3 \end{array}$$

$C_4H_{12}O_3NP$ <u>DL-1-Amino-2-methylpropylphosphonic acid</u> H_2L

Metal ion	Equilibrium	Log K 25°, 0.1
H^+	HL/H.L	10.31
	$H_2L/HL.H$	5.78
Cu^{2+}	ML/M.L	9.17
	$ML_2/M.L^2$	$(17.0)^a$
	MHL/ML.H	4.53
	$MHL_2/ML_2.H$	5.4
	$MH_2L_2/MHL_2.H$	5.1

---

[a] Optical isomerism not stated.

Bibliography: 72WN

$$\begin{array}{c} NH_2 \\ | \\ CH_3CPO_3H_2 \\ | \\ CH_3 \end{array}$$

$C_3H_{10}O_3NP$ <u>2-Amino-2-propylphosphonic acid</u> H_2L

Metal ion	Equilibrium	Log K 25°, 0.1
H^+	HL/H.L	10.30 ±0.01
	$H_2L/HL.H$	5.82 ±0.03
	$H_3L/H_2L.H$	1.7
Be^{2+}	ML/M.L	4.6
	$M_2L_2/M^2.L^2$	12.6
Mn^{2+}	ML/M.L	4.03
	$ML_2/M.L^2$	7.43
	MHL/ML.H	9.22
Fe^{2+}	ML/M.L	4.68
	$ML_2/M.L^2$	8.61
	MHL/ML.H	8.70
Co^{2+}	ML/M.L	5.19
	$ML_2/M.L^2$	10.27
	MHL/ML.H	8.23
Ni^{2+}	ML/M.L	5.65
	$ML_2/M.L^2$	10.98
	MHL/ML.H	7.68
Cu^{2+}	ML/M.L	8.95 -0.5
	$ML_2/M.L^2$	16.5 -1
	MHL/ML.H	4.68 +1
	$MHL_2/ML_2.H$	5.8
	$MH_2L_2/MHL_2.H$	5.1
Zn^{2+}	ML/M.L	6.13
	$ML_2/M.L^2$	11.97
	MHL/ML.H	7.62

B. AMINOPHOSPHONIC ACIDS

2-Amino-2-propylphosphonic acid   (continued)

Metal ion	Equilibrium	Log K 25°, 0.1
Al^{3+}	$ML/M.L$	11.42
	$ML_2/M.L^2$	19.53
	$MHL/M.HL$	5.59
	$M(HL)_2/M.(HL)^2$	9.61
	$M(HL)_3/M.(HL)^3$	13.07

Bibliography:

H^+ 68DM,72WN

Be^{2+} 68DM

Mn^{2+}-Ni^{2+},Zn^{2+},Al^{3+} 69DM

Cu^{2+} 69DM,72WN

$$H_2NCH_2CH_2CH_2PO_3H_2$$

$C_3H_{10}O_3NP$

3-Aminopropylphosphonic acid

H_2L

Metal ion	Equilibrium	Log K 25°, 0.1
H^+	$HL/H.L$	11.00
	$H_2L/HL.H$	6.87
Cu^{2+}	$ML/M.L$	7.65
	$MHL/ML.H$	6.03
	$M(HL)_2/M.(HL)^2$	27.3
	$M(HL)_2/MHL_2.H$	6.4

Bibliography: 72WNa

$$\begin{array}{c} NH_2 \\ | \\ H_2O_3PCPO_3H_2 \end{array}$$

$C_7H_{11}O_6NP_2$ Amino(phenyl)methylenediphosphonic acid H_4L

Metal ion	Equilibrium	Log K 25°, 0.1
H^+	HL/H.L	10.29
	H_2L/HL.H	8.17
	H_3L/H_2L.H	5.29
	H_4L/H_3L.H	1.6
Be^{2+}	ML/M.L	16.20
	MHL/ML.H	4.50
	M_2L/ML.H	7.21
	M_2HL/M_2L.H	4.00
Mg^{2+}	ML/M.L	7.39
	MHL/ML.H	8.36
Ca^{2+}	ML/M.L	6.56
	MHL/ML.H	8.55
Sr^{2+}	ML/M.L	5.54
	MHL/ML.H	9.30
Ba^{2+}	ML/M.L	5.16
	MHL/ML.H	9.49
Mn^{2+}	ML/M.L	9.93
	MHL/ML.H	8.65
	M_2L/ML.M	5.48
Fe^{2+}	ML/M.L	10.40
	MHL/ML.H	7.26
	M_2L/ML.M	5.22
Co^{2+}	ML/M.L	10.63
	MHL/ML.H	7.02
	M_2L/ML.M	5.09

Amino(phenyl)methylenediphosphonic acid   (continued)

Metal ion	Equilibrium	Log K 25°, 0.1
Ni^{2+}	ML/M.L	12.39
	MHL/ML.H	5.97
	$M_2L/ML.M$	4.48
Cu^{2+}	ML/M.L	15.63
	MHL/ML.H	4.67
	$M_2L/ML.M$	3.41
Fe^{3+}	ML/M.L	20.15
	$ML_2/M.L^2$	27.52
	MHL/M.HL	15.08
	$M(HL)_2/M.(HL)^2$	22.75
Zn^{2+}	ML/M.L	11.64
	MHL/ML.H	7.14
	$M_2L/ML.M$	4.67
Al^{3+}	ML/M.L	18.79
	$ML_2/M.L^2$	26.31
	MHL/M.HL	12.50
	$M(HL)_2/M.(HL)^2$	19.68
	$M_2L/ML.M$	4.30

Bibliography: 69DM

$$H_2NCH_2CH_2NHCH_2PO_3H$$

$C_3H_{11}O_3N_2P$ 2-Aminoethylaminomethylphosphonic acid H_2L

(ethylenediamine–N–methylenephosphonic acid)

Metal ion	Equilibrium	Log K 25°, 0.1
H^+	HL/H.L	10.3
	H_2L/HL.H	7.35
	H_3L/H_2L.H	4.85
Mn^{2+}	ML/M.L	5.2
	MHL/ML.H	7.3
Co^{2+}	ML/M.L	8.0
	ML_2/M.L^2	13.4
	MHL/ML.H	5.5
Ni^{2+}	ML/M.L	9.6
	ML_2/M.L^2	16.3
	MHL/ML.H	5.0
Cu^{2+}	ML/M.L	14.8
	ML_2/M.L^2	20.9
	MHL/ML.H	3.7
Cd^{2+}	ML/M.L	8.2
	ML_2/M.L^2	13.8
	MHL/ML.H	5.6

Bibliography: 72AU

$$H_2O_3PCH_2NHCH_2CH_2NHCH_2PO_3H_2$$

$C_4H_{14}O_6N_2P_2$ <u>Ethylenebis(iminomethylenephosphonic acid)</u> H_4L

Metal ion	Equilibrium	Log K 25°, 0.1
H^+	HL/H.L	10.60 ±0.1
	H_2L/HL.H	7.72 +0.3
	H_3L/H_2L.H	5.74 −0.09
	H_4L/H_3L.H	4.58 ±0.1
Be^{2+}	MH_2L/M.H_2L	8.76
	M_2H_2L/M^2.H_2L	11.4
Y^{3+}	MH_2L/M.H_2L	8.79
La^{3+}	MH_2L/M.H_2L	7.49
Pr^{3+}	MH_2L/M.H_2L	7.98
Nd^{3+}	MH_2L/M.H_2L	8.31
Sm^{3+}	MH_2L/M.H_2L	8.43
Eu^{3+}	MH_2L/M.H_2L	8.54
Gd^{3+}	MH_2L/M.H_2L	8.31
Dy^{3+}	MH_2L/M.H_2L	8.79
Mn^{2+}	ML/M.L	7.25 +0.3
Co^{2+}	ML/M.L	10.23 +0.6
	MHL/ML.H	5.98 −0.3
	MH_2L/MHL.H	5.33 ±0.0
Ni^{2+}	ML/M.L	11.70 +0.3
	MHL/ML.H	5.53 +0.3
	MH_2L/MHL.H	4.99 −0.1
Cu^{2+}	ML/M.L	17.52 +1
	MHL/ML.H	4.72 −0.2
	MH_2L/MHL.H	3.80 −0.2

Ethylenebis(iminomethylenephosphonic acid)   (continued)

Metal ion	Equilibrium	Log K 25°, 0.1
Cd^{2+}	ML/M.L	10.9
	MHL/ML.H	5.6
	MH_2L/MHL.H	(5.8)

Bibliography:

H^+ 65DK,65UA,71MMa,72AU Mn^{2+}-Cd^{2+} 65DK,71MMa,72AU

Be^{2+} 68DMa   Other reference: 67MR

Y^{3+}-Dy^{3+} 65DK

$$\begin{array}{cc}
CH_3 & CH_3 \\
| & | \\
H_2O_3PCNHCH_2CH_2NHCPO_3H_2 \\
| & | \\
CH_3 & CH_3
\end{array}$$

$C_8H_{22}O_6N_2P_2$ Ethylenebis[imino(dimethyl)methylenephosphonic acid] H_4L

Metal ion	Equilibrium	Log K 25°, 0.1
H^+	HL/H.L	11.68
	H_2L/HL.H	8.55
	H_3L/H_2L.H	6.00
	H_4L/H_3L.H	4.95
Be^{2+}	$MH_2L/M.H_2L$	7.65
	M_2H_2L/MH_2L.M	3.68
Y^{3+}	ML/M.L	12.87
	$MH_2L/ML.H^2$	13.84
La^{3+}	ML/M.L	10.13
	$MH_2L/ML.H^2$	15.47
Nd^{3+}	ML/M.L	11.60
	$MH_2L/ML.H^2$	14.45
Sm^{3+}	ML/M.L	12.56
	$MH_2L/ML.H^2$	13.87
Gd^{3+}	ML/M.L	12.27
	$MH_2L/ML.H^2$	14.12
Dy^{3+}	ML/M.L	12.89
	$MH_2L/ML.H^2$	13.74
Er^{3+}	ML/M.L	13.39
	$MH_2L/ML.H^2$	13.72
Lu^{3+}	ML/M.L	13.37
	$MH_2L/ML.H^2$	(14.53)
Mn^{2+}	ML/M.L	8.00
	$MH_2L/ML.H^2$	15.80
Co^{2+}	ML/M.L	11.39
	$MH_2L/ML.H^2$	12.88

Ethylenebis[imino(dimethyl)methylenephosphonic acid]   (continued)

Metal ion	Equilibrium	Log K 25°, 0.1
Ni^{2+}	ML/M.L	11.13
	$MH_2L/ML.H^2$	12.94
Cu^{2+}	ML/M.L	20.35
	$MH_2L/ML.H^2$	8.71
Zn^{2+}	ML/M.L	13.38
	$MH_2L/ML.H^2$	11.66

Bibliography:
$H^+, Y^{3+} - Zn^{2+}$ 65DK
Be^{2+} 68DMa

Other references: 67MR,68LS

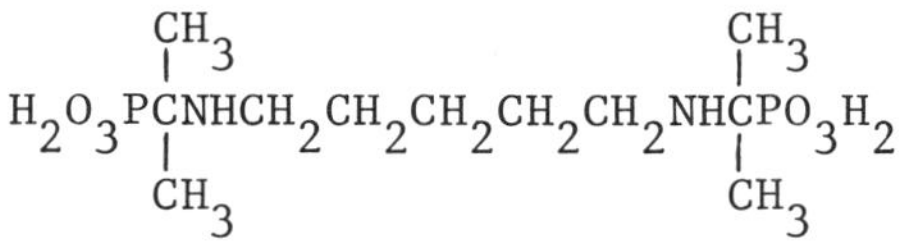

$$\underset{\substack{| \\ CH_3}}{\overset{\substack{CH_3 \\ |}}{H_2O_3PCNHCH_2CH_2CH_2CH_2CH_2NHCPO_3H_2}} \quad \overset{\substack{CH_3 \\ |}}{\underset{\substack{| \\ CH_3}}{}}$$

$C_{11}H_{28}O_6N_2P_2$ <u>Pentamethylenebis[imino(dimethyl)methylenephosphonic acid]</u> H_4L

Metal ion	Equilibrium	Log K 25°, 0.1
H^+	$H_3L/H_2L.H$	6.43
	$H_4L/H_3L.H$	5.34
Be^{2+}	$MH_2L/M.H_2L$	6.15
	$M_2H_2L/MH_2L.M$	4.06
La^{3+}	$MH_2L/M.H_2L$	4.44
Nd^{3+}	$MH_2L/M.H_2L$	4.96
Sm^{3+}	$MH_2L/M.H_2L$	5.49
Gd^{3+}	$MH_2L/M.H_2L$	5.49
Dy^{3+}	$MH_2L/M.H_2L$	5.49
Lu^{3+}	$MH_2L/M.H_2L$	(6.90)

Bibliography:

H^+ 67KM,68DMa $La^{3+}-Lu^{3+}$ 67KM

Be^{2+} 68DMa

$$H_2O_3PCHNHCH_2CH_2NHCHPO_3H_2$$

$C_{16}H_{22}O_6N_2P_2$ Ethylenebis[imino(phenyl)methylenephosphonic acid] H_4L

Metal ion	Equilibrium	Log K 25°, 0.1	Log K 25°, 0.5
H^+	HL/H.L	9.69	10.05
	$H_2L/HL.H$	7.44	7.25
	$H_3L/H_2L.H$	5.56	5.46
	$H_4L/H_3L.H$	4.18	4.31
Co^{2+}	ML/M.L		10.5
	MHL/ML.H		5.17
	$MH_2L/MHL.H$		4.73
Ni^{2+}	ML/M.L	11.78	11.3
	MHL/ML.H	5.07	5.63
	$MH_2L/MHL.H$	4.99	4.50
Cu^{2+}	ML/M.L	16.55	18.5
	MHL/ML.H	4.92	4.35
	$MH_2L/MHL.H$	4.05	3.50
Fe^{3+}	ML/M.L	15.90	
	$MH_2L/ML.H^2$	9.66	
Zn^{2+}	ML/M.L	12.18	12.4
	MHL/ML.H	5.33	5.22
	$MH_2L/MHL.H$	4.75	
Cd^{2+}	ML/M.L		10.7
	MHL/ML.H		5.52
	$MH_2L/MHL.H$		4.77

Bibliography:
H^+,Ni^{2+},Cu^{2+},Zn^{2+} 68MRD,72GT Fe^{3+} 68MRD
Co^{2+},Cd^{2+} 72GT

$$H_2O_3PCHNHCH_2CH_2NHCHPO_3H_2$$

HO—⬡   ⬡—OH

$C_{16}H_{22}O_8N_2P_2$ Ethylenebis[imino(2-hydroxyphenyl)methylenephosphonic acid] H_6L

Metal ion	Equilibrium	Log K 25°, 0.1	Log K 25°, 0.5
H^+	HL/H.L	11.19	(13.1)
	H_2L/HL.H	9.94	(12.1)
	H_3L/H_2L.H	8.77	9.12
	H_4L/H_3L.H	6.64	6.61
	H_5L/H_4L.H	5.41	5.20
	H_6L/H_5L.H	4.37	4.23
Ni^{2+}	ML/M.L	12.26	(19.1)
	MHL/ML.H	9.63	
	MH_2L/MHL.H	8.23	
	MH_3L/MH_2L.H	4.37	
Cu^{2+}	ML/M.L	18.58	(23.3)
	MH_2L/ML.H^2	17.59	
	MH_3L/MH_2L.H	1.96	
Fe^{3+}	ML/M.L	25	
	MH_2L/M.H_2L	17.45	
Zn^{2+}	ML/M.L	14.50	
	MHL/ML.H	9.25	
	MH_2L/MHL.H	7.94	
	MH_3L/MH_2L.H	3.14	

Bibliography:
H^+-Cu^{2+} 68MRD,72GT Fe^{3+},Zn^{2+} 68MRD

$$H_2O_3\overset{\overset{\displaystyle CH_3}{|}}{\underset{\underset{\displaystyle CH_3}{|}}{P}}CNHCH_2CH_2OCH_2CH_2NH\overset{\overset{\displaystyle CH_3}{|}}{\underset{\underset{\displaystyle CH_3}{|}}{C}}PO_3H_2$$

$C_{10}H_{26}O_7N_2P_2$ Oxybis[ethyleneimino(dimethyl)methylenephosphonic acid] H_4L

Metal ion	Equilibrium	Log K 25°, 0.1
H^+	HL/H.L	11.61
	H_2L/HL.H	10.67
	H_3L/H_2L.H	6.45
	H_4L/H_3L.H	5.16
Be^{2+}	MH_2L/M.H_2L	7.34
	M_2H_2L/MH_2L.M	5.12
Mg^{2+}	ML/M.L	3.65
Ca^{2+}	ML/M.L	5.61
Sr^{2+}	ML/M.L	3.97
Ba^{2+}	ML/M.L	3.41
La^{3+}	ML/M.L	11.06
	MH_2L/ML.H^2	16.36
Nd^{3+}	ML/M.L	(11.04)
	MH_2L/ML.H^2	(16.71)
Sm^{3+}	ML/M.L	12.02
	MH_2L/ML.H^2	16.03
Gd^{3+}	ML/M.L	(13.37)
	MH_2L/ML.H^2	(14.68)
Dy^{3+}	ML/M.L	12.04
	MH_2L/ML.H^2	16.01
Lu^{3+}	ML/M.L	(13.36)
	MH_2L/ML.H^2	15.79
Mn^{2+}	ML/M.L	7.45
Co^{2+}	ML/M.L	9.02
Ni^{2+}	ML/M.L	(8.84)

Oxybis[ethyleneimino(dimethyl)methylenephosphonic acid]   (continued)

Metal ion	Equilibrium	Log K 25°, 0.1
Cu^{2+}	ML/M.L	16.18
	$MH_2L/ML.H^2$	12.16
Zn^{2+}	ML/M.L	11.17
	$MH_2L/ML.H^2$	14.63
Cd^{2+}	ML/M.L	12.04
	$MH_2L/ML.H^2$	14.96
Pb^{2+}	ML/M.L	11.68
	$MH_2L/ML.H^2$	16.01

Bibliography:

H^+ 67KM,68DMa

Be^{2+} 68DMa

$Ca^{2+}-Pb^{2+}$ 67KM

Other reference: 67MR

$$H_2O_3PCNHCH_2CH_2SCH_2CH_2NHCPO_3H_2$$

with CH_3 groups substituted above and below each carbon center.

$C_{10}H_{26}O_6N_2P_2S$ Thiobis[ethyleneimino(dimethyl)methylenephosphonic acid] H_4L

Metal ion	Equilibrium	Log K 25°, 0.1
H^+	HL/H.L	11.41
	$H_2L/HL.H$	10.42
	$H_3L/H_2L.H$	6.34
	$H_4L/H_3L.H$	5.21
Be^{2+}	$MH_2L/M.H_2L$	7.15
	$M_2H_2L/MH_2L.M$	4.82
La^{3+}	ML/M.L	10.01
	$MH_2L/ML.H^2$	16.65
Nd^{3+}	ML/M.L	10.91
	$MH_2L/ML.H^2$	16.65
Sm^{3+}	ML/M.L	11.80
	$MH_2L/ML.H^2$	15.95
Gd^{3+}	ML/M.L	11.80
	$MH_2L/ML.H^2$	15.95
Dy^{3+}	ML/M.L	11.80
	$MH_2L/ML.H^2$	15.95
Lu^{3+}	ML/M.L	(13.03)
	$MH_2L/ML.H^2$	15.26
Mn^{2+}	ML/M.L	8.57
	$MH_2L/ML.H^2$	16.37
Co^{2+}	ML/M.L	10.71
	$MH_2L/ML.H^2$	13.86
Ni^{2+}	ML/M.L	11.39
Cu^{2+}	ML/M.L	15.90
	$MH_2L/ML.H^2$	11.89
Zn^{2+}	ML/M.L	12.86
	$MH_2L/ML.H^2$	12.72

Thiobis[ethyleneimino(dimethyl)methylenephosphonic acid]   (continued)

Metal ion	Equilibrium	Log K 25°, 0.1
Cd^{2+}	ML/M.L	12.70
Pb^{2+}	ML/M.L	14.39
	$MH_2L/ML.H^2$	12.84

Bibliography:

H^+ 67KM,68DMa La^{3+}-Pb^{2+} 67KM

Be^{2+} 68DMa

$$\begin{array}{ccc}
& CH_3 & CH_3 \\
& | & | \\
H_2O_3PCNHCH_2CH_2NHCH_2CH_2NHCPO_3H_2 & & \\
& | & | \\
& CH_3 & CH_3
\end{array}$$

$C_{10}H_{27}O_6N_3P_2$ <u>Iminobis[ethyleneimino(dimethyl)methylenephosphonic acid]</u> H_4L

Metal ion	Equilibrium	Log K 25°, 0.1
H^+	HL/H.L	11.20
	H_2L/HL.H	10.40
	H_3L/H_2L.H	6.55
	H_4L/H_3L.H	5.39
La^{3+}	ML/M.L	(9.89)
	MH_2L/ML.H^2	17.21
Nd^{3+}	ML/M.L	12.01
	MH_2L/ML.H^2	15.55
Sm^{3+}	ML/M.L	14.01
	MH_2L/ML.H^2	14.09
Mn^{2+}	ML/M.L	9.82
	MH_2L/ML.H^2	15.21
Co^{2+}	ML/M.L	13.86
	MH_2L/ML.H^2	12.19
Ni^{2+}	ML/M.L	14.36
	MH_2L/ML.H^2	12.22
Cu^{2+}	ML/M.L	19.85
	MH_2L/ML.H^2	11.30
Zn^{2+}	ML/M.L	15.60
	MH_2L/ML.H^2	11.40
Cd^{2+}	ML/M.L	13.34
	MH_2L/ML.H^2	12.91
Pb^{2+}	ML/M.L	17.68
	MH_2L/ML.H^2	10.22

Bibliography: 67KM   Other reference: 72GL

$$CH_3CH_2 \diagdown$$
$$\qquad NCH_2PO_3H_2$$
$$CH_3CH_2 \diagup$$

$C_5H_{14}O_3NP$ N,N-Diethylaminomethylphosphonic acid H_2L

Metal ion	Equilibrium	Log K 25°, 1.0
H^+	HL/H.L	12.3
	H_2L/HL.H	5.8
Mg^{2+}	MHL/M.HL	1.3
Ca^{2+}	ML/M.L	1.3

Bibliography: 67CC

$$HOCH_2CH_2 \diagdown$$
$$\qquad NCH_2PO_3H_2$$
$$HOCH_2CH_2 \diagup$$

$C_5H_{14}O_5NP$ N,N-Bis(2-hydroxyethyl)aminomethylphosphonic acid H_2L

Metal ion	Equilibrium	Log K 20°, 0.1
H^+	HL/H.L	9.46
	H_2L/HL.H	4.98
Ni^{2+}	ML/M.L	6.78
	$ML_2/M.L^2$	9.46
Cu^{2+}	ML/M.L	9.52
	MHL/ML.H	3.37
	MOHL/ML.OH	6.92
Fe^{3+}	M(OH)HL/M.OH.HL	19.19
	MOHL/M.OH.L	24.36
	$M(OH)_3L/MOHL.(OH)^2$	14.21

Bibliography: 70KM

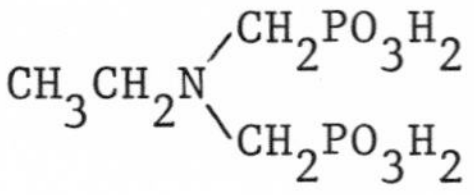

$$CH_3CH_2N \overset{\displaystyle CH_2PO_3H_2}{\underset{\displaystyle CH_2PO_3H_2}{<}}$$

$C_4H_{13}O_6NP_2$ N-Ethyliminodimethylenediphosphonic acid H_4L

Metal ion	Equilibrium	Log K 25°, 1.0
H^+	HL/H.L	12.42
	H_2L/HL.H	5.92
	H_3L/H_2L.H	4.70
Mg^{2+}	ML/M.L	4.42
	MHL/ML.H	10.33
	MH_2L/MHL.H	5.5
Ca^{2+}	ML/M.L	3.36
	MHL/ML.H	10.35

Bibliography: 67CC

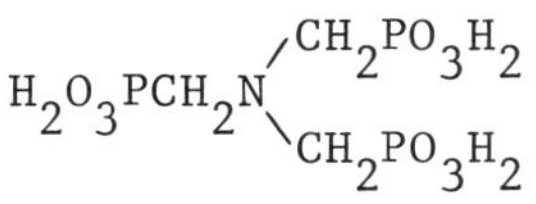

$$H_2O_3PCH_2N \begin{matrix} CH_2PO_3H_2 \\ CH_2PO_3H_2 \end{matrix}$$

$C_3H_{12}O_9NP_3$ <u>Nitrilotris(methylenephosphonic acid)</u> H_6L

Metal ion	Equilibrium	Log K 25°, 1.0
H^+	HL/H.L	12.3
	$H_2L/HL.H$	6.66
	$H_3L/H_2L.H$	5.46
	$H_4L/H_3L.H$	4.30
Mg^{2+}	ML/M.L	6.49
	MHL/ML.H	9.1
	$MH_2L/MHL.H$	6.2
	$MH_3L/MH_2L.H$	4.6
Ca^{2+}	ML/M.L	6.68
	MHL/ML.H	8.5
	$MH_2L/MHL.H$	6.1
	$MH_3L/MH_2L.H$	4.9

Bibliography: 67CC

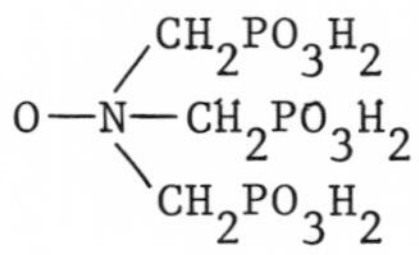

$$
\begin{array}{c}
\diagup \mathrm{CH_2PO_3H_2} \\
\mathrm{O-N-CH_2PO_3H_2} \\
\diagdown \mathrm{CH_2PO_3H_2}
\end{array}
$$

$\mathrm{C_3H_{12}O_{10}NP_3}$ <u>Nitrilotris(methylenephosphonic acid) N-oxide</u> $\mathrm{H_6L}$

Metal ion	Equilibrium	Log K 25°, 1.0
$\mathrm{H^+}$	HL/H.L	12.1
	$\mathrm{H_2L/HL.H}$	6.95
	$\mathrm{H_3L/H_2L.H}$	5.26
	$\mathrm{H_4L/H_3L.H}$	3.28
$\mathrm{Mg^{2+}}$	ML/M.L	8.3
	MHL/ML.H	7.4
	$\mathrm{MH_2L/MHL.H}$	5.5
	$\mathrm{MH_3L/MH_2L.H}$	4.2
$\mathrm{Ca^{2+}}$	ML/M.L	5.7
	MHL/ML.H	9.3
	$\mathrm{MH_2L/MHL.H}$	5.8

Bibliography: 67CCa

$$H_2O_3PCH_2 \diagdown NCH_2CH_2N \diagup CH_2PO_3H_2$$

$C_{18}H_{26}O_8N_2P_2$ N,N'-Bis(2-hydroxybenzyl)ethylenedinitrilo-N,N'-bis(methylenephosphonic acid)

H_6L

Metal ion	Equilibrium	Log K 25°, 0.1
H^+	HL/H.L	13.57
	H_2L/HL.H	11.48
	H_3L/H_2L.H	9.67
	H_4L/H_3L.H	7.24
	H_5L/H_4L.H	5.31
	H_6L/H_5L.H	3.61
Mg^{2+}	ML/M.L	7.95
	MH_2L/M.H_2L	3.04
	MHL/ML.H	11.05
	MH_2L/MHL.H	9.10
Ca^{2+}	ML/M.L	8.36
	MH_2L/M.H_2L	3.61
	MHL/ML.H	10.72
	MH_2L/MHL.H	9.59
Co^{2+}	ML/M.L	18.0
	MH_2L/M.H_2L	9.58
	MHL/ML.H	9.88
	MH_2L/MHL.H	6.70
	MH_3L/MH_2L.H	5.09
	MH_4L/MH_3L.H	5.28
Ni^{2+}	ML/M.L	17.9
	MH_2L/M.H_2L	10.83
	MHL/ML.H	10.02
	MH_2L/MHL.H	7.95
	MH_3L/MH_2L.H	5.53
	MH_4L/MH_3L.H	4.69

N,N'-Bis(2-hydroxybenzyl)ethylenedinitrilo-N,N'-bis(methylenephosphonic acid)   (continued)

Metal ion	Equilibrium	Log K 25°, 0.1
Cu^{2+}	$ML/M.L$	24.0
	$MH_2L/M.H_2L$	16.09
	$MHL/ML.H$	9.39
	$MH_2L/MHL.H$	7.75
	$MH_3L/MH_2L.H$	4.74
	$MH_4L/MH_3L.H$	3.41
Zn^{2+}	$MHL/ML.H$	10.27

Bibliography: 73MMa

$$\begin{array}{c} H_2O_3PCH_2 \diagdown \qquad \diagup CH_2PO_3H_2 \\ \qquad\qquad NCH_2CH_2N \\ H_2O_3PCH_2 \diagup \qquad \diagdown CH_2PO_3H_2 \end{array}$$

$C_6H_{20}O_{12}N_2P_4$ <u>Ethylenedinitrilotetrakis(methylenephosphonic acid)</u> H_8L

Metal ion	Equilibrium	Log K 25°, 0.1	Log K 25°, 3.0
H^+	HL/H.L	$(13.14)^a$	12.36
	H_2L/HL.H	10.01 ±0.2	9.36
	H_3L/H_2L.H	8.13 −0.3	7.60
	H_4L/H_3L.H	6.57 −0.2	6.05
	H_5L/H_4L.H	5.26 −0.04	4.86
	H_6L/H_5L.H	3.15 −0.1	2.98
	H_7L/H_6L.H	$(1.37)^a$+0.1	1.24
Co^{2+}	ML/M.L	17.36 −2	
	MHL/ML.H	8.48 −0.08	
	MH_2L/MHL.H	6.61 +0.3	
	MH_3L/MH_2L.H	5.27 +0.4	
	MH_4L/MH_3L.H	4.86 −0.2	
Ni^{2+}	ML/M.L	16.95 −0.6	
	MHL/ML.H	9.03 −0.2	
	MH_2L/MHL.H	7.47 −0.2	
	MH_3L/MH_2L.H	5.51 +0.2	
	MH_4L/MH_3L.H	4.57 −0.08	
Cu^{2+}	ML/M.L	23.01 −4	
	MHL/ML.H	7.99 −0.02	
	MH_2L/MHL.H	6.22 +0.3	
	MH_3L/MH_2L.H	4.81 +0.4	
	MH_4L/MH_3L.H	3.86 +0.04	
Zn^{2+}	ML/M.L	19.52 −1	
	MHL/ML.H	8.34 +0.2	
	MH_2L/MHL.H	6.05 +0.5	
	MH_3L/MH_2L.H	4.93 +0.2	
	MH_4L/MH_3L.H	4.46 −0.2	

---

[a] Estimated value calculated from values at 3.0 ionic strength.

Ethylenedinitrilotetrakis(methylenephosphonic acid)   (continued)

Bibliography:

H^+ 65UA,67KD,71MMb   Other references: 65WR,70Ta,72S

Co^{2+}-Zn^{2+} 67KD,71MMb,73MMa

$$H_2O_3PCH_2 \diagdown \quad \quad \overset{\displaystyle CH_2PO_3H_2}{\overset{\displaystyle |}{\diagup CH_2PO_3H_2}}$$
$$\quad \quad \quad NCH_2CH_2NCH_2CH_2N$$
$$H_2O_3PCH_2 \diagup \quad \quad \quad \diagdown CH_2PO_3H_2$$

$C_9H_{28}O_{15}N_3P_5$ <u>Diethylenetrinitrilopentakis(methylenephosphonic acid)</u> $H_{10}L$

Metal ion	Equilibrium	Log K 25°, 0.1
H^+	HL/H.L	12.04
	H_2L/HL.H	10.10
	H_3L/H_2L.H	8.15
	H_4L/H_3L.H	7.17
	H_5L/H_4L.H	6.38
	H_6L/H_5L.H	5.50
	H_7L/H_6L.H	4.45
	H_8L/H_7L.H	2.8
Mg^{2+}	ML/M.L	6.40
	MHL/ML.H	11.04
	MH_2L/MHL.H	9.40
	MH_3L/MH_2L.H	7.39
	MH_4L/MH_3L.H	6.36
	MH_5L/MH_4L.H	5.61
Ca^{2+}	ML/M.L	7.11
	MHL/ML.H	10.35
	MH_2L/MHL.H	9.17
	MH_3L/MH_2L.H	7.70
	MH_4L/MH_3L.H	6.24
	MH_5L/MH_4L.H	5.56
Mn^{2+}	ML/M.L	11.15
	MHL/ML.H	9.30
	MH_2L/MHL.H	8.00
	MH_3L/MH_2L.H	7.18
	MH_4L/MH_3L.H	6.48
	MH_5L/MH_4L.H	5.67
	MH_6L/MH_5L.H	4.20

Diethylenetrinitrilopentakis(methylenephosphonic acid)   (continued)

Metal ion	Equilibrium	Log K 25°, 0.1
Co^{2+}	$ML/M.L$	15.73
	$MHL/ML.H$	8.38
	$MH_2L/MHL.H$	7.20
	$MH_3L/MH_2L.H$	6.33
	$MH_4L/MH_3L.H$	5.56
	$MH_5L/MH_4L.H$	4.94
	$MH_6L/MH_5L.H$	4.30
Cu^{2+}	$ML/M.L$	19.47
	$MHL/ML.H$	8.45
	$MH_2L/MHL.H$	7.60
	$MH_3L/MH_2L.H$	6.37
	$MH_4L/MH_3L.H$	5.23
	$MH_5L/MH_4L.H$	4.20
	$MH_6L/MH_5L.H$	3.24
	$MH_7L/MH_6L.H$	2.42
Zn^{2+}	$ML/M.L$	16.45
	$MHL/ML.H$	8.95
	$MH_2L/MHL.H$	7.15
	$MH_3L/MH_2L.H$	6.18
	$MH_4L/MH_3L.H$	5.50
	$MH_5L/MH_4L.H$	4.84
	$MH_6L/MH_5L.H$	4.18
Cd^{2+}	$ML/M.L$	13.37
	$MHL/ML.H$	8.93
	$MH_2L/MHL.H$	7.52
	$MH_3L/MH_2L.H$	6.83
	$MH_4L/MH_3L.H$	6.14
	$MH_5L/MH_4L.H$	5.45
	$MH_6L/MH_5L.H$	4.80
	$MH_7L/MH_6L.H$	2.75

Bibliography: 67KD   Other reference: 68T

For other aminophosphonic acids see:

    Volume 1: pp. 103,157,158,159

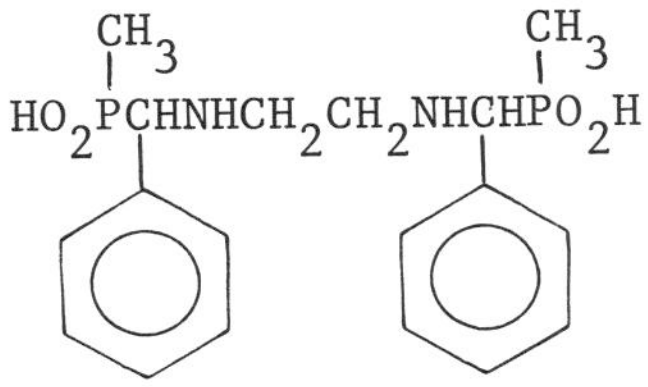

$$\begin{array}{cc} & CH_3 & CH_3 \\ & | & | \\ HO_2PCHNHCH_2 & CH_2NHCHPO_2H \end{array}$$

$C_{18}H_{26}O_4N_2P_2$ Ethylenebis[imino(phenyl)methylene(methyl)phosphinic acid] H_2L

Metal ion	Equilibrium	Log K 25°. 0.1
H^+	HL/H.L	7.84
	H_2L/HL.H	4.61
Ni^{2+}	ML/M.L	6.91
Cu^{2+}	ML/M.L	10.32

Bibliography: 70DM

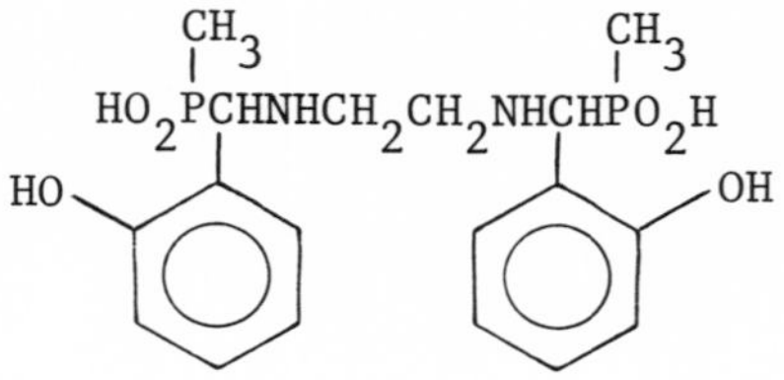

$$\text{HO}_2\text{PCHNHCH}_2\text{CH}_2\text{NHCHPO}_2\text{H}$$

$C_{18}H_{26}O_6N_2P_2$ Ethylenebis[imino(2-hydroxyphenyl)methylene(methyl)phosphinic acid] H_4L

Metal ion	Equilibrium	Log K 25°, 0.1
H^+	HL/H.L	11.58
	H_2L/HL.H	10.56
	H_3L/H_2L.H	7.55
	H_4L/H_3L.H	4.78
Ni^{2+}	ML/M.L	15.39
	MHL/ML.H	7.75
	MH_2L/MHL.H	6.06
Cu^{2+}	ML/M.L	20.14
	MHL/ML.H	8.18
	MH_2L/MHL.H	4.80
Fe^{3+}	ML/M.L	31.25
Al^{3+}	ML/M.L	20
	MHL/M.HL	15.36

Bibliography: 70DM

$$H_2NCH_2CH_2OPO_3H_2$$

$C_2H_8O_4NP$ 2-Aminoethyl-dihydrogenphosphate   (0-phosphorylethanolamine) H_2L

Metal ion	Equilibrium	Log K 25°, 0.1	Log K 25°, 0	ΔH 25°, 0	ΔS 25°, 0
H^+	HL/H.L	10.15 ±0.05	10.638	-11.0	12
	H_2L/HL.H	5.58 ±0.02	5.838	0.5	28
Mg^{2+}	ML/M.L	1.70[a]			
	MHL/ML.H	9.65[a]			
Ca^{2+}	ML/M.L	1.57[a]			
	MHL/ML.H	9.70[a]			
Mn^{2+}	ML/M.L	2.55[a]			
	MHL/ML.H	9.29[a]			
Cu^{2+}	ML/M.L	6.42 ±0.03			
	ML_2/M.L^2	12.42 ±0.03			
	MHL/ML.H	5.8 ±0.2			
	MHL_2/ML_2.H	6.0 ±0.4			
	MH_2L_2/MHL_2.H	5.9[a]			

[a] 25°, 0.15

Bibliography:
H^+ 55CD,59FO,600,62DG,620,72WNa Cu^{2+} 600, 72WNa
Mg^{2+},Ca^{2+},Mn^{2+} 620   Other references: 59DG,65HF

$C_6H_8O_4NP$ 2-Pyridylmethyl-dihydrogenphosphate   (O-phosphorylpicolinol) H_2L

Metal ion	Equilibrium	Log K 25°, 0.1
H^+	HL/H.L	6.30
	H_2L/HL.H	4.42
	H_3L/H_2L.H	1.8
Mg^{2+}	ML/M.L	1.7
Mn^{2+}	ML/M.L	2.44
Co^{2+}	ML/M.L	2.27
Ni^{2+}	ML/M.L	2.85
Cu^{2+}	ML/M.L	4.44
	MHL/ML.H	4.09
Zn^{2+}	ML/M.L	2.83

Bibliography:

H^+ 66MT,68MT Mg^{2+}-Zn^{2+} 68MT

For other amino-dihydrogenphosphates see:

    Volume 1: pp. 29,303,312,320,321,329

    Volume 2: pp. 182,266,267,268,269,270,273

VI. PROTONATION VALUES FOR OTHER LIGANDS

A. Amines:

1. Primary amines: $R-NH_2$

Ligand	Log K 25°, 0	ΔH 25°, 0	ΔS 25°, 0	Bibliography
2-Propylamine (isopropylamine), L	10.67	-13.95±0.03	2.0	68OW,69CI
DL-2-Butylamine (s-butylamine), L	10.56	-14.03	1.3	69CI
2-Methyl-2-propylamine (t-butylamine), L	10.685	-14.39±0.04	0.6	62HR,68OW,69CI
3-Methylbutylamine (isopentylamine), L	10.64	-14.03	1.8	32HS,69CI
Cyclopropylamine, L	9.10	-11.7	2	69CI
3-Methylaniline (m-toluidine)[*], L	4.72 ±0.01 4.96[a]	-7.5 ±0.1	-4	61BR,66D,67BH,68O, 73LP
3-Methoxyaniline (m-anisidine), L	4.22 ±0.02 4.42[a]	-7.0 -0.1	-4	61BR,66D,68BH, 73LP

$HO_3SO-R-NH_2$

	Log K 25°, 0	ΔH 25°, 0	ΔS 25°, 0	Bibliography
2-Aminoethyl hydrogensulfate, HL	9.182	-11.0	5	62DG, Other reference: 59DG

$HO_3S-R-NH_2$

	Log K 25°, 0	ΔH 25°, 0	ΔS 25°, 0	Bibliography
2-Aminobenzenesulfonic acid, HL	2.458	-2.45±0.02	3.0	65CS,69CI
3-Amino-4-methylbenzenesulfonic acid, HL	3.633	-5.1	0	65CS
4-Amino-3-methylbenzenesulfonic acid, HL	3.124	-4.5	-1	65CS

[a] 25°, 0.1; [*] Metal constants were also reported but are not included in the compilation of selected constants.

Primary amines (continued)

Ligand	Log K 25°, 0	ΔH 25°, 0	ΔS 25°, 0	Bibliography
Tyrosine methyl ester, HL	9.89[a]			67HP
Tyrosinamide, HL	9.77[b]			58MEW
	7.36[b]			
2-Aminophenol[*], HL	9.97[c]			61P, Other
	4.78[c]			reference: 59Sa
3-Amino-4-hydroxybenzenesulfonic acid[*], H_2L	9.15[c]			61P
	4.12[c]			
4-Amino-5-hydroxynaphthalene-2,7-disulfonic acid[*], H_3L	8.83[c]			61P
	3.63[c]			
2-(4-Hydroxyphenyl)ethylamine (tyramine), HL	10.56[d]			74GS
	9.42[d]			
2-(3,4-Dihydroxyphenyl)ethylamine (dopamine)[*], H_2L	11.90[e]			73AW,74GS, Other
	10.31[a]			reference: 63Ha
	10.45[e]			
	8.86[a]			
	8.91[e]			
L-2-Amino-1-(3,4-dihydroxyphenyl)ethanol (noradrenaline)[*], H_3L	(13)[a]			65JN,71AV,
	(13)[a]			Other references:
	9.68[a]±0.03			62AL,62Ha,65JNa,
	8.63[a]±0.01			66JN,74GS

$$HO-R-NH_2$$

Ligand	Log K 25°, 0	ΔH 25°, 0	ΔS 25°, 0	Bibliography
DL-2-Aminobutanol[*], L	9.516	-12.46	1.8	68TE, Other reference: 72VE
2-Amino-2-methylpropanol[*], L	9.694	-12.91±0.02	1.1	68OW,68TE, Other
	9.72[a]			reference: 71HS

---

[a] 25°, 0.1;   [b] 25°, 0.16;   [c] 20°, 0;   [d] 20°, 0.5;   [e] 20°, 0.37;   [*] Metal constants were also reported but are not included in the compilation of selected constants.

A. AMINES

Primary amines   (continued)

$$\begin{array}{c} O \\ \| \\ R'-OC-R-NH_2 \end{array}$$

Ligand	Log K 25°, 0	ΔH 25°, 0	ΔS 25°, 0	Bibliography
Glycine butyl ester, L	7.78[f]			68HA
Alanine methyl ester, L	7.74[a]+0.1	(-11)[g]	(-2)[a]	67HP,69LA,70HM
Alanine ethyl ester, L	7.91[f]			68HA
2-Aminobutanoic acid ethyl ester, L	7.64[a]	(-11)[g]	(-2)[a]	70HM
Norvaline methyl ester, L	7.66[a]			70HM
Norleucine methyl ester, L	7.68[a]			70HM
Valine methyl ester, L	7.48[a]+0.01	(-11)[g]	(-3)[a]	67HP,70HM
Valine ethyl ester, L	7.75[f]			68HA
Leucine methyl ester, L	7.62[a]-0.09			66HP,67HP,69LA
Leucine ethyl ester, L	7.64[a]±0.02	(-11)[g]	(-2)[a]	66HP,67HP
Isoleucine methyl ester, L	7.54[a]±0.01			67HP,70HM
Phenylalanine methyl ester, L	7.05[a]	(-11)[g]	(-5)[a]	67HP
Phenylalanine ethyl ester, L	7.12[a]±0.01			66HP,67HP
O-Methyltyrosine ethyl ester, L	7.19[b]			58MEW
Glutamic acid dimethyl ester, L	7.03[a]±0.01			66HP,67HP
Glutamic acid diethyl ester, L	7.04			36N
3-Aminopropanoic acid methyl ester, L	9.17[a]	(-13)[g]	(-2)[a]	70HM
3-Aminopropanoic acid ethyl ester, L	9.23[a]-0.1			68HA,71AA
4-Aminobutanoic acid methyl ester, L	9.84[a]	(-13)[g]	(1)[a]	70HM
4-Aminobutanoic acid ethyl ester, L	9.71			36N
5-Aminopentanoic acid ethyl ester, L	10.15			36N
2-Aminobenzoic acid methyl ester, L	2.34	-4.6	-5	67CW
3-Aminobenzoic acid methyl ester, L	3.60	-6.5	-5	67CW
4-Aminobenzoic acid methyl ester, L	2.45 ±0.05	-5.1	-6	48KH,67CW

$$\begin{array}{c} O \\ \| \\ R'-NHC-R-NH_2 \end{array}$$

Ligand	Log K 25°, 0	ΔH 25°, 0	ΔS 25°, 0	Bibliography
β-Alanyl-β-alaninamide, L	9.21[a]			73YN
Triglycine ethyl ester, L	7.91[h]			41GH
Tetraglycine ethyl ester, L	7.81[h]			41GH
Hexaglycine ethyl ester, L	7.86[h]			41GH

---

[a] 25°, 0.1;   [b] 25°, 0.16:   [f] 25°, 0.05;   [g] 25-50°, 0.1:   [h] 20°, 0.05

Primary amines   (continued)

$$HS-R-NH_2$$

Ligand	Log K 25°, 0	ΔH 25°, 0	ΔS 25°, 0	Bibliography
Cysteine ethyl ester, HL	8.94[b]			65CMa,65FCW
	6.53[b]			

$$R'-S-R-NH_2$$

S–Methylcysteine methyl ester, L	6.70[a]	(−11)[g]	(−6)[a]	67HP
Methionine methyl ester, L	7.1[a]			67HP

$$H_2N-R-NH_2$$

Hexamethylenediamine, L	10.930	−13.91	3.4	52EP,70BP
	10.97[a]			
	11.02[i]	−13.9[i]	4[i]	
	9.830	−13.82	−1.4	
	10.09[a]			
	10.24[i]	−13.7[i]	1[i]	
2,4-Diaminobutanoic acid methyl ester, L	8.66[a]			72HM
	6.10[a]			
Ornithine methyl ester, L	(8.32)[a]			72HM
	6.73[a]			
Lysine methyl ester, L	10.25[a]	(−13)[g]	(3)[a]	72HM
	7.19[a]	(−12)[g]	(−7)[a]	

---

[a] 25°, 0.1;   [b] 25°, 0.16;   [g] 25-50°, 0.1;   [i] 25°, 0.5

2. Secondary amines: R-NH-R'

Ligand	Log K 25°, 0	ΔH 25°, 0	ΔS 25°, 0	Bibliography
N-Methylbutylamine, L	10.90	-12.7 ± 0.2	7	64WB,69CI
Dipropylamine, L	11.00	-13.2	6	69CI
Di-2-propylamine (diisopropylamine), L	11.20	-13.6	6	69CI
Dibutylamine, L	11.25	-13.7	6	69CI
DL-Di-2-butylamine (di-s-butylamine), L	11.91	-14.0	3	69CI
Bis(2-methylpropyl)amine (diisobutyl-amine), L	11.82	-13.4	4	32HS,69CI
Bis(3-methylbutyl)amine (diisopentyl-amine), L	10.92	-13.4	5	69CI
Dicyclopentylamine, L	10.93	-14.2	2	69CI
Dicyclohexylamine, L	11.25	-14.2	4	69CI
Diprop-2-enylamine (diallylamine), L	9.29	-11.7	3	69CI
Dibenzylamine, L	8.52	-11.3	1	69CI
Dimethyleneimine (aziridine), L	8.04	-9.02	6.5	56ST,71CC
Trimethyleneimine (azetidine), L	11.29	-12.52	9.7	56ST,71CC
2,2,6,6-Tetramethylpiperidine, L	11.09^i	-14.2^i	3^i	74BEB
Hexamethyleneimine, L	11.12	-13.01	7.2	63RR,71CC
Heptamethyleneimine, L	11.1	-12.86	7.7	71CC

L-2-(Methylamino)-1-(3,4-dihydroxyphenyl)- ethanol (adrenaline)[*], H_3L	$(13)^a$ $(13)^a$ $9.90^a \pm 0.05$ $8.65^a \pm 0.01$	65JN,66AT, Other references: 62AL,62Hb,65JNa, 66JN,69CA,74GS
DL-2-(1-Methylethylamino)-1-(3,4-di- hydroxyphenyl)ethanol (isoprenaline), H_2L	$--$ 9.90^a 8.65^a	73AW

---

[a] 25°, 0.1; [i] 25°, 0.5; [*] Metal constants were also reported but are not included in the compilation of selected constants.

Secondary amines   (continued)

$$H_3C \quad CH_3$$
$$HON=C-CNHR$$
$$CH_3$$

Ligand	Log K 25°, 0	ΔH 25°, 0	ΔS 25°, 0	Bibliography
3-Methylamino-3-methylbutan-2-one oxime, HL		-10.1		64WB
3-Ethylamino-3-methylbutan-2-one oxime, HL		-10.4		64WB
3-Butylamino-3-methylbutan-2-one oxime, HL		-10.8		64WB
3-(1-Methylethylamino)-3-methylbutan-2-one oxime, HL		-10.4		64WB
3-Allylamino-3-methylbutan-2-one oxime, HL		-10.6		64WB

HO-R-NHR'

Ligand	Log K 25°, 0	ΔH 25°, 0	ΔS 25°, 0	Bibliography
L-erythro-2-(Methylamino)-1-phenylpropanol ((-)-ephedrine)[*], L	9.544 / 9.57[a]	-10.8	7	58EH, Other references: 61S,64AR,69CA
L-threo-2-(Methylamino)-1-phenylpropanol ((+)-ψ-ephedrine), L	9.706	-11.0	8	58EH
DL-2-(2-Hydroxypropylamino)ethanol, L	8.81	-9.8	7	59SG

$$O$$
$$\|$$
$$R'OCCH_2NHR$$

Ligand	Log K 25°, 0	Bibliography
Sarcosine ethyl ester, L	8.12[f]	69LA

R-NHCH$_2$CH$_2$NH-R

Ligand	Log K 25°, 0	ΔH 25°, 0	ΔS 25°, 0	Bibliography
N,N'-Di-2-propylethylenediamine, L	10.25[i] / 7.44[i]			53BM
2,6-Dimethylpiperazine, L	9.57[a] / 5.40[a]	-10.47[a] / -6.65[a]	9.2[a] / 2.9[a]	74BEH
cis-2,5-Dimethylpiperazine, L	9.76[a] / 5.32[a]	-10.76[a] / -7.40[a]	9.1[a] / 0.0[a]	74BEH
trans-2,5-Dimethylpiperazine, L	9.58[a] / 5.37[a]	-10.37[a] / -7.32[a]	9.6[a] / 0.5[a]	74BEH

[a] 25°, 0.1; [f] 25°, 0.05; [i] 25°, 0.5; [*] Metal constants were also reported but are not included in the compilation of statility constants.

3. <u>Tertiary amines</u>:

$$R-N\overset{\displaystyle R'}{\underset{\displaystyle R''}{}}$$

Ligand	Log K 25°, 0	ΔH 25°, 0	ΔS 25°, 0	Bibliography
Trimethylamine[*], L	9.800-0.05	-8.83-0.03	15.2	28HR,30HO,41EW,
	9.85[j]			56CG,65PS,69CI,
				Other reference:
				35BW
Tripropylamine, L	10.66	-10.5	15	69CI
N,N-Dimethylcyclohexylamine, L	10.72	-10.1	15	69CI
Triprop-2-enylamine (triallylamine), L	8.31	-8.8	9	69CI
N,N-Diethylbenzylamine, L	9.44	-9.9	10	69CI
N-Methylpyrrolidine, L	10.46	-9.05	17.5	56ST,71CC

$$HO-R-N\overset{\displaystyle R'}{\underset{\displaystyle R''}{}}$$

Ligand	Log K 25°, 0	ΔH 25°, 0	ΔS 25°, 0	Bibliography
DL-1-(Dimethylamino)-2-propanol, L	9.418			73NKa
DL-1-(Diethylamino)-2-propanol, L	10.064			73NK
3-(Diethylamino)propanol, L	10.096			73NK
2,2-Bis(hydroxymethyl)nitrilotri-2-ethanol, L	6.483	-6.75	7.0	70PB

$$\overset{\displaystyle R'}{\underset{\displaystyle R'}{}}N-R-N\overset{\displaystyle R''}{\underset{\displaystyle R'''}{}}$$

Ligand	Log K 25°, 0	ΔH 25°, 0	ΔS 25°, 0	Bibliography
N,N-Dimethyltrimethylenediamine, L	10.18			71NT
	10.22[a]			
	10.41[i]			
	7.95			
	8.23[a]			
	8.56[i]			

---

[a] 25°, 0.1; [i] 25°, 0.5; [j] 23°, 0.2; [*] Metal constants were also reported but are not included in the compilation of stability constants.

Tertiary amines   (continued)

Ligand	Log K 25°, 0	ΔH 25°, 0	ΔS 25°, 0	Bibliography
N,N-Dibutyltrimethylenediamine, L	10.57			72KK
	10.63[a]			
	10.83[i]			
	8.30			
	8.59[a]			
	8.90[i]			
N,N,N'-Triethylethylenediamine, L	10.03[a]			72TR
	6.98[a]			
N,N,N',N'-Tetraethylethylenediamine, L	9.67			69J,72TR
	9.68[a]±0.05			
	10.00[i]			
	6.13			
	6.44[a]±0.01			
	6.79[i]			

R-N s N-R'

N,N'-Dimethylpiperazine, L	8.13[a]	−6.66[a]	15.4[a]	74BEH
	4.18[a]	−3.76[a]	6.5[a]	
Piperazine-N,N'-bis(ethylenesulfonic acid), H_2L	6.93[a]	−2.72[a]	22.6[a]	71BS
N-(2-Hydroxyethyl)piperazine-N'-ethylene-sulfonic acid, HL	5.01[a]	−4.9[a]	7[a]	71BS

$$CH_3CH_2\!\!\diagdown\atop CH_3CH_2\!\!\diagup N CH_2CH_2 \overset{\overset{\displaystyle CH_3}{|}}{N} CH_2CH_2 N \!\!{\diagup CH_2CH_3 \atop \diagdown CH_2CH_3}$$

6-Methyl-3,9-diethyl-3,6,9-triazaundecane,	9.71[k]			71AA
(N,N,N'',N''-tetraethyl-N'-methyldi-	9.02[k]			
ethylenetriamine), L	2.29[k]			

---

[a] 25°, 0.1;   [i] 25°, 0.5;   [k] 25°, 0.13

B. Azoles:

Ligand	Log K 25°, 0	ΔH 25°, 0	ΔS 25°, 0	Bibliography
2-Phenylimidazole, L	6.39			38KN
4-Phenylimidazole, L	6.00			38KN
1-Methyl-5-(2-oxo-3-ethyl-4-furanylmethyl)- imidazole, L	6.98 7.12[a] 7.36[i]			68LP
2,2'-Dimidazolyl, L	5.01[a]	-6.1[a]	2[a]	67HW

Ligand	Log K 25°, 0	ΔH 25°, 0	ΔS 25°, 0	Bibliography
2-Methylbenzimidazole, L	6.17[b]	(-10)[1]	(-5)[b]	60LQ
2-Ethylbenzimidazole, L	6.15[b]	(-9)[1]	(-2)[b]	60LQ
2-(Hydroxymethyl)benzimidazole[*], L	5.28[b]			60LQ
1-Methyl-2-(hydroxymethyl)benzimidazole[*], L	5.43[b]			60LQ

---

[a] 25°, 0.1; [b] 25°, 0.16; [i] 25°, 0.5; [1] 5-35°, 0.16; [*] Metal constants were also reported but are not included in the compilation of stability constants.

C. <u>Azines</u>:

1. <u>Pyridines</u>:

R_4 R_3 R_2 R_6 N

Ligand	Log K 25°, 0	ΔH 25°, 0	ΔS 25°, 0	Bibliography
2-Ethylpyridine, L	5.89			54AC,73BEM
	6.22[i]	-6.48[i]	6.7[i]	
2-Propylpyridine, L	6.14[i]	-6.83[i]	5.2[i]	73BEM
3-Ethylpyridine, L	5.56			54AC
2,3,6-Trimethylpyridine (2,3,6-collidine), L	7.44[i]	-8.33[i]	6.1[i]	73BEM
2-Chloropyridine, L	0.49	-0.02	2.2	72CS
3-Chloropyridine, L	2.83	-2.11	5.9	72CS
2-Bromopyridine, L	0.69	-0.08	2.9	72CS
3-Bromopyridine*, L	2.88	-1.35	8.7	72CS, Other reference: 67N
2-Hydroxypyridine, L	1.24	0.07	5.4	72CS
3-Hydroxypyridine, L	4.77	-4.01	8.4	72CS
4-Hydroxypyridine, L	3.23	-1.49	9.8	72CS
Pyridine-3-carboxaldehyde oxime, HL	10.32			58C,73P
	10.37[c]			
	10.15[d]			
	3.94			
	3.94[c]			
	4.14[d]			
Pyridine-4-carboxaldehyde oxime, HL	9.91			58C
	4.58			
2-(Hydroxymethyl)-6-methylpyridine, L	5.43			67TT
2-(2-Hydroxyethyl)pyridine, L	5.31			67TT
2-(2-Hydroxyethyl)-6-methylpyridine, L	6.10			67TT
2-(3-Hydroxypropyl)pyridine, L	5.61			67TT
3-(3-Hydroxypropyl)pyridine, L	5.47			67TT
4-(2-Hydroxyethyl)pyridine, L	5.60			67TT
4-(3-Hydroxypropyl)pyridine, L	5.84			67TT

---

[c] 20°, 0; [d] 20°, 0.5; [i] 25°, 0.5; * Metal constants were also reported but are not included in the compilation of selected constants.

Pyridines   (continued)

Ligand	Log K 25°, 0	ΔH 25°, 0	ΔS 25°, 0	Bibliography
2,6-Bis(hydroxymethyl)pyridine[*], H_2L	$(14.6)^d$			72PF,72PP
	$(13.5)^d$			
	4.39^d			
Picolinic acid methyl ester, L	2.21^m			56GT
Nicotinic acid methyl ester, L	3.13^m			56GT
Isonicotinic acid methyl ester, L	3.26^m			56GT
2-Acetylpyridine, L	2.66	−3.38	0.8	65CC
3-Acetylpyridine, L	3.28	−3.13	4.4	65CC
4-Acetylpyridine, L	3.51	−3.68	3.7	65CC

3,5,6,8-Tetramethyl-1,10-phenanthroline, L	5.80^n	61JW

Ligand	Log K 25°, 0	ΔH 25°, 0	ΔS 25°, 0	Bibliography
4,4'-Bipyridyl, L	4.77^a	$(-5)^o$	$(5)^a$	71U
	2.69^a	$(-4)^o$	$(-1)^a$	
4,4'-Dimethyl-2,2'-bipyridyl[*], L	5.32^a			56YS
Iminobis(methylene-3-pyridine), L	7.15^a			68RB
	3.90^a			
	2.95^a			

---

[a] 25°, 0.1;   [d] 20°, 0.5;   [m] 22°, 0;   [n] 25°, 0.3;   [o] 19-40°, 0.1;   [*] Metal constants were also reported but are not included in the compilation of selected constants.

2. <u>Pyrimidines</u>:

Ligand	Log K 25°, 0	ΔH 25°, 0	ΔS 25°, 0	Bibliography
Deoxycytidine-5'-(dihydrogenphosphate), H_3L		-4.3^a		60RS
2,4-Dioxopyrimidine (uracil), HL	9.49 ±0.03 9.34^f	-7.85	17.1	61NS,67CR,69RW
1-(β-D-Ribofuranosyl)uracil (uridine)*, H_2L	12.59 9.30 9.18^f 9.06^p	-10.6 -7.24	21 18.3	65FB,67CR,69RWa, Other reference: 67Wa
2,4,6-Trioxopyrimidine (barbituric acid), L	4.06 ±0.02 3.98^q	-0.06	18.4	56B,59NH,65MA
1-Methylbarbituric acid, L	4.35			56B
1,3-Dimethylbarbituric acid, L	4.68	0.05	21.6	56B,65MA
5-Methylbarbituric acid, L	4.40	1.28	24.4	65MA
5-(2-Propyl)barbituric acid, L	4.94			56B
5,5-Diethylbarbituric acid (barbital), L	7.980±0.02 7.78^a	-5.81	17.0	52MS,56B,59NH,65MA
5-Butyl-5-ethylbarbituric acid, L	7.98			56B
5-Ethyl-5-pentylbarbituric acid, L		-5.22		65MA
5-Ethyl-5-(2-methylbutyl)barbituric acid, L	7.96	-5.81	16.9	56B,65MA
5-Allylbarbituric acid, L	4.78	1.89	28.2	65MA
5-Allyl-5-(1-methylethyl)barbituric acid, L	7.99			56B
5-Allyl-5-(2-methylpropyl)barbituric acid, L	7.79			56B
5-Allyl-5-(2-methylbutyl)barbituric acid, L	7.96	-5.00	19.7	56B,65MA
5,5-Diallylbarbituric acid, L	7.77	-4.92	19.1	56B,65MA
5-Methyl-5-phenylbarbituric acid, L	7.73			56B

---

^{a} 25°, 0.1; ^{f} 25°, 0.05; ^{p} 20°, 1.0; ^{q} 25°, 1.0; * Metal constants were also reported but are not included in the compilation of selected constants.

Pyrimidines   (continued)

Ligand	Log K 25°, 0	ΔH 25°, 0	ΔS 25°, 0	Bibliography
5-Ethyl-5-phenylbarbituric acid, L	7.45	−4.60	18.6	56B,65MA
5-Oxobarbituric acid 5-oxime (violuric acid), L	4.58			59NH
	4.26^q			

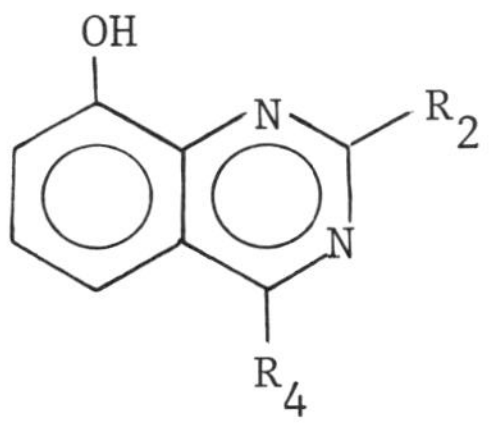

8-Hydroxybenzo[d]-1,3-diazine	8.54		55IR, Other
(8-hydroxyquinazoline)[*], L	8.35^n		reference: 54AH
	3.36		
	3.55^n		
2,4-Dimethyl-8-hydroxyquinazoline, L	9.41		55IR
	9.22^n		
	3.79		
	3.98^n		

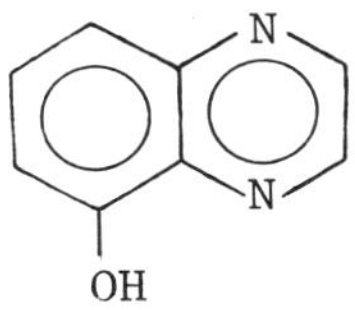

5-Hydroxybenzo[b]-1,4-diazine	8.75		55IR, Other
(5-hydroxyquinoxaline)[*], L	8.56^n		reference: 54AH
	0.8		
	0.97^n		

---

[n] 25°, 0.3;   [q] 25°, 1.0;   [*] Metal constants were also reported but are not included in the compilation of selected constants.

3. Purines:

Ligand	Log K 25°, 0	ΔH 25°, 0	ΔS 25°, 0	Bibliography
1,3-Diazolo[4,5-d]1,3-diazine (purine)*,	8.66[f]			69RW
HL	2.41[f]			
6-Chloropurine*, HL	7.5[r]			71TK
	1.1[r]			
6-Mercaptopurine*, H_2L	9.1[r]			71TK, Other
	6.9[r]			reference: 69GK
	2.2[r]			
2-Amino-6-mercaptopurine*, H_2L	9.8[r]			73TK
	7.9[r]			
	3.2[r]			
6-(Diethylamino)purine*, HL	9.65[f]			69RW
	4.53[f]			
6-Amino-7-methylpurine (7-methyladenine)*,	4.13[f]			69RW
L				
9-Methyladenine*, L	3.90[f]			69RW
2-Mercaptoadenine*, H_2L	9.6[r]			73TK
	7.7[r]			
	3.8[r]			
8-Azaadenine*, H_2L	5.8[r]			73TK
	2.4[r]			
9-(β-D-Xylofuranosyl)adenine ((trans-	12.34	-8.4	28	66CR
hydroxy)adenosine), HL				
Deoxyadenosine, HL	3.62[f]	-3.9[a]	4[f]	60RS,69RWa
Deoxyadenosine-5'-(dihydrogenphosphate),H_3L		-2.6[a]		60RS
Adenosine-5'-(dihydrogenphosphate) N(1)-	12.49[a]			64SB,67SPb
oxide*, H_3L	6.12[a]			
	2.58[a]			

---

[a] 25°, 0.1; [f] 25°, 0.05; [r] 45°, 0.1; * Metal constants were also reported but are not
included in the complilation of selected constants.

Purines (continued)

Ligand	Log K 25°, 0	ΔH 25°, 0	ΔS 25°, 0	Bibliography
2,6-Diaminopurine*, HL	10.37[f]			69RW,73TK
	9.8[r]			
	(4.89)[f]			
	(5.1)[r]			
2-Amino-6-oxopurine (guanine), HL	9.31[a]	(-10)[s]	(9)[a]	64SPB
9-(β-D-Ribofuranosyl)guanine (guanosine)*, H_2L	12.33	-10.9	20	60RS,64SPB,65FB,
	9.25	-7.7	17	66BS,70CR, Other
	9.13[a]			reference: 53A
	9.10[p]			
	1.9	-3.2	-2	
	1.8[a]±0.1	-1.0[a]	5[a]	
	2.1[p]			
Deoxyguanosine*, HL	9.11[f]			60RS,69RWa, Other
	2.12[f]	-1.9[a]	3[f]	reference: 65RV
Deoxyguanosine-5'-(dihydrogenphosphate), H_3L		-0.1[a]		60RS
6-Oxopurine (hypoxanthine)*, H_2L	12.04 ±0.04	-9.8 ±0.3	22	64SPB,69RW,70CR,
	8.85 ±0.06	-7.9 ±0.1	14	70WW,71TK, Other
	8.7[a] ±0.1			reference: 53A
	8.79[f]			
	1.85 ±0.05	-2.7 ±0.2	-1	
7-Methylhypoxanthine*, HL	9.01[f]			69RW
9-Methylhypoxanthine*, HL	9.16[f]			69RW
9-(β-D-Ribofuranosyl)hypoxanthine (inosine)*, HL	12.36	-10.65	20.9	64SPB,70CRa,
	8.96	-6.5	19	Other references:
	8.8[f]			53A,67Wa
Inosine-5'-(dihydrogenphosphate) (IMP), HL	6.64 ±0.02	(2)[t]	(37)	65PE,71MM
Inosine-5'-(trihydrogendiphosphate) (IDP), H_3L	7.18			65PE
2,6-Dioxopurine (xanthine)*, H_2L	11.84	-9.61	22.0	69RW,70CRa
	7.53	-6.3	13	
	7.43[f]			
1,3-Dimethylxanthine*, HL	8.68[f]			69RW
9-(β-D-Ribofuranosyl)xanthine (xanthosine)*, HL	12.00	-10.9	18	70CRa, Other
	5.67	-3.74	13.4	reference: 53A

[a] 25°, 0.1; [f] 25°, 0.05; [p] 20°, 1.0; [r] 45°, 0.1; [s] temperature range not given, 0.1: [t] 10-40°, 0; * Metal constants were also reported but are not included in the compilation of selected constants.

Purines (continued)

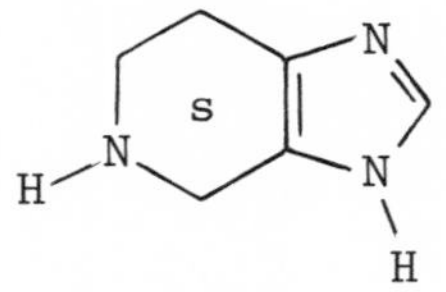

Ligand	Log K 25°, 0	ΔH 25°, 0	ΔS 25°, 0	Bibliography
4,5,6,7-Tetrahydroimidazolo[4,5-c]pyridine	8.90[a]	(-10)[u]	(7)[a]	73BD
(spinaceamine), L	4.90[a]	(-6)[u]	(2)[a]	

D. Aminophosphorus acids:

1. Aminophosphonic acids: $\quad H_2N-R-PO_3H_2$

	Log K	ΔH	ΔS	
1-Aminobutylphosphonic acid, H_2L	10.24[a]			72WN
	5.66[a]			
1-Aminopentylphosphonic acid, H_2L	10.25[a]			72WN
	5.69[a]			
2-Aminoethylphosphonic acid, H_2L	11.499	-14.0	6	73U
	6.505	0.3	31	
	1.30	2.0	13	
Phenyl(methylamino)methylphosphonic acid, H_2L	10.12[i]			72GT
	5.43[i]			
2-Hydroxyphenyl(methylamino)methyl-phosphonic acid, H_3L	12.65[i]			72GT
	9.34[i]			
	5.03[i]			
Ethylenediamine-N,N,N'-tris(methylene-phosphonic acid), H_6L	10.9[a]			65UA
	8.76[a]			
	6.70[a]			
	5.48[a]			
	3.03[a]			

For other protonation values of aminophosphonic acids see: Volume 1, p. 396.

2. Amino-dihydrogenphosphates : $\quad H_2N-R-OPO_3H_2$

	Log K			
O-Phosphoryl-DL-serine methyl ester, H_2L	7.83[b]			59FO
	5.33[b]			

For other protonation values of amino-dihydrogenphosphates see: Volume 1: p. 401 and
Volume 2: pp. 344,345.

[a] 25°, 0.1; [b] 25°, 0.16; [i] 25°, 0.5; [u] 5-35°, 0.1

VII. LIGANDS CONSIDERED BUT NOT INCLUDED

A. Amines:

1. Primary amines:

Ligand	Bibliography
Heptylamine	64SM
2,6-Dimethylaniline (2,6-xylidine)	52G
3,5-Dimethylaniline (3,5-xylidine)	52G
1-(3,4-Dihydroxyphenyl)-2-aminopropanol (corbadrine)	62Hc
2-Amino-4,5-dimethylphenol	59Sa
1-Amino-8-naphthol-3,6-disulfonic acid	63RS
7-Amino-1-naphthol-3,6-disulfonic acid	63SD,68AS,68BD
8-Amino-2-(2-hydroxyphenylazo)-1-naphthol-3,6-disulfonic acid	60DE
8-Amino-2-(2-hydroxyphenylazo)-1-naphthol-5,7-disulfonic acid	60DE
4,5-Diamino-1,8-dihydroxyanthraquinone (diaminochrysazin)	64Ba
Aminoacethydroxamic acid (glycine hydroxamic acid)	56CD,66MPa,71KM
6-Aminohexanol	72VE
3-Amino-1,2-propanediol	72VE
L-Leucyl-L-phenylalaninamide	73KK
2-Aminoethylthiourea	61KP
3-Aminopropanethiol	61KPa
1,1-Diethylethylenediamine	69KJ
L-Leucinamide	59DL

2. Secondary amines:

N-Methylaniline	52G
1-(3,4-Dihydroxyphenyl)-2-(propylamino)ethanol	63H
2-(3-Hydroxypropylamino)ethanol	71HS
N-Methylglucamine	59J
erythro-Phenyl(piperidyl)methanol	65T
threo-Phenyl(piperidyl)methanol	65T
Sarcosinamide	59DL

Secondary amines   (continued)

Ligand	Bibliography
Sarcosine-N-methylamide	59DL
Sarcosine-N,N-dimethylamide	59DL
Iminodiacetic acid diamide	68P,68Pa
Iminodithioacetic acid diamide	68P,68Pa
2,2'-Bipiperidyl	64SM
N-Benzylethylenediamine	61H
N,N'-Dibenzylethylenediamine	64SM
Ethylenediiminobis(ethylenesulfonic acid)	71ES,71IP
3-(2-Aminoethylamino)propanol (N-(3-hydroxypropyl)ethylenediamine)	50E
DL-N-(2-Amino-?-methylethylamino)ethanol	71KP
Ethylenediiminodi(4-pent-3-en-2-ol)	61AV
N-(2-Furylmethyl)ethylenediamine	58H,64SM
N-(3-Furylmethyl)ethylenediamine	61H
N,N'-Bis(2-furylmethyl)ethylenediamine	56H,58H,64SM
3,7-Diazanonanedioic acid diamide	74KZ
3,7-Diazanonanedioic acid bis(ethylamide)	74KZ
N-(2-Thienylmethyl)ethylenediamine	61H
N,N'-Bis(2-thienylmethyl)ethylenediamine	64SM
1,4,7-Triazacyclononane	73AH
1,4,7,10,13,16-Hexaazahexadecane (pentaethylenehexamine)	57JB,62JS
6-(2-Aminoethylaminomethyl)-6-ethyl-1,4,8,11-tetraazaundecane	63GC

3.   Tertiary amines:

Ligand	Bibliography
N-Ethyliminodi-2-ethanol	67FH
N-Butyliminodi-2-ethanol	66SKa
N,N-Dimethyl-2-(diphenylmethoxy)ethylamine (benadryl)	61S,62AL,64AR,69CA
Nitrilotris(acetamide)	70P
2-(Dimethylamino)ethanethiol	73MS
2-[Bis(1-methylethyl)amino]ethanethiol	73SC
Ethylenediamine-N,N ,N'-tris(ethylenesulfonic acid)	71EI
4-Aminobenzoic acid 2-(diethylamino)ethyl ester (procaine)	61GK
N-[2-(Dimethylamino)ethyl]oxamide	68KZ
N,N'-Bis[2-(dimethylamino)ethyl]oxamide	68ZK
Ethylenedinitrilotetrakis(acetamide)	70P
Hexamethylenetetramine	24P,28J

B. Azoles:

1. 1,2-Diazoles:

Ligand	Bibliography
1,2-Diazole (pyrazole)	63AR,65C,66C,68CW
3-Methylpyrazole	74CA
3-(2-Aminoethyl)pyrazole (betazole)	60HJ
1-Phenyl-2,3-dimethyl-4-dimethylamino-1,2-diazol-5-one (pyramidone)	68Sb
3,6-Bis(4-antipyrylazo)-4,5-dihydroxynaphthalene-2,7-disulfonic acid	68BB,74Ba

2. 1,3-Diazoles:

4-Aminomethyl-2-methylimidazole	60HJ
3-Methylhistamine	60LR
4,4'-Methylenediimidazole (di-4-imidazolylmethane)	65DF
4-Aminobenzimidazole	69RW
2-(2-Hydroxyphenyl)imidazoline	56AR
2-(3-Hydroxy-2-naphthyl)imidazoline	56AR
2-(N-Benzyl-N-phenylaminomethyl)imidazoline (antistine)	61S,62AL,64AR,69CA

3. Thiazoles:

4-(2-Aminoethyl)thiazole	60HJ
2,4'-Thiazolylazoresorcinol (TAR)	66HS,67SI
1-(2-Thiazolylazo)-2-naphthol	64N,66N,67Na
1-(2-Thiazolylazo)-2-naphthol-6-sulfonic acid	67NP

4. Triazoles:

2-(2-Hydroxy-5-methylphenyl)benzotriazole	68NL

C. Azines:

1. Pyridines:

Ligand	Bibliography
2,4,6-Trimethylpyridine (2,4,6-collidine)	73SB
2-{2-[(Dimethylamino)ethyl]benzyl}pyridine (trimeton)	62AL
2-[3-(Dimethylamino)-1-(4-chlorophenyl)propyl]pyridine (chlortrimeton)	62AL
2-[N-Benzyl-N-(2-dimethylaminoethyl)amino]pyridine (pyribenzamine)	69CA
2-Pyridyl(hydroxy)methanesulfonic acid	64BG
6-Methyl-2-pyridyl(hydroxy)methanesulfonic acid	64BG
3-Pyridyl(hydroxy)methanesulfonic acid	64BG
Pyridine-3-carbohydroxamic acid (nicotinylhydroxamic acid)	64RMa
2-Acetylpyridine oxime	60BT
1-(2-Pyridylazo)-2-naphthol (PAN)	63BF,66N,67Na
1-(2-Pyridylazo)-4-naphthol	63BT
2-Hydroxypyridine 1-oxide	56AR
2-(2-Hydroxyphenyl)pyridine	56AR
Phenyl(2-pyridyl)methanol	57PT
2-(1-Hydroxyethyl)pyridine	55LFa
Pyridine-3-carboxylic acid ethyl ester (ethyl nicotinate)	67N
N-(2-Pyridylmethyl)glycinamide	71KZ
N-(2-Pyridylmethyl)glycine-N-ethylamide	71KZ
N-(2-Pyridyl)-3-oxobutanoic acid amide	67H
N,N'-Bis(2-pyridylmethyl)oxamide	68GF
N-Picolinoylethylenediamine	71KZ
2-(N-Glycylaminomethyl)pyridine	71KZ
N-(2-Pyridylmethyl)ethylenediamine	60H,69BZ
N-(3-Pyridylmethyl)ethylenediamine	60H
N-(4-Pyridylmethyl)ethylenediamine	60H
2-(2-Piperidyl)pyridine	64SM
2-Mercaptopyridine 1-oxide	56AR

2. Benzo[b]pyridines:

Ligand	Bibliography
8-Methoxyquinoline	64SM
2-Methyl-8-methoxyquinoline	64SM
5-Chloroquinoline-7-sulfonic acid	70BB
5-Bromoquinoline-7-sulfonic acid	70BB
5-Iodoquinoline-7-sulfonic acid	70BB
3,3'-Dimethylene-4,4'-diphenyl-2,2'-biquinolyldisulfonic acid	67UH
3,3'-Trimethylene-4,4'-diphenyl-2,2'-biquinolyldisulfonic acid	67UH
4-Methyl-8-hydroxyquinoline	65CF
2-Hydroxyquinoline 1-oxide	56AR
5-Chloro-8-hydroxyquinoline	68CF
5-Bromo-8-hydroxyquinoline	68CF
5-Iodo-8-hydroxyquinoline	68CF
5-Nitro-8-hydroxyquinoline	68CF
2-Methyl-7-bromo-8-hydroxyquinoline	63FF
2-Methyl-7-nitro-8-hydroxyquinoline	63FF
5,7-Dibromo-8-hydroxyquinoline	65NK,68RS,69SR
5,7-Diiodo-8-hydroxyquinoline	68RS
8-Hydroxyquinoline-7-sulfonic acid	70BB
7-Chloro-8-hydroxyquinoline-5-sulfonic acid	70ABa,72AB
7-Bromo-8-hydroxyquinoline-5-sulfonic acid	70ABa,71AB,72AB,73M
5-Iodo-2-hydroxyquinoline-7-sulfonic acid	72BBa
2-Hydrazino-4-methylquinoline	57FE
8-Amino-2-methylquinoline (8-aminoquinaldine)	64SM

3. Dipyridines:

Ligand	Bibliography
1,2-Di-2-pyridylethane-1,2-diol	69MG
1,2-Bis(6-methyl-2-pyridyl)ethane-1,2-diol	69MG
Terpyridine	54BW,54ML,61JW,63OG, 66HH,69PP,73YB
Quaterpyridine	64BO

4. 1,10-Phenanthrolines:

5-Phenyl-1,10-phenanthroline	52BG
4-Chloro-1,10-phenanthroline	61HD
4-Hydroxy-1,10-phenanthroline	61HD

5. 1,2-Diazines:

8-Hydroxybenzo[c]-1,2-diazine (8-hydroxycinnoline)	54AH
8-Hydroxy-4-methylcinnoline	54AH
3-Hydrazinobenzo[d]-1,2-diazine (1-hydrazinophthalazine)	57FE

6. 1,3-Diazines:

2,2'-Bipyrimidyl	63BMa
8-Hydroxy-4-methylquinazoline	54AH
2-[N-(2-Dimethylaminoethyl)-N-(4-methoxybenzyl)amino]pyrimidine (neohetramine)	61S,62AL,64AR

7. <u>Triazines</u>:

<u>Ligand</u>	<u>Bibliography</u>
2,4-Diamino-6-phenyl-1,3,5-triazine	63MB
2,4,6-Tripyridyl-1,3,5-triazine	66BC,71LP

8. <u>Purines</u>:

Adenine N(1)-oxide	60P
Adenosine N(1)-oxide	60P

9. <u>Other azines</u>:

4-Hydroxypyridino[2,3-e]pyridine (4-hydroxy-1,5-naphthyridine)	54AH
8-Hydroxy-1,6-naphthyridine	54AH
8-Hydroxy-1,7-naphthyridine	54AH
4-Hydroxypyridino[2,3-d]-1,3-diazine	54AH
4'-Hydroxypyridino[2,3-b]-1,4-diazine	54AH
2'-Hydroxypyridino[2,3-b]-1,4-diazine	54AH
4-Hydroxy-1,3-diazino[4,5-b]-1,4-diazine (4-hydroxypteridine)	53A
2,4-Dihydroxypteridine	53A
6-Anilinomethyl-4-hydroxypteridine	53A
6-(4-Sulfoanilinomethyl)-4-hydroxypteridine	53A
6,7-Dimethyl-9-D-1'-ribitylbenzo[b]-1,4-diazino[2,3-d]-1,3-diazin-2,4-dione (riboflavin)	53A,72ND

D.   Aminophosphorus acids:

1.   Aminophosphonic acids:

Ligand	Bibliography
1,2-Cyclohexylenedinitrilotetrakis(methylenephosphonic acid)	59BY
Ethylenediamine-N,N'-bis(ethylenesulfonic)-N-methylenephosphonic acid	71IP

VIII. BIBLIOGRAPHY

Russian translations have the page of the original in parentheses.

03BE   G. Bodlander and W. Eberlein, Chem. Ber., 1903, 36, 3945

03E   H. von Euler, Chem. Ber., 1903, 36, 2878

04E   H. von Euler, Chem. Ber., 1904, 37, 2768

24P   F. G. Pawelka, Z. Elektrochem., 1924, 30, 180

28HR   H. S. Harned and R. A. Robinson, J. Amer. Chem. Soc., 1928, 50, 3157

28J   P. Job, Ann. Chim. (France), 1928, 9, 113

30HO   H. S. Harned and B. B. Owen, J. Amer. Chem. Soc., 1930, 52, 5079

30K   F. K. V. Koch, J. Chem. Soc., 1930, 2053

32HS   N. F. Hall and M. R. Sprinkle, J. Amer. Chem. Soc., 1932, 54, 3469

33AT   M. Aumeras and A. Tamisier, Bull. Soc. Chim. France, 1933, 53, 97

33T   A. Tamisier, Bull. Soc. Chim. France, 1933, 53, 157

35BW   H. T. S. Britton and W. C. Williams, J. Chem. Soc., 1935, 796

36BW   H. T. S. Britton and W. C. Williams, J. Chem. Soc., 1936, 96

36N   A. Neuberger, Biochem. J., 1936, 30, 2085

37N   A. Neuberger, Proc. Roy. Soc. (London), 1937, A158, 68

37P   K. J. Pedersen, Kgl. Danske Vid. Sel., Math.-Fys. Medd., 1937, 14, No. 9

38KN   A. H. M. Kirby and A. Neuberger, Biochem. J., 1938, 32, 1146

41B   J. Bjerrum, Metal-Amine Formation, P. Haase and Son, Copenhagen, 1941

41EW   D. H. Everett and W. F. K. Wynne-Jones, Proc. Roy. Soc. (London), 1941, 177A, 499

41GH   S. Glastone and E. F. Hammel, Jr., J. Amer. Chem. Soc., 1941, 63, 243

43VC   W. C. Vosburgh and S. A. Cogswell, J. Amer. Chem. Soc., 1943, 65, 2412

45BA   J. Bjerrum and P. Anderson, Kgl. Danske Vid. Sel., Math.-Fys. Medd., 1945, 22, No. 7

45CM   G. A. Carlson, J. P. McReynolds, and F. H. Verhoek, _J. Amer. Chem. Soc._,
        1945, _67_, 1334

46DN   F. P. Dwyer and R. S. Nyholm, _J. Proc. Roy. Soc. New South Wales_, 1946, _80_, 28

47BR   J. Bjerrum and S. Refn, _Nordiska Kemistmote_, Lund, 1947, 277 (see 64SM)

47DM   F. P. Dwyer and H. A. McKenzie, _J. Proc, Roy. Soc. New South Wales_, 1947, _81_, 97

47SW   G. Schwarzenbach, A. Willi, and R. O. Bach, _Helv. Chim. Acta_, 1947, _30_, 1303

48BN   J. Bjerrum and E. J. Nielsen, _Acta Chem. Scand._, 1948, _2_, 297

48BV   R. J. Bruehlman and F. H. Verhoek, _J. Amer. Chem. Soc._, 1948, _70_, 1401

48KH   S. Kilpi and P. Harjanne, _Suomen Kem._, 1948, _B21_, 14

48LK   T. S. Lee, I. M. Kolthoff, and D. L. Leussing, _J. Amer. Chem. Soc._, 1948,
        _70_, 2348, 3596

48MM   D. P. Mellor and C. Maley, _Nature_, 1948, _161_, 436

49AS   H. Ackerman and G. Schwarzenbach, _Helv. Chim. Acta_, 1949, _32_, 1543

49BP   R. G. Bates and G. D. Pinching, _J. Res. Nat. Bur. Standards_, 1949, _43_, 519

49HJ   E. B. Hughes, H.H.G. Jellinek, and B. A. Ambrose, _J. Phys. Chem._, 1949, _53_, 410

49K   P. Krumholz, _J. Amer. Chem. Soc._, 1949, _71_, 3654

49LM   D. L. Levi, W. S. McElwan, and J. H. Wolfenden, _J. Chem. Soc._, 1949, **760**

49LO   H. A. Latiner, E. L. Onstatt, J. C. Bailer, Jr., and S. Swann, Jr.,
        J. Amer. Chem. Soc., 1949, _71_, 1550

49MM   L. E. Maley and D. P. Mellor, _Aust. J. Sci. Res._, 1949, _A2_, 579

49SS   E. B. Sandell and D. C. Spindler, _J. Amer. Chem. Soc._, 1949, _71_, 3806

50A   A. Albert, _Biochem. J._, 1950, _47_, 531

50B   J. Bjerrum, _Chem. Rev._, 1950, _46_, 381

50BG   J. H. Baxendale and P. George, _Trans. Faraday Soc._, 1950, _46_, 55

50BL   J. Bjerrum and C. G. Lamm, _Acta Chim. Scand._, 1950, _4_, 997

50DL   B. E. Douglas, H. A. Laitiner, and J. C. Bailar, Jr., _J. Amer. Chem. Soc._,
        1950, _72_, 2484

50E   L. J. Edwards, Diss., University of Michigan, 1950 (see 64SM)

50JLM   H. B. Jonassen, R. B. LeBlanc, A. W. Meibohm, and R. M. Rogan, _J. Amer. Chem. Soc._,
        1950, _72_, 2430

50JLR   H. B. Jonassen, R. B. LeBlanc, and R. M. Rogan, _J. Amer. Chem. Soc._, 1950, _72_, 4968

50K   P. Krumholz, _An. Acad. Brasil. Cien._, 1950, _22_, No. 3

50KL   I. M. Kolthoff, D. L. Leussing, and T. S. Lee, J. Amer. Chem. Soc., 1950, 72, 2173

50OL   E. I. Onstott and H. A. Laitiner, J. Amer. Chem. Soc., 1950, 72, 4724

50PS   J. E. Prue and G. Schwarzenbach, Helv. Chim. Acta, 1950, 33, 963

50PSa   J. E. Prue and G. Schwarzenbach, Helv. Chim. Acta, 1950, 33, 985

50PSb   J. E. Prue and G. Schwarzenbach, Helv. Chim. Acta, 1950, 33, 995

50S   G. Schwarzenbach, Helv. Chim. Acta, 1950, 33, 974

51AS   R. A. Alberty, R. M. Smith, and R. M. Bock, J. Biol. Chem., 1951, 193, 425

51BP   R. G. Bates and G. D. Pinching. J. Res. Nat. Bur. Standards, 1951, 56, 349

51CL   C. H. Cook, Jr. and F. A. Long, J. Amer. Chem. Soc., 1951, 73, 4119

51EH   A. G. Evans and S. D. Hamann, Trans. Faraday Soc., 1951, 47, 34

51G   E. Gonik, Diss., Penn. State Coll., 1951

51JM   H. B. Jonassen and A. W. Meibohm, J. Phys. Chem., 1951, 55, 726

51JW   H. H. G. Jellinek and M. G. Wayne, J. Phys. Chem., 1951, 55, 173

51KL   I. M. Kolthoff, D. L. Leussing, and T. S. Lee, J. Amer. Chem. Soc., 1951, 73, 390

51N   R. Nasanen, Acta Chem. Scand., 1951, 5, 1293

51NL   R. Nasanen, P. Lumme, and A. L. Mukula, Acta Chem. Scand., 1951, 5, 1199

52A   A. Albert, Biochem. J., 1952, 50, 690

52BG   W. W. Brandt and D. K. Gullstrom, J. Amer. Chem. Soc., 1952, 74, 3532

52BM   F. Basolo and R. K. Murmann, J. Amer. Chem. Soc., 1952, 74, 2373

52BMa   F. Basolo and R. K. Murmann, J. Amer. Chem. Soc., 1952, 74, 5243

52BR   J. Bjerrum and S. E. Rasmussen, Acta Chem. Scand., 1952, 6, 1265

52D   D. Dyrssen, Svensk. Kem. Tidskr., 1952, 64, 213

52EP   D. H. Everett and B. R. W. Pinsent, Proc. Roy. Soc., 1952, A215, 416

52F   W. S. Fyfe, Nature, 1952, 169, 69

52G   C. Golumbic, J. Amer. Chem. Soc., 1952, 74, 5777

52H   G. B. Hares, Diss., Penn. State Coll., 1952

52Ha   G. B. Hares, unpublished data (see G. H. McIntyre, Diss., Penn. State Coll., 1953)

52J   H. B. Jonassen, G. G. Hurst, R. B. LeBlanc, and A. W. Meibohm, J. Phys. Chem.,
           1952, 56, 16

52MS   G. G. Manov, K. E. Schuette, and F. S. Kirk, J. Res. Nat. Bur. Standards, 1952, 48, 84

52N   R. Nasanen, Acta Chem. Scand., 1952, 6, 352

52NE   R. Nasanen and A. Ekman, Acta Chem. Scand., 1952, 6, 1384

52NP   R. Nasanen and V. Penttinen, Acta Chem. Scand., 1952, 6, 837

52SA   G. Schwarzenbach, H. Ackerman, B. Maissen, and G. Anderegg, Helv. Chim. Acta, 1952, 35, 2337

52SM   G. Schwarzenbach, B. Maissen, and H. Ackerman, Helv. Chim. Acta, 1952, 35, 2333

53A   A. Albert, Biochem. J., 1953, 54, 646

53Aa   A. Albert, Experientia, 1953, 9, 370

53BH   J. M. Booling and J. L. Hall, J. Amer. Chem. Soc., 1953, 75, 3953

53BM   F. Basolo, R. K. Murmann, and Y. T. Chen, J. Amer. Chem. Soc., 1953, 75, 1478

53CG   A. K. Chakraburtty, N. N. Ghosh, and P. Ray, J. Indian Chem. Soc., 1953, 30, 185

53D   D. Dyrssen, Svensk. Kem. Tidskr., 1953, 65, 43

53DD   D. Dyrssen and V. Dahlberg, Acta Chem. Scand., 1953, 7, 1186

53DN   V. Di Stefano and W. F. Neuman, J. Biol. Chem., 1953, 200, 759

53EN   A. Ekman and R. Nasanen, Acta Chem. Scand., 1953, 7, 1261

53IC   H. M. Irving, J. Cabell, and D. H. Mellor, J. Chem. Soc., 1953, 3417

53K   E. J. King, J. Amer. Chem. Soc., 1953, 75, 2204

53KL   J. M. Klotz and W. C. Loh Ming, J. Amer. Chem. Soc., 1953, 75, 4159

53L   J. Lotz, unpublished data (see G. H. McIntyre, Diss., Penn. State Coll., 1953)

53N   R. Nasanen, Suomen Kem., 1953, B26, 11

53Na   R. Nasanen, Suomen Kem., 1953, B26, 69

53Nb   C. J. Nyman, J. Amer. Chem. Soc., 1953, 75, 3575

53P   D. J. Perkins, Biochem. J., 1953, 55, 649

53SM   G. Schwarzenbach and P. Moser, Helv. Chim. Acta, 1953, 36, 581

53SP   C. G. Spike and R. W. Parry, J. Amer. Chem. Soc., 1953, 75, 2726

53SPa   C. G. Spike and R. W. Parry, J. Amer. Chem. Soc., 1953, 75, 3770

53TW   C. Tanford and M. L. Wagner, J. Amer. Chem. Soc., 1953, 75, 434

53WS   E. J. Wheelright, F. H. Spedding, and G. Schwarzenbach, J. Amer. Chem. Soc., 1953, 75, 4196

54AC   R. J. L. Andon, J. D. Cox, and E. F. G. Harington, <u>Trans. Faraday Soc.</u>,
              1954, <u>50</u>, 918

54AH   A. Albert and A. Hampton, <u>J. Chem. Soc.</u>, 1954, 505

54BC   F. Basolo, Y. T. Chen, and R. K. Murmann, <u>J. Amer. Chem. Soc.</u>, 1954, <u>76</u>, 956

54BM   F. Basolo and R. K. Murmann, <u>J. Amer. Chem. Soc.</u>, 1954, <u>76</u>, 211

54BW   W. W. Brandt and J. P. Wright, <u>J. Amer. Chem. Soc.</u>, 1954, <u>76</u>, 3082

54D   D. Dryssen, <u>Svensk. Kem. Tidskr.</u>, 1954, <u>66</u>, 234

54DS   T. Davies, S. S. Singer, and L. A. K. Staveley, <u>J. Chem. Soc.</u>, 1954, 2304

54EF   J. T. Edsall, G. Felsenfeld, D. S. Goodman, and F. R. N. Gurd, <u>J. Amer. Chem. Soc.</u>
              1954, <u>76</u>, 3054

54GF   E. Gonick, W. C. Fernelius, and B. E. Douglas, <u>J. Amer. Chem. Soc.</u>, 1954, <u>76</u>, 4671

54GFa   E. Gonick, W. C. Fernelius, and B. E. Douglas, <u>J. Amer. Chem. Soc.</u>, 1954,
              <u>76</u>, 5354

54IG   H. Irving and J. M. M. Griffiths, <u>J. Chem. Soc.</u>, 1954, 213

54IW   H. Irving, R. J. P. Williams, D. J. Ferrett, and A. E. Williams, <u>J. Chem. Soc.</u>,
              1954, 3494

54JU   H. H. G. Jellinek and J. R. Urwin, <u>J. Phys. Chem.</u>, 1954, <u>58</u>, 548

54L   J. R. Lotz, Diss., Penn. State Coll., 1954

54LW   N. C. Li, J. M. White, and E. Doody, <u>J. Amer. Chem. Soc.</u>, 1954, <u>76</u>, 6219

54M   N. C. Melchior, <u>J. Biol. Chem.</u>, 1954, <u>208</u>, 615

54NU   R. Nasanen and E. Uusitalo, <u>Acta Chem. Scand.</u>, 1954, <u>8</u>, 112

54PB   R. T. Pflaum and W. W. Brandt, <u>J. Amer. Chem. Soc.</u>, 1954, <u>76</u>, 6215

54S   J. Schubert, <u>J. Amer. Chem. Soc.</u>, 1954, <u>76</u>, 3442

54SS   E. Scrocco and O. Salvetti, <u>Boll. Sci. Fac. Chim. Ind. Bologna</u>, 1954, <u>12</u>, 98

54WN   V. R. Williams and J. B. Neilands, <u>Arch. Biochem. Biophys.</u>, 1954, <u>53</u>, 56

55BM   H. C. Brown and X. R. Mihm, <u>J. Amer. Chem. Soc.</u>, 1955, <u>77</u>, 1723

55CD   H. B. Clarke, S. P. Datta, and B. R. Rabin, <u>Biochem. J.</u>, 1955, <u>59</u>, 209

55CH   F. A. Cotton and F. E. Harris, <u>J. Phys. Chem.</u>, 1955, <u>59</u>, 1203

55D   D. Dryssen, <u>Svensk. Kem. Tidskr.</u>, 1955, <u>67</u>, 711

55F   W. S. Fyfe, <u>J. Chem. Soc.</u>, 1955, 1347

55FK   R. J. Flannery, B. Ke, M. W. Greib, and D. Trivich, <u>J. Amer. Chem. Soc.</u>,
              1955, <u>77</u>, 2996

55FR   E. Felder, C. Rescigno, and C. Radice, Gazz. Chim. Ital., 1955, 85, 453

55GF   E. Gonick, W. G. Fernelius, and B. E. Douglas, J. Amer. Chem. Soc., 1955, 77, 6506

55IM   H. Irving and D. H. Mellor, J. Chem. Soc., 1955, 3457

55IR   H. Irving, H. S. Rossotti, and G. Harris, Analyst, 1955, 80, 83

55JR   H. B. Jonassen, R. E. Reeves, and L. Sogal, J. Amer. Chem. Soc., 1955, 77, 2748

55LC   N. C. Li, T. L. Chu, C. T. Fujii, and J. M. White, J. Amer. Chem. Soc., 1955, 77, 859

55LF   P. G. Langer, S. Fallab, and H. Erlenmeyer, Helv. Chim. Acta, 1955, 38, 92

55LFa   W. D. Luz, S. Fallab, and H. Erlenmeyer, Helv. Chim. Acta, 1955, 38, 1114

55LM   N. C. Li and R. A. Manning, J. Amer. Chem. Soc., 1955, 77, 5225

55M   H. A. McKenzie, Aust. J. Chem., 1955, 8, 569

55MA   B. L. Michel and A. C. Andrews, J. Amer. Chem. Soc., 1955, 77, 5291

55MB   R. K. Murmann and F. Basolo, J. Amer. Chem. Soc., 1955, 77, 3484

55MBa   R. R. Miller and W. Brandt, J. Amer. Chem. Soc., 1955, 77, 1384

55MS   D. E. Metzler and E. E. Snell, J. Amer. Chem. Soc., 1955, 77, 2431

55N   R. Nasanen, Suomen Kem., 1955, B28, 161

55Na   R. Nasanen, Suomen Kem., 1955, B28, 123

55NM   C. J. Nyman, E. W. Muhrbach, and G. B. Millard, J. Amer. Chem. Soc., 1955, 77, 4194

55NR   C. J. Nyman, D. K. Roe, D. B. Masson, J. Amer. Chem. Soc., 1955, 77, 4191

55NU   R. Nasanen and E. Uusitalo, Suomen Kem., 1955, B28, 17

55PB   I. Poulsen and J. Bjerrum, Acta Chem. Scand., 1955, 9, 1407

55SK   K. Sune, P. Krumholz, and H. Stammrich, J. Amer. Chem. Soc., 1955, 77, 777

56A   A. Albert, Nature, 1956, 177, 525

56AR   A. Albert, C. W. Reese, and A. J. H. Tomlinson, Brit. J. Exp. Path., 1956, 37, 500

56B   A. I. Biggs, J. Chem. Soc., 1956, 2485

56BB   R. G. Bates and V. E. Bower, J. Res. Nat. Bur. Standards, 1956, 57, 153

56BF   C. R. Bertsch, W. C. Fernelius, and B. P. Block, J. Phys. Chem., 1956, 60, 384

56BL   R. Boek, N. Ling, S. Morell, and S. Lipton, Arch. Biochem. Biophys., 1956, 62, 253

56BR   J. Bjerrum and S. Refn, Suomen Kem., 1956, B29, 68

56CD   V. Cieleszky, A. Dines, and E. Sandi, Acta Chim. Acad. Sci. Hung., 1956, 9, 381

56CG   J. Chatt and G. A. Gamleni, J. Chem. Soc., 1956, 2371

56DD   D. Dyrssen, M. Dyrssen, and E. Johansson, Acta Chem. Scand., 1956, 10, 341

56DR   S. P. Datta and B. R. Rabin, Trans. Faraday Soc., 1956, 52, 1117, 1123, 1130

56GT   R. W. Green and H. K. Tong, J. Amer. Chem. Soc., 1956, 78, 4896

56H   E. Hoyer, Z. Anorg. Allg. Chem., 1956, 288, 36

56HF   G. B. Hares, W. C. Fernelius, and B. E. Douglas, J. Amer. Chem. Soc., 1956,
        78, 1816

56M   K. Morinaga, Bull. Chem. Soc. Japan, 1956, 29, 793

56MB   D. W. Margerum, R. I. Bystroff, and C. V. Banks, J. Amer. Chem. Soc., 1956,
        78, 4211

56ML   R. B. Martin and J. A. Lissfelt, J. Amer. Chem. Soc., 1956, 78, 938

56MS   A. E. Martell and G. Schwarzenbach, Helv. Chim. Acta, 1956, 39, 653;
        J. Phys. Chem., 1958, 62, 886

56NU   R. Nasanen and E. Uusitalo, Suomen Kem., 1956, B29, 11

56RK   R. A. Robinson and A. K. Kiang, Trans. Faraday Soc., 1956, 52, 327

56SA   R. M. Smith and R. A. Alberty, J. Phys. Chem., 1956, 60, 180

56SAa   R. M. Smith and R. A. Alberty, J. Amer. Chem. Soc., 1956, 78, 2376

56SB   G. Schwarzenbach and R. Bauer, Helv. Chim. Acta , 1956, 39, 722

56ST   S. Searles, M. Tamres, F. Block, and L. A. Quaterman, J. Amer. Chem. Soc.,
        1956, 78, 4917

56WI   H. Walba and R. W. Isensee, J. Org. Chem., 1956, 21, 702

56WM   J. I. Watters and J. G. Mason, J. Amer. Chem. Soc., 1956, 78, 285

56WML   J. M. White, R. A. Manning, and N. C. Li, J. Amer. Chem. Soc., 1956, 78, 2367

56YS   M. Yasuda, K. Sone, and K. Yamasaki, J. Phys. Chem., 1956, 60, 1667

56YY   K. Yamasaki and M. Yasuda, J. Amer. Chem. Soc., 1956, 78, 1324

57B   W. E. Bennett, J. Amer. Chem. Soc., 1957, 79, 1290

57BI   E. Baciocchi and G. Illuminati, Gazz. Chim. Ital., 1957, 87, 981

57FE   S. Fallab and H. Erlenmeyer, Helv. Chim. Acta, 1957, 40, 363

57GM   R. L. Gustafson and A. E. Martell, Arch. Biochem. Biophys., 1957, 68, 485

57HJ   J. L. Hall, F. R. Jones, C. E. Delchamps, and C. W. Williams, J. Amer. Chem. Soc.,
        1957, 79, 3361

57JB   H. B. Jonassen, J. A. Bertrand, F. R. Groves, Jr., and R. I. Steams, J. Amer. Chem. Soc., 1957, 79, 4279

57JF   H. B. Jonassen, F. W. Frey, and A. Schaafsma, J. Phys. Chem., 1957, 61, 504

57JW   H. B. Jonassen and L. Westerman, J. Phys. Chem., 1957, 61, 1006

57JWa   H. B. Jonassen and L. Westerman, J. Amer. Chem. Soc., 1957, 79, 4275

57LD   N. C. Li, E. Doody, and J. M. White, J. Amer. Chem. Soc., 1957, 79, 5859

57LH   D. L. Leussing and R. C. Hansen, J. Amer. Chem. Soc., 1957, 79, 4270

57MC   A. E. Martell, S. Chaberk, S. Westerback, and H. Hyytianen. J. Amer. Chem. Soc., 1957, 79, 3036

57N   L. B. Nanninga, J. Phys. Chem., 1957, 61, 1144

57NG   Y. Nozaki, F. R. N. Gurd, R. F. Chen, and J. T. Edsall, J. Amer. Chem. Soc., 1957, 79, 2123

57PB   P. L. Peczok and J. Bjerrum, Acta Chem. Scand., 1957, 11, 1419

57PT   J. C. Pariaud and C. Tissler, J. Chim. Phys., 1957, 54, 533, 544

57RS   C. N. Reilly and R. W. Schmid, J. Elisha Mitchell Sci. Soc., 1957, 73, 279

57S   R. S. Subrahmanya, Proc. Indian Acad. Sci. (A), 1957, 46, 377

57SA   G. Schwarzenbach and G. Anderegg, Helv. Chim. Acta, 1957, 40, 1229

57TB   R. Trujillo and F. Brito, Anales Real. Soc. Espan., Fis. Quim., 1957, 53B, 249

57TBa   R. Trujillo and F. Brito, Anales Real. Soc. Espan., Fis. Quim., 1957, 53B, 313

57V   H. Vink, Arkiv. Kemi., 1957, 11, No. 3, 9

57WM   P. E. Wenger, D. Monnier, and I. Kapetanidis, Helv. Chim. Acta, 1957, 40, 1456

57WS   K. Wallenfels and H. Sund, Biochem. Z., 1957, 329, 41

58AC   S. Ahrland, J. Chatt, N. R. Davies, and A. A. Williams, J. Chem. Soc., 1958, 276

58ACa   S. Ahrland, J. Chatt, N. R. Davies, and A. A. Williams, J. Chem. Soc., 1958, 1403

58AS   D. J. Alner and A. G. Smeeth, J. Chem. Soc., 1958, 4207

58BB   C. R. Bertsch, B. P. Block, and W. C. Fernelius, J. Phys. Chem., 1958, 62, 503

58BF   C. R. Bertsch, W. C. Fernelius, and B. P. Block, J. Phys. Chem., 1958, 62, 444

58BG   M. T. Beck and S. Gorog, Acta Phys. Chem. Szeged, 1958, 4, 59

58C   P. Cecchi, Ric. Sci., 1958, 28, 2526

58CS   S. Canani and E. Scrocco, J. Inorg. Nucl. Chem., 1958, 8, 332

58EH   D. H. Everett and J. B. Hyne, J. Chem. Soc., 1958, 1636

58FO   B. F. Freasier, A. G. Oberg, and W. W. Wendlandt, J. Phys. Chem., 1958, 62, 700

58H   E. Hoyer, Z. Anorg. Allg. Chem., 1958, 297, 167

58HD   J. L. Hall and W. G. Dean, J. Amer. Chem. Soc., 1958, 80, 4183

58HF   T. R. Harkins and H. Freiser, J. Amer. Chem. Soc., 1958, 80, 1132

58JS   H. B. Jonassen, A. Schaafsma, and L. Westerman, J. Phys. Chem., 1958, 62, 1022

58KD   W. L. Koltun, R. N. Dexter, R. E. Clark, and F. R. N. Gurd, J. Amer. Chem. Soc.,
        1958, 80, 4188

58LC   N. C. Li and M. C. M. Chen, J. Amer. Chem. Soc., 1958, 80, 5678

58LD   N. C. Li, E. Doody, and J. M. White, J. Amer. Chem. Soc., 1958, 80, 5901

58M   R. K. Murmann, J. Amer. Chem. Soc., 1958, 80, 4174

58ME   R. B. Martin and J. T. Edsall, J. Amer. Chem. Soc., 1958, 80, 5033

58MEW   R. B. Martin, J. T. Edsall, D. B. Wetlaufer, and B. R. Hollingsworth,
        J. Biol. Chem., 1958, 233, 1429

58RH   C. N. Relly and J. H. Holloway, J. Amer. Chem. Soc., 1958, 80, 2917

58SL   J. L. Schubert, E. L. Ling, W. M. Westfall, R. Pleger, and N. C. Li,
        J. Amer. Chem. Soc., 1958, 80, 4799

58V   S. N. R. von Schalien, Suomen Kem., 1958, B31, 372

58W   E. Walaas, Acta Chem. Scand., 1958, 12, 528; 1957, 11, 1082

58WM   P. E. Wenger, D. Monnier, and I. Kapetanidis, Helv. Chim. Acta, 1958, 41, 1548

58WS   G. Weitzel and T. Speer, Z. Physiol. Chem., 1958, 313, 212

59A   G. Anderegg, Helv. Chim. Acta, 1959, 42, 344

59B   K. Burton, Biochem. J., 1959, 71, 388

59BB   C. V. Banks and R. C. Bystroff, J. Amer. Chem. Soc., 1959, 81, 6153

59BY   C. V. Banks and R. E. Yerick, Anal. Chim. Acta, 1959, 20, 301

59CG   R. C. Courtney, R. L. Gustafson, S. Chaberek, and A. E. Martell, J. Amer Chem.
        Soc., 1959, 81, 519

59DG   S. P. Datta and A. K. Grzybowski, J. Chem. Soc., 1959, 1091

59DL   S. P. Datta, R. Leberman, and B. R. Rabin, Trans. Faraday Soc., 1959, 55,
        1982, 2141

59FO   G. Folsch and R. Osterberg, J. Biol. Chem., 1959, 234, 2298

59GF   D. E. Goldberg and W. C. Fernelius, J. Phys. Chem., 1959, 63, 1246

59GFa   D. E. Goldberg and W. C. Fernelius, J. Phys. Chem., 1959, 63, 1328

59GM   R. L. Gustafson and A. E. Martell, J. Amer. Chem. Soc., 1959, 81, 525

59J   R. S. Juvet, J. Amer. Chem. Soc., 1959, 81, 1796

59K   D. A. Keyworth, Talanta, 1959, 2, 383

59KC   W. L. Koltun, R. E. Clark, R. N. Dexter, P. G. Katsoyannis, and F. R. N. Gurd,
        J. Amer. Chem. Soc., 1959, 81, 295

59KL   H. Kristiansen and F. J. Langmyhr, Acta Chem. Scand., 1959, 13, 1473

59LB   J. R. Lotz, B. P. Block, and W. C. Fernelius, J. Phys. Chem., 1959, 63, 541

59MB   G. H. McIntyre, Jr., B. P. Block, and W. C. Fernelius, J. Amer. Chem. Soc.,
        1959, 81, 529

59ML   R. Mathur and H. Lal, J. Phys. Chem., 1959, 63, 439

59MP   P. K. Migal and A. N. Pushnyak, Russ. J. Inorg. Chem., 1959, 4, 601 (1336)

59NH   R. Nasanen and T. Heikkila, Suomen Kem., 1959, B32, 163

59RG   C. F. Richard, R. L. Gustafson, and A. E. Martell, J. Amer. Chem. Soc.,
        1959, 81, 1033

59S   J. A. Swisher, M. S. Thesis, West Virginia Univ., 1959

59Sa   P. Sims, J. Chem. Soc., 1959, 3648

59SG   K. Schwabe, W. Graichen, and D. Spiethoff, Z. Phys. Chem. (Frankfurt), 1959,
        20, 68

59V   S. N. R. von Schalien, Suomen Kem., 1959, B32, 148

60AS   A. Albert and E. P. Serjeant, Biochem. J., 1960, 76, 621

60BA   R. G. Bates and G. F. Allen, J. Res. Nat. Bur. Standards, 1960, 64A, 343

60BD   P. Brooks and N. Davidson, J. Amer. Chem. Soc., 1960, 82, 2118

60BH   R. G. Bates and H. B. Hetzer, J. Res. Nat. Bur. Standards, 1960, 64A, 427

60BT   D. Banerjee and K. K. Tripathi, Anal. Chem., 1960, 32, 1196

60C   J. C. Colleter, Ann. Chim., 1960, 415

60CP   M. Ciampolini, P. Paoletti, and L. Sacconi, J. Chem. Soc., 1960, 4553

60DE   H. Diehl and J. Ellingboe, Anal. Chem., 1960, 32, 1120

60GG   D. H. Gold and H. P. Gregor, J. Phys. Chem., 1960, 64, 1461

60H   E. Hoyer, Chem. Ber., 1960, 93, 2475

60HD   J. L. Hall, W. G. Dean, and E. A. Pacofsky, J. Amer. Chem. Soc., 1960, 82, 3303

60HJ   F. Holmes and F. Jones, J. Chem. Soc., 1960, 2398

60LQ   T. J. Lane and K. P. Quinlan, J. Amer. Chem. Soc., 1960, 82, 2994

60LR   R. Leberman and B. R. Rabin, Nature, 1960, 185, 768

60MC   R. B. Martin, M. Chamberlin, and J. T. Edsall, J. Amer. Chem. Soc., 1960, 82, 495

60MP   P. K. Migal and A. N. Pushnyak, Russ. J. Inorg. Chem., 1960, 5, 293 (610)

60O   R. Osterberg, Acta Chem. Scand., 1960, 14, 471

60Oa   J. Olivard, Arch. Biochem. Biophys., 1960, 88, 382

60P   D. D. Perrin, J. Amer. Chem. Soc., 1960, 82, 5642

60R   J. Rydberg, Acta Chem. Scand., 1960, 14, 157

60RS   M. Rawitscher and J. M. Sturtevant, J. Amer. Chem. Soc., 1960, 82, 3739

60S   S. C. Srivastava, Doctor's Thesis, Allahabad Univ., India, 1960 (see 64SM)

60SP   L. Sacconi, P. Paoletti, and M. Ciampolini, J. Amer. Chem. Soc., 1960, 82, 3831

60V   S. N. R. von Schalien, Suomen Kem., 1958, B33, 5

61AL   D. J. Alner and R. C. Lansbury, J. Chem. Soc., 1961, 3169

61ALa   V. Armeanu and C. Luca, Z. Phys. Chem. (Leipzig), 1961, 217, 389

61AV   K. V. Astakhov, V. B. Verenikin, and V. I. Zimin, Russ. J. Inorg. Chem.,
          1961, 6, 1062 (2077)

61B   H. Brintzinger, Helv. Chim. Acta, 1961, 44, 935, 1199

61BH   R. G. Bates and H. B. Hetzer, J. Phys. Chem., 1961, 65, 667

61BM   J. A. Broomhead, H. A. McKenzie, and D. P. Mellor, Aust. J. Chem., 1961, 14, 649

61BR   A. I. Biggs and R. A. Robinson, J. Chem. Soc., 1961, 388, 2572

61BS   P. K. Bhattacharya, M. C. Saxena, and S. N. Banerji, J. Indian Chem. Soc.,
          1961, 38, 801

61CP   M. Ciampolini and P. Paoletti, J. Phys. Chem., 1961, 65, 1224

61CPS   M. Ciampolini, P. Paoletti, and L. Sacconi, J. Chem. Soc., 1961, 2994

61CS   G. Curthoys and D. A. J. Swinkels, Anal. Chim. Acta, 1961, 24, 589

61DK   A. G. Desai and M. B. Kabadi, J. Indian Chem. Soc., 1961, 38, 805

61ES   J. M. Essery and K. Schofield, J. Chem. Soc., 1961, 3939

61FB   R. Ferreira, E. Ben-Zvi, T. Yamane, J. Vasileuskis, and N. Davidson, Adv. in
          Chem. of Coord. Compds., S. Kirschner, Ed., Macmillan, New York, N. Y.,
          1961, 457

61GF   R. W. Green and I. R. Freer, J. Phys. Chem., 1961, 65, 2211

61GG   A. A. Grinberg and M. I. Gelfman, Proc. Acad. Sci. USSR, 1961, 137, 257 (87)

61GK   A. A. Grinberg and Kh. Kh. Khakimov, Russ. J. Inorg. Chem., 1961, 6, 71 (144)

61H   E. Hoyer, Z. Anorg. Allg. Chem., 1961, 312, 282

61HB   K. Hotta, J. Brahms, and M. Morales, J. Amer. Chem. Soc., 1961, 83, 997

61HD   C. J. Hawkins, H. Duewell, and W. F. Pickering, Anal. Chim. Acta, 1961, 25, 257

61HS   M. Hnilickova and L. Sommer, Coll. Czech. Chem. Comm., 1961, 26, 2189

61I   T. Iwamoto, Bull. Chem. Soc. Japan, 1961, 34, 605

61JW   B. R. James and R. J. P. Williams, J. Chem. Soc., 1961, 1007

61KM   F. Ya. Kulba, Ya. A. Makashev, and V. E. Mironov, Russ. J. Inorg. Chem., 1961,
           6, 321 (630)

61KP   E. C. Knoblock and W. C. Purdy, J. Electroanal. Chem., 1961, 2, 493

61KPa   E. C. Knoblock and W. C. Purdy, Radiation Res., 1961, 15, 94

61LL   C. H. Liu and C. F. Liu, J. Amer. Chem. Soc., 1961, 83, 4169

61LS   F. J. Langmyhr and A. R. Storm, Acta Chem. Scand., 1961, 15, 1461

61M   R. B. Martin, Fed. Proc., 1961, 20, No. 3, Suppl. 10, 54

61N   R. Nasanen, Suomen Kem., 1961, B34, 4

61Na   L. B. Nanninga, Biochem. Biophys. Acta, 1961, 54, 330

61NF   W. C. Nicholas and W. C. Fernelius, J. Phys. Chem., 1961, 65, 1047

61NM   R. Nasanen and P. Merilainen, Suomen Kem., 1961, B34, 127

61NS   K. Nakanishi, N. Suzuki, and F. Yamazaki, Bull. Chem. Soc. Japan, 1961, 34, 53

61OP   W. J. O'Sullivan and D. D. Perrin, Biochem. Biophys. Acta, 1961, 52, 612

61P   D. D. Perrin, J. Chem. Soc., 1961, 2244

61PG   J. M. Pagano, D. E. Goldberg, and W. C. Fernelius, J. Phys. Chem., 1961, 65, 1062

61RF   R. E. Reichard and W. C. Fernelius, J. Phys. Chem., 1961, 65, 380

61RM   D. K. Roe, D. B. Masson, and C. J. Nyman, Anal. Chem., 1961, 33, 1464

61S   I. C. Smith, Diss., Kansas State Univ., 1961

61SB   T. G. Spiro and C. J. Ballhausen, Acta Chem. Scand., 1961, 15, 1707

61SL   A. R. Strom and F. J. Langmyhr, Acta Chem. Scand., 1961, 15, 1765

61SP   L. Sacconi, P. Paoletti, and M. Ciampolini, J. Chem. Soc., 1961, 5115

61SS   W. Schaeg and F. Schneider, Z. Physiol. Chem., 1961, 326, 40

61TD   E. R. Tucci, E. Doody, and N. C. Li, J. Phys. Chem., 1961, 65, 1570

62A   G. Anderegg, Helv. Chim. Acta, 1962, 45, 1303

62Aa   G. Anderegg, Helv. Chim. Acta, 1962, 45, 1643

62AB   G. Atkinson and J. E. Bauman, Jr., Inorg. Chem., 1962, 1, 900

62AL   A. C. Andrews, T. D. Lyons, and T. D. O'Brien, J. Chem. Soc., 1962, 1776

62AM   H. Asai and M. Morales, Arch. Biochem. Biophys., 1962, 99, 383

62BE   S. Bolton and R. E. Ellin, J. Pharm. Sci., 1962, 51, 533

62BR   B. E. Bower, R. A. Robinson, and R. G. Bates, J. Res. Nat. Bur. Standards,
        1962, 66A, 71

62BS   P. A. Brauner and G. Schwarzenbach, Helv. Chim. Acta, 1962, 45, 2030

62CI   J. J. Christensen and R. M. Izatt, J. Phys. Chem., 1962, 66, 1030

62CL   S. Cabani and M. Landucci, J. Chem. Soc., 1962, 278

62CM   S. Cabani, G. Moretti, and E. Scrocco, J. Chem. Soc., 1962, 88

62CW   L. Cockerell and H. F. Walton, J. Phys. Chem., 1962, 66, 75

62DG   S. P. Datta and A. K. Grzybowski, J. Chem. Soc., 1962, 3068

62DP   G. Douheret and J. C. Pariand, J. Chim. Phys., 1962, 59, 1021

62FH   J. F. Fisher and J. L. Hall, Anal. Chem., 1962, 34, 1094

62GN   W. J. Geary, G. Nickless, and F. H. Pollard, Anal. Chim. Acta, 1962, 27, 71

62H   J. L. Hall, Proc. W. Va. Acad. Sci., 1962, 35, 104

62Ha   J. Halmekoski, Suomen Kem., 1962, B35, 209

62Hb   J. Halmekoski, Suomen Kem., 1962, B35, 238

62Hc   J. Halmekoski, Suomen Kem., 1962, B35, 241

62HB   H. B. Hetzer and R. G. Bates, J. Phys. Chem., 1962, 66, 308

62HBa   U. Handschin and H. Brintzinger, Helv. Chim. Acta, 1962, 45, 1037

62HI   G. I. H. Hanania and D. H. Irvine, J. Chem. Soc., 1962, 2745

62HJ   F. Holmes and F. Jones, J. Chem. Soc., 1962, 2818

62HP   C. J. Hawkins and D. D. Perrin, J. Chem. Soc., 1962, 1351

62HR   H. B. Hetzer, R. A. Robinson, and R. G. Bates, J. Phys. Chem., 1962, 66, 2696

62HS   J. L. Hall, J. A. Swisher, D. G. Brannon, and T. M. Liden, Inorg. Chem.,
        1962, 1, 409

62IC   R. M. Izatt and J. J. Christensen, J. Phys. Chem., 1962, 66, 359

62IJ   Y. Israeli and E. Jungreis, Bull. Res. Council Israel, 1962, 11A, 121

62IM   H. Irving and D. H. Miller, J. Chem. Soc., 1962, 5222

62IMa   H. Irving and D. H. Mellor, J. Chem. Soc., 1962, 5237

62JS   E. Jacobsen and K. Schroder, Acta Chem. Scand., 1962, 16, 1393

62K   B. Kirson, Bull. Soc. Chim. France, 1962, 1030

62KI   B. Kirson and J. Israeli, Bull. Soc. Chim. France, 1962, 1572

62KM   F. Ya. Kulba, Yu. A. Makashev, B. D. Culler, and C. V. Kiselev, Russ. J.
        Inorg. Chem., 1962, 7, 351 (689)

62M   R. K. Murmann, J. Amer. Chem. Soc., 1962, 84, 1349

62MB   W. A. E. McBryde, D. A. Brisbin, and H. Irving, J. Chem. Soc., 1962, 5245

62MS   P. K. Migal and G. F. Serova, Russ. J. Inorg. Chem., 1962, 7, 827 (1601)

62NMB   R. Nasanen, P. Merilainen, and O. Butkewitsch, Suomen Kem., 1962, B35, 219

62NMH   R. Nasanen, P. Merilainen, and E. Heinanen, Suomen Kem., 1962, B35, 15

62NMK   R. Nasanen, P. Merilainen, and M. Koskinen, Suomen Kem., 1962, B35, 59

62O   R. Osterberg, Acta Chem. Scand., 1962, 16, 2434

62SG   A. Ya. Sychev and A. P. Gerbeleu, Russ. J. Inorg. Chem., 1962, 7, 138 (269)

62SH   L. Sommer and M. Hnilickova, Anal. Chim. Acta, 1962, 27, 241

63HS   M. Hnilickova and L. Sommer, Z. Anal. Chem., 1963, 193, 171

62SS   G. Schwarzenback and I. Szilard, Helv. Chim. Acta, 1962, 45, 1222

62ST   J. G. Schmidt and R. F. Trimble, J. Phys. Chem., 1962, 66, 1063

62TA   P. Teyssie, G. Anderegg, and G. Schwarzenbach, Bull. Soc. Chim. Belges,
        1962, 71, 177

62TM   M. M. Taqui Khan and A. E. Martell, J. Phys. Chem., 1962, 66, 10

62TMa   M. M. Taqui Khan and A. E. Martell, J. Amer. Chem. Soc., 1962, 84, 3037

62W   I. Wadso, Acta Chem. Scand., 1962, 16, 479

63A   G. Anderegg, Helv. Chim. Acta, 1963, 46, 2397

63Aa   G. Anderegg, Helv. Chim. Acta, 1963, 46, 2813

63AB   G. Atkinson and J. E. Bauman, Jr., Inorg. Chem., 1963, 2, 64

63AR   A. C. Andrews and J. K. Romary, Inorg. Chem., 1963, 2, 1060

63BB   F. Becker, J. Barthel, N. G. Schmahl, and H. M. Luschow, Z. Phys. Chem.
         (Frankfurt), 1963, 37, 52

63BF   D. Betteridge, Q. Fernando, and H. Freiser, Anal. Chem., 1963, 35, 294

63BFa   S. P. Bag, Q. Fernando, and H. Freiser, Anal. Chem., 1963, 35, 719

63BM   D. A. Brisbin and W. A. E. McBryde, Canad. J. Chem., 1963, 41, 1135

63BMa   D. D. Bly and M.G. Mellon, Anal. Chem., 1963, 35, 1386

63BT   D. Betteridge, P. K. Todd, Q. Fernando, and H. Freiser, Anal. Chem., 1963,
         35, 729

63C   M. Cadiot-Smith, J. Chim. Phys., 1963, 60, 957, 976, 991

63Ca   C. Caullet, Bull. Soc. Chim. France, 1963, 688

63CC   A. Chakravorty and F. A. Cotton, J. Phys. Chem., 1963, 67, 2878

63CO   E. Campi, G. Ostacoli, and A. Vanni, Ric. Sci., 1963, 33 (II-A), 1073

63DB   J. M. Dale and C. V. Banks, Inorg. Chem., 1963, 2, 591

63DG   S. P. Datta, A. K. Grzybowski, and B. A. Weston, J. Chem. Soc., 1963, 792

63DK   A. G. Desai and M. B. Kabadi, Curr. Sci. (India), 1963, 32, 15

63FF   J. Fresco and H. Freiser, Inorg. Chem., 1963, 2, 82

63GC   R. W. Green, K. W. Catchpole, A. T. Phillip, and F. Lions, Inorg. Chem.,1963,2,597

63GP   P. George, R. Phillips, and R. Rutman, Biochemistry, 1963, 2, 501

63GPa   P. George,R. Phillips, and R. Rutman, Biochemistry, 1963, 2, 508

63H   J. Halmekoski, Suomen Kem., 1963, B36, 40

63Ha   J. Halmekoski, Suomen Kem., 1963, B36, 55

63HB   H. B. Hetzer, R. G. Bates, and R. A. Robinson, J. Phys. Chem., 1963, 67, 1124

63HG   R. P. Held and D. E. Goldberg, Inorg. Chem., 1963, 2, 585

63HP   C. J. Hawkins and D. D. Perrin, Inorg. Chem., 1963, 2, 843

63KV   D. Kourad and A. A. Vlcek, Coll. Czech. Chem. Comm., 1963, 28, 595

63L   D. L. Leussing, Inorg. Chem., 1963, 2, 77

63LM   D. L. Lewis and R. K. Murmann, J. Inorg. Nucl. Chem., 1963, 25, 1431

63MB   C. E. Meloan and J. Butel, Anal. Chem., 1963, 35, 768

63MR   D. W. Margerum, D. B. Rorabacher, and J. F. G. Clarke, Jr., Inorg. Chem.
         1963, 2, 667

63MS   S. Mahapatra and R. S. Subrahmanya, Proc. Indian Acad. Sci. (A), 1963, 58, 161

63NK   K. Nagano, H. Kinoshita, and Z. Tamura, Chem. Pharm. Bull., 1963, 11, 999

63NM   R. Nasanen, P. Merilainen, and M. Koskinen, Suomen Kem., 1963, B36, 9

63NMa   R. Nasanen and P. Merilainen, Suomen Kem., 1963, B36, 97; Acta Chem. Scand.,
          1964, 18, 1337

63NMb   R. Nasamen, P. Merilainen, and M. Koskinen, Suomen Kem., 1963, B36, 110

63NMc   R. Nasanen and P. Merilainen, Suomen Kem., 1963, B36, 205

63NML   R. Nasanen, P. Merilainen, and S. Lukkari, Suomen Kem., 1963, B36, 135

63NT   K. Nagano, H. Tsukahara, H. Kinoshita, and Z. Tamura, Chem. Pharm. Bull.,
          1963, 11, 797

63OG   P. O'D. Offenhartz, P. George, and G. P. Haight, Jr., J. Phys. Chem., 1963,
          67, 116

63PC   P. Paoletti and M. Ciampolini, Ric. Sci., 1963, 3 (II-A), 399

63PCa   P. Paoletti and M. Ciampolini, Ric. Sci., 1963, 3 (II-A), 405

63PCS   P. Paoletti, M. Ciampolini, and L. Sacconi, J. Chem. Soc., 1963, 3589;
          Ric. Sci., 1960, 30, 1791

63PCV   P. Paoletti, M. Ciampolini, and A. Vacca, J. Phys. Chem., 1963, 67, 1065

63PL   G. Popa, C. Luca, and V. Magearu, J. Chim. Phys., 1963, 60, 355

63PV   P. Paoletti, A. Vacca, and I Guisti, Ric. Sci., 1963, 33 (II-A), 523

63RR   R. Reynaud and P. Rumpf, Bull. Soc. Chim. France, 1963, 1805

63RS   M. L. N. Reddy and U. V. Seshaiah, Indian J. Chem., 1963, 1, 536

63S   A. R. Storm, Acta Chem. Scand., 1963, 17, 667

63Sa   P. E. Sturrock, Anal. Chem., 1963, 35, 1092

63SB   H. Sigel and H. Brintzinger, Helv. Chim. Acta, 1963, 46, 701

63SBE   H. Sigel, H. Brintzinger, and H. Erlenmeyer, Helv. Chim. Acta, 1963, 46, 712

63SD   R. L. Seth and A. K. Dey, Z. Anal. Chem., 1963, 194, 271

63SG   A. Ya. Sychev, A. P. Gerbeleu, and P. K. Migal, Russ. J. Inorg. Chem., 1963,
          8, 1081 (2070)

63TA   Y. Tsuchitani, T. Ando, and K. Ueno, Bull. Chem. Soc. Japan, 1963, 36, 1534

63TN   Z. Tamura and K. Nagano, Chem. Pharm. Bull., 1963, 11, 793

63TY   R. S. Tobias and M. Yasuda, Inorg. Chem., 1963, 2, 1307

63YT   M. Yasuda and R. S. Tobias, Inorg. Chem., 1963, 2, 207

64AK   D. J. Alner and M. A. A. Kahn, J. Chem. Soc., 1964, 5265

64AM   F. J. Anderson and A. E. Martell, J. Amer. Chem. Soc., 1964, 86, 715

64AR   A. C. Andrews and J. Romary, J. Chem. Soc., 1964, 405

64B   J. Bjerrum, Acta Chem. Scand., 1964, 18, 843

64Ba   R. S. Brown, Canad. J. Chem., 1964, 42, 2635

64BG   L. Banford and W. J. Geary, J. Chem. Soc., 1964, 378

64BO   A. Bergh, P. O'D. Offenhartz, P. George, and G. P. Haight, Jr., J. Chem. Soc.
         1964, 1533

64BW   J. E. Bauman, Jr. and J. C. Wang, Inorg. Chem., 1964, 3, 368

64DC   M. A. Doran, S. Chaberek, and A. E. Martell, J. Amer. Chem. Soc., 1964, 86, 2129

64FB   R. Fischer and J. Bye, Bull. Soc. Chim. France, 1964, 2920

64FF   J. Fresco and H. Freiser, Anal. Chem., 1964, 36, 372

64GHI   P. George, G. I. H. Hanania, D. H. Irvine, and I. Abu-Issa, J. Chem. Soc.,
         1964, 5689

64GHL   R. W. Green, P. S. Hallman, and F. Lions, Inorg. Chem., 1964, 3, 376

64IN   R. J. Irving, L. Nelander, and I. Wadso, Acta Chem. Scand., 1964, 18, 769

64KS   K. Kahmann, H. Sigel, and H. Erlenmeyer, Helv. Chim. Acta, 1964, 47, 1754

64LA   S. C. Lahiri and S. Aditya, Z. Phys. Chem. (Frankfort), 1964, 41, 173

64LAa   S. C. Lahiri and S. Aditya, Z. Phys. Chem. (Frankfort), 1964, 43, 282

64LM   R. G. Lacoste and A. E. Martell, Inorg. Chem., 1964, 3, 881

64MS   S. Mahapatra and R. S. Subrahmanya, Proc. Indian Acad. Sci. (A), 1964, 59, 299

64MSa   P. K. Mical and G. F. Serova, Russ. J. Inorg. Chem., 1964, 9, 978 (1806)

64N   O. Navratil, Coll. Czech. Chem. Comm., 1964, 29, 2490

64NA   L. V. Nazarova, A. V. Ablov, and V. A. Dagaer, Russ. J. Inorg. Chem.,
         1964, 9, 1150 (2129)

64NK   R. Nasanen and M. Koskinen, Acta Chem. Scand., 1964, 18, 1337

64NM   R. Nasanen and P. Merilainen, Suomen Kem., 1964, B37, 54

64NMK   R. Nasanen, P. Merilainen, and M. Koskinen, Suomen Kem., 1964, B37, 41

64NML   R. Nasanen, P. Merilainen, and S. Lukkari, Suomen Kem., 1964, B37, 1

64OM   M. Ohta, H. Matsukawa, and R. Tsuchiya, Bull. Chem. Soc. Japan, 1964, 37, 692
         (see also J. Bjerrum and O. Monsted, ibid, 1971, 44, 3492

64OP   W. J. O'Sullivan and D. D. Perrin, Biochemistry, 1964, 3, 18

64P   F. Pantani, <u>Ric. Sci.</u>, 1964, <u>34</u> (II-A-<u>6</u>), 417

64PG   K. H. Pearson and K. H. Gayer, <u>Inorg. Chem.</u>, 1964, <u>3</u>, 476

64PV   P. Paoletti and A. Vacca, <u>J. Chem. Soc.</u>, 1964, 5051

64RM   R. Rowland and C. E. Meloan, <u>Anal. Chem.</u>, 1964, <u>36</u>, 1997

64SB   H. Sigel and H. Brintzinger, <u>Helv. Chim. Acta</u>, 1964, <u>47</u>, 1701

64SBE   P. W. Schneider, H. Brintzinger, and H. Erlenmeyer, <u>Helv. Chim. Acta</u>,
            1964, <u>47</u>, 992

64SD   R. L. Seth and A. K. Dey, <u>Indian J. Chem.</u>, 1964, <u>2</u>, 291

64SM   Unpublished values quoted in L. G. Sillen and A. E. Martell, <u>Stability</u>
            <u>Constants of Metal-Ion Complexes</u>, Special Publication No. 17,
            The Chemical Soc., London, 1964

64SPB   B. T. Suchorukow, U. I. Poltew, and L. Blumenfeld, <u>Abh. Deut. Akad. Wiss. Berlin,</u>
            <u>Kl. Med.</u>, 1964, 381

64SPC   L. Sacconi, P. Paoletti, and M. Ciampolini, <u>J. Chem. Soc.</u>, 1964, 5046

64W   R. G. Wilkins, <u>Inorg. Chem.</u>, 1964, <u>3</u>, 520

64WB   J. C. Wang, J. E. Bauman, Jr., and R. K. Murmann, <u>J. Phys. Chem.</u>, 1964, <u>68</u>, 2296

64WD   T. H. Wirth and N. Davidson, <u>J. Amer. Chem. Soc.</u>, 1964, <u>86</u>, 4314

65AZ   A. C. Andrews and D. M. Zebolsky, <u>J. Chem. Soc.</u>, 1965, 742

65BC   J. Botts, A. Chashin, and H. L. Young, <u>Biochemistry</u>, 1965, <u>4</u>, 1788

65BY   H. Berge and P. Yeroschewski, <u>Z. Phys. Chem.</u> (Leipzig), 1965, <u>228</u>, 239

65C   D. R. Crow, <u>J. Polarog. Soc.</u>, 1965, <u>11</u>, 22, 67

65CC   S. Cabani and G. Conti, <u>Gazz. Chim. Ital.</u>, 1965, <u>95</u>, 533

65CF   F. C. Chou, Q. Fernando, and H. Freiser, <u>Anal. Chem.</u>, 1965, <u>37</u>, 361

65CJ   W. A. Connor. M. M. Jones, and D. L. Tuleen, <u>Inorg. Chem.</u>, 1965, <u>4</u>, 1129

65CM   H. L. Conley and R. B. Martin, <u>J. Phys. Chem.</u>, 1965, <u>69</u>, 2914

65CMa   H. L. Conley and R. B. Martin, <u>J. Phys. Chem.</u>, 1965, <u>69</u>, 2923

65CS   P. J. Conn and D. F. Swinehart, <u>J. Phys. Chem.</u>, 1965, <u>69</u>, 2653

65D   G. Douheret, <u>Bull. Soc. Chim. France</u>, 1965, 2915

65DD   R. L. Davies and K. W. Dunning, <u>J. Chem. Soc.</u>, 1965, 4168

65DF   C. N. C. Drey and J. S. Fruton, <u>Biochemistry</u>, 1965, <u>4</u>, 1258

65DK   N. M. Dyatlova, M. I. Kabachnik, T. Ya. Medved, M. V. Rudomino, and
            Ya. F. Belugin, <u>Doklady Chem.</u>, 1965, <u>161</u>, 307 (607)

65FB   A. M. Fiskin and M. Beer, Biochemistry, 1965, 4, 1289

65FCW   M. Friedman, J. F. Cavins, and J. S. Wall, J. Amer. Chem. Soc., 1965, 87, 3672

65H   J. Hala, J. Inorg. Nucl. Chem., 1965, 27, 2659

65HF   H. S. Hendrickson and J. G. Fullington, Biochemistry, 1965, 4, 1599

65HI   G. I. H. Hanania, D. H. Irvine, and F. Shurayh, J. Chem. Soc., 1965, 1149

65HWH   H. P. Hopkins, C. Wu, and L. G. Hepler, J. Phys. Chem., 1965, 69, 2244

65I   Y. J. Israeli, J. Inorg. Nucl. Chem., 1965, 27, 2271

65JN   R. F. Jameson and W. F. S. Neillie, J. Chem. Soc., 1965, 2391

65JNa   R. F. Jameson and W. F. S. Neillie, J. Inorg. Nucl. Chem., 1965, 27, 2623

65KSD   A. F. Krivis, G. R. Supp, and R. L. Doerr, Anal. Chem., 1965, 37, 52

65KSE   K. Kahmann, H. Sigel, and H. Erlenmeyer, Helv. Chim. Acta, 1965, 48, 295

65MA   F. J. Millero, J. C. Ahluwalia, and L. G. Hepler, J. Chem. Eng. Data, 1965,
            10, 199

65MB   E. Mario and S. M. Bolton, Anal. Chem., 1965, 37, 165

65MP   P. K. Migal and K. I. Ploae, Russ. J. Inorg. Chem., 1965, 10, 1368 (2517)

65MT   Y. Murakami and M. Takagi, Bull. Chem. Soc. Japan, 1965, 38, 828

65N   L. V. Nazarova, Russ.J. Inorg. Chem., 1965, 10, 1364

65NK   O. Navratil and J. Kotas, Coll. Czech. Chem. Comm., 1965, 30, 1824

65NKK   R. Nasanen, M. Koskinen, and K. Kajander, Suomen Kem., 1965, B38, 103

65NKS   R. Nasanen, M. Koskinen, R. Salonen, and A. Kiiski, Suomen Kem., 1965, B38, 81

65NKT   R. Nasanen, M. Koskinen, and P. Tilus, Suomen Kem., 1965, B38, 125

65PE   R. Phillips, P. Eisenberg, P. George, and R. J. Rutman, J. Biol. Chem., 1965,
            240, 4393

65PG   R. L. Pecsok, R. A. Garber, and L. D. Shields, Inorg. Chem., 1965, 4, 447

65PL   G. Popa, C. Luca, and V. Magearu, J. Chim. Phys., 1965, 62, 449, 853

65PS   P. Paoletti, J. H. Stern, and A. Vacca, J. Phys. Chem., 1965, 69, 3759

65R   M. Robson, Nature, 1965, 208, 265

65RV   C. Ropars and R. Viovy, J. Chim. Phys., 1965, 62, 408

65S   H. Sigel, Helv. Chim. Acta, 1965, 48, 1513

65Sa   H. Sigel, Helv. Chim. Acta, 1965, 48, 1519

65Sb   A. P. Savostin, Russ. J. Inorg. Chem., 1965, 10, 1394 (2565)

65SG   V. K. Sharma and J. M. Gaur, J. Electroanal. Chem., 1965, 9,321

65SS   G. Schwarzenbach and M. Schellenberg, Helv. Chim. Acta, 1965, 48, 28

65T   C. Tissier, Bull. Soc. Chim. France, 1965, 124

65TS   M. L. Tomlinson, M. L. R. Sharp, and H. M. N. H. Irving, J. Chem. Soc.,
          1965, 603

65UA   E. Uhling and W. Achilles, Z. Chem., 1965, 3, 109

65WB   J. C. Wang and J. E. Bauman, Inorg. Chem., 1965, 4, 1613

65WR   S. Westerback, K. S. Ragan, and A. E. Martell, J. Amer. Chem. Soc., 1965, 87,2567

65ZK   Yu. A. Zolotov and N.M. Kuzmin, J. Anal. Chem. USSR, 1965, 20, 442 (476)

54ZL   Yu. A. Zolotov and V. G. Lambrev, J. Anal. Chem. USSR, 1965, 20, 1204 (1153)

66AT   P. J. Antikainen and K. Tevanen, Suomen Kem., 1966, B39, 247, 285

66BC   E. B. Buchanan, Jr., D. Crichton, and J. R. Bacon, Talanta, 1966, 13, 903

66BE   K. Burger, I. Egyed, and I. Raff, J. Inorg. Nucl. Chem., 1966, 28, 139

66BS   L. G. Bunville and S. J. Schwalbe, Biochemistry, 1966, 5, 3521

66BV   A. K. Babko, A. I. Volkova, and T. E. Getman, Russ. J. Inorg. Chem., 1966,
          11, 203 (374)

66C   D. R. Crow, J. Polarog. Soc., 1966, 12, 101

66CR   J. J. Christensen, J. H. Rytting, and R. M. Izatt, J. Amer. Chem. Soc.,
          1966, 88, 5105

66D   A. de Courville, Comp. Rend. Acad. Sci. Paris, Ser. C, 1966, 262, 1196

66DG   S. P. Datta and A. K. Grzybowski, J. Chem. Soc. (B), 1966, 136

66DGa   S. P. Datta and A. K. Grzybowski, J. Chem. Soc. (A), 1966, 1059

66DK   A. G. Desai and M. B. Kabadi, J. Inorg. Nucl. Chem., 1966, 28, 1279

66DM   C. D. Dwivedi, K. N. Munshi, and A. K. Dey, J. Inorg. Nucl. Chem., 1966, 28, 245

66DT   E. Doody, E. R. Tucci, R. Scruggs, and N. C. Li, J. Inorg. Chem., 1966, 28, 833

66ES   F. P. Emmenegger and G. Schwarzenbach, Helv. Chim. Acta, 1966, 49, 625

66FB   M. J. Fahsel and C. V. Banks, J. Amer. Chem. Soc., 1966, 88, 878

66FL   Ya. D. Fridman, M. G. Levina, and R. I. Sorochan, Russ. J. Inorg. Chem.,
          1966, 11, 877 (1641)

66GC   S. L. Gupta and M. K. Chatterjee, Indian J. Chem., 1966, 4, 22

66GD   D. W. Gruenwedel and N. Davidson, J. Mol. Biol., 1966, 21, 129

66HB   H. B. Hetzer, R. G. Bates, and R. A. Robinson, J. Phys. Chem., 1966, 70, 2869

66HE   D. N. Hague and M. Eigen, Trans.Faraday Soc., 1966, 62, 1236

66HH   R. H. Holyer, C. D. Hubbard, S. F. A. Kettle, and R. G. Wilkins, Inorg. Chem.,
        1966, 5, 622

66HJ   J. E. Hix, Jr. and M. M. Jones, Inorg. Chem., 1966, 5, 1863

66HP   R. W. Hay, L. J. Porter, and P. J. Morris, Aust. J. Chem., 1966, 19, 1197

66HS   M. Hnilickova and L. Sommer, Talanta, 1966, 13, 667

66HW   D. P. Hanlon, D. S. Watt, and E. W. Westhead, Anal. Biochem., 1966, 16, 225

66IR   R. M. Izatt, J. H. Rytting, L. D. Hansen, and J. J. Christensen, J. Amer.
        Chem. Soc., 1966, 88, 2641

66JN   R. F. Jameson and W. F. S. Neillie, J. Inorg. Nucl. Chem., 1966, 28, 2667

66KZ   T. Kaden and A. Zuberbuhler, Helv. Chim. Acta, 1966, 49, 2189

66LA   K. Lal and R. P. Agarwal, J. Indian Chem. Soc., 1966, 43, 169

66LB   S. Lewin and M. A. Barnes, J. Chem. Soc. (B), 1966, 478

66LBH   J. W. Larson, G. L. Bertrand, and L. G. Hepler, J. Chem. Eng. Data, 1966,
        11, 595

66LH   D. L. Leussing and N. Hug, Anal. Chem., 1966, 38, 1388

66LHa   S. Lewin and D. A. Humphries, J. Chem. Soc. (B), 1966, 210

66LK   B. I. Lobov, F. Ya. Kulba, and V. E. Mironov, Russ. J. Phys. Chem., 1966,
        40, 1353 (2527)

66LM   G. A. L'Heureux and A. E. Martell, J. Inorg. Nucl. Chem., 1966, 28, 481

66MM   G. E. Mont and A. E. Martell, J. Amer. Chem. Soc., 1966, 88, 1387

66MPa   J. Majer, Z. Pikylikova, and V. Springer, Acta Fac. Pharm. Bohemoslov,
        1966, 12, 131

66MT   Y. Murakami, M. Takagi, and H. Nishi, Bull. Chem. Soc. Japan, 1966, 39, 1197

66N   O. Navratil, Coll. Czech. Chem. Comm., 1966, 31, 2492

66NK   R. Nasanen, M. Koskinen, L Anttila, and M. L. Korvola, Suomen Kem., 1966,
        B39, 122

66NT   R. Nasanen, P. Tilus, and A. M. Rinne, Suomen Kem., 1966, B39, 45

66OC   G. Ostacoli, E. Campi, and M. C. Gennaro, Gazz. Chim. Ital., 1966, 96, 741

66PA   O. N. Puplikova, L. N. Akimova, and I. A. Savich, Moscow Univ. Chem. Bull.,
        1966, 21, 237 (106)

66PC   H. K. J. Powell and N. F. Curtis, J. Chem. Soc. (B), 1966, 1205

66PCI   J. A. Partridge, J. J. Christensen, and R. M. Izatt, J. Amer. Chem. Soc.,
        1966, 88, 1649

66PG   R. C. Phillips, P. George, and R. J. Rutman, J. Amer. Chem. Soc., 1966, 88,
        2631

66PN   P. Paoletti, F. Nuzzi, and A. Vacca, J. Chem. Soc. (A), 1966, 1385

66PS   D. D. Perrin and V. S. Sharma, J. Inorg. Nucl. Chem., 1966, 28, 1271

66PSa   D. D. Perrin and V. S. Sharma, Biochem. Biophys. Acta, 1966, 127, 35

66PV   P. Paoletti, A. Vacca, and D. Arenare, J. Phys. Chem., 1966, 70, 193

66RG   T. I. Romantseva, M. I. Gromova, and V. M. Peshkova, Russ. J. Inorg. Chem.,
        1966, 11, 935 (1748)

66SK   E. V. Sklenskaya and M. Kh. Karapetyants, Russ. J. Inorg. Chem., 1966, 11,
        1102, (2061)

66SKa   E. V. Sklenskaya and M. Kh. Karapetyants, Russ. J. Inorg. Chem., 1966, 11,
        1478 (2749)

66TM   M. M. Taqui Khan and A. E. Martell, J. Amer. Chem. Soc., 1966, 88, 668

66VA   A. Vacca, D. Arenare, and P. Paoletti, Inorg. Chem., 1966, 5, 1384

66Z   J. Zarembowitch, J. Chim. Phys., 1966, 63, 420

66ZB   L. J. Zampa and R. F. Bogucki, J. Amer. Chem. Soc., 1966, 88, 5186

67AD   B. V. Agarwalla and A. K. Dey, Curr. Sci. (India), 1967, 36, 544

67AW   G. Anderegg and F. Wenk, Helv. Chim. Acta, 1967, 50, 2330

67BH   P. D. Bolton and F. M. Hall, Aust. J. Chem., 1967, 20, 1797

67CC   R. P. Carter, R. L. Carroll, and R. R. Irani, Inorg. Chem., 1967, 6, 939

67CCa   R. P. Carter, M. M. Crutchfield, and R. R. Irani, Inorg. Chem., 1967, 6, 943

67CR   J. J. Christensen, J. H. Rytting, and R. M. Izatt, J. Phys. Chem., 1967,
        71, 2700

67CW   J. J. Christensen, D. P. Wrathall, R. M. Izatt, and D. O. Talman, J. Phys.
        Chem., 1967, 71, 3001

67EH   W. J. Eilbeck and F. Holmes. J. Chem. Soc. (A), 1967, 1777

67EHP   W. J. Eilbeck, F. Holmes, G. G. Phillips, and A. E. Underhill, J. Chem.
        Soc. (A), 1967, 1161

67FH   J. F. Fisher and J. L. Hall, Anal. Chem., 1967, 39, 1550

67FL   Ya. D. Fridman and M. G. Levina, Russ. J. Inorg. Chem., 1967, 12, 1425 (2704)

67H   H. J. Harries, J. Inorg. Nucl. Chem., 1967, 29, 2484

67HM   R. W. Hay and P. J. Morris, Chem. Comm., 1967, 23

67HP   R. W. Hay and L. J. Porter, J. Chem. Soc. (B), 1967, 1261

67HV   F. R. Hartley and L. M. Venanzi, J. Chem. Soc. (A), 1967, 333

67HW   F. Holmes and D. R. Williams, J. Chem Soc. (A), 1967, 1256

67HWa   F. Holmes and D. R. Williams, J. Chem. Soc. (A), 1967, 1702

67KD   M. I. Kubachnik, N. M. Dyatlova, T. Ya. Medved, Yu. F. Begulin, and
        V. V. Sidorenko, Doklady Chem., 1967, 175, 621 (351)

67KM   M. I. Kabachnik, T. Ya. Medved, N. M. Dyatlova, M. N. Rusina, and M. V.
        Rudomino, Bull. Acad. Sci. USSR, Div. Chem. Sci., 1967, 1450 (1501)

67L   C. Luca, Bull. Soc. Chim. France, 1967, 2556

67LA   S. C. Lahiri and S. Aditya, Z. Phys. Chem. (Frankfurt), 1967, 55, 6

67LAa   S. C. Lahiri and S. Aditya, J. Indian Chem. Soc., 1967, 44, 9

67LAb   K. Lal and R. P. Agarwal, J. Less-Common Metals, 1967, 12, 269

67LK   B. I. Lobov, F. Ya. Kulba, and V. E. Mironov, Russ. J. Inorg. Chem., 1967,
        12, 176 (341)

67LR   P. Lingaiah, J. M. Rao, and U. V. Seshaiah, Curr. Sci. (India), 1967, 36, 197

67MB   J. N. Mathur and S. N. Banerji, J. Indian Chem. Soc., 1967, 44, 513

67MR   T. Ya. Meduel, M. V. Rudomino, E. A. Mironova, V. S. Balabukha, and M. I.
        Kabachnik, Bull. Acad. Sci. USSR, Div. Chem. Sci., 1967, 332 (351)

67N   L. V. Nazarova, Russ. J. Inorg. Chem., 1967, 12, 1620 (3062)

67Na   O. Navratil, Coll. Czech. Chem. Comm., 1967, 32, 2004

67NKA   R. Nasanen, M. Koskinen, M. L. Alatalo, L. Adler, and S. Koskela, Suomen Kem.,
        1967, B40, 124

67NKJ   R. Nasanen, M. Koskinen, R. Jarvinen, and R. Penttonen, Suomen Kem., 1967,
        B40, 25

67NP   G. Nickless, F. H. Pollard, and T. J. Samuelson, Anal. Chim. Acta, 1967, 39, 37

67PC   H. K. J. Powell and N. F. Curtis, J. Chem. Soc. (A), 1967, 1441

67PM   G. Popa and V. Magearu, Rev. Roum. Chim., 1967, 12, 1107

67PS   D. D. Perrin, I. G. Sayce, and V. S. Sharma, J. Chem. Soc. (A), 1967, 1755

67R   C. G. Regardh, Acta Pharm. Suecica, 1967, 4, 335

67RB   P. S. Relan and P. K. Bhattacharya, J. Indian Chem. Soc., 1967, 44, 536

67RBB   J. K. Romary, J. E. Bunds, and J. D. Barger, J. Chem. Eng. Data, 1967, 12, 224

67SB   M. S. Sun and D. G. Brewer, Canad. J. Chem., 1967, 45, 2729

67SBM   H. Sigel, K. Becker, and D. B. McCormick, Biochim. Biophys. Acta, 1967, 148, 655

67SI   L. Sommer and V. M. Ivanov, Talanta, 1967, 14, 171

67SIN   L. Sommer, V. M. Ivanov, and H. Novotna, Talanta, 1967, 14, 329

67SN   L. Sommer and H. Novotna, Talanta, 1967, 14, 457

67SP   J. P. Scharff and M. R. Paris, Compt. Rend Acad. Sci. Paris, Sec. C, 1967, 265, 488

67SPa   J. P. Scharff and M. R. Paris, Bull. Soc. Chim. France, 1967, 1782

67SPb   H. Sigel and B. Prijs, Helv. Chim. Acta, 1967, 50, 2357

67SS   R. Sundaresan, S. C. Saraiya, and A. K. Sundaram, Proc. Indian Acad. Sci.,
         1967, 66A, 120

67SSa   R. Sundaresan, S. C. Saraiya, and A. K. Sundaram, Proc. Indian Acad. Sci.,
         1967, 66A, 246

67TK   E. G. Timofeeva, A. A. Knyazeva, and I. I. Kalinichenko, Russ. J. Inorg. Chem.,
         1967, 12, 1117 (2119)

67TM   M. M. Taqui Khan and A. E. Martell, J. Amer. Chem. Soc., 1967, 89, 5585

67TMC   J. Tsau, S. Matsono, P. Clerc, and R. Benoit, Bull. Soc. Chim. France, 1967, 1039

67TT   M. Tissier and C. Tissier, Bull. Soc. Chim. France, 1967, 3155

67UH   E. Uhlemann and U. Hammerschick, Z. Anorg. Allg. Chem., 1967, 352, 53

67VA   A. Vacca and D. Arenare, J. Phys. Chem., 1967, 71, 1495

67W   M. R. Wright, J. Chem. Soc. (B), 1967, 1265

67Wa   U. Weser, Z. Naturforsch., 1967, 22B, 457

67ZF   A. Zuberbuhler and S. Fallab, Helv. Chim. Acta, 1967, 50, 889

68AL   D. J. Alner, R. C. Lansbury, and A. G. Smeeth, J. Chem. Soc. (A), 1968, 417

68AS   B. V. Agarwala, S. P. Sangal, and A. K. Dey, Mikrochim. Acta, 1968, 442

68BB   A. Bezdekova and B. Budesinsky, Coll. Czech. Chem. Comm., 1968, 33, 4178

68BD   A. Banerjee and A. K. Dey, Anal. Chim. Acta, 1968, 42, 473

68BH   P. D. Bolton and F. M. Hall, Aust. J. Chem., 1968, 21, 939

68CE   M. C. Cox, D. H. Everett, D. A. Landsman, and R. J. Munn, J. Chem. Soc. (B),
         1968, 1373

68CF   F. C. Chou and H. Freiser, Anal. Chem., 1968, 40, 34

68CW   D. R. Crow and J. V. Westwood, J. Inorg. Nucl. Chem., 1968, 30, 179

68CWI   J. J. Christensen, D. P. Wrathall, and R. M. Izatt, Anal. Chem., 1968, 40, 175

68DM   N. M. Dyatlova, U. V. Medyntsev, T. Ya. Medved, and M. I. Kabachnik,
          J. Gen. Chem. USSR, 1968, 38, 1025 (1065)

68DMa   N. M. Dyatlova, U. V. Medyntsev, T. Ya. Medved, and M. I. Kabachnik,
          J. Gen. Chem. USSR, 1968, 38, 1035 (1076)

68DP   A. Dei, P. Paoletti, and A. Vacca, Inorg. Chem., 1968, 7, 865

68EH   W. J. Eilbeck, F. Holmes, T. W. Thomas, and G. Williams, J. Chem. Soc. (A),
          1968, 2348

68FD   E. Fischerova, O. Dracka, and M. Meloun, Coll. Czech. Chem. Comm., 1968, 33, 473

68G   D. W. Gruenwedel, Inorg. Chem., 1968, 7, 495

68GF   G. Guntnikov and H. Freiser, Anal. Chem., 1968, 40, 39

68GFa   R. Griesser and S. Fallab, Chimia (Switz.), 1968, 22, 90

68GG   R. W. Green and W. G. Goodwin, Aust. J. Chem., 1968, 21, 1165

68GS   J. N. Gaur and V. K. Sharma, Acta Chim. Acad. Sci. Hung., 1968, 55, 255

68HA   D. Hopgood and R. J. Angelici, J. Amer. Chem. Soc., 1968, 90, 2508

68HG   J. L. Hall and W. B. Glenn, Jr., Proc. W. Va. Acad. Sci., 1968, 40, 270

68HR   H. B. Hetzer, R. A. Robinson, and R. G. Bates, J. Phys. Chem., 1968, 72, 2081

68IE   R. M. Izatt, D. Eatough, R. L. Snow, and J. J. Christensen, J. Phys. Chem.,
          1968, 72, 1208

68JK   R. F. Jameson and I. A. Khan, J. Chem. Soc. (A), 1968, 921

68KD   P. F. Knowles and H. Diebler, Trans. Faraday Soc., 1968, 64, 977

68KZ   T. Kaden and A. Zuberbuhler, Helv. Chim. Acta, 1968, 51, 1797

68L   P. Lanza, Electroanal. Chem., 1968, 19, 289

68LC   T. T. Lai and J. Y. Chen, J. Electroanal. Chem., 1968, 16, 413

68LP   S. Lukkari and M. Palonen, Suomen Kem., 1968, B41, 225

68LS   I. A. Lebedev and A. B. Shalinets, Soviet Radiochem., 1968, 10, 218 (233)

68MP   D. W. Margerum, B. L. Powell, and J. A. Luthy, Inorg. Chem., 1968, 7, 800

68MRD   T. Ya. Medved, M. V. Rudomino, N. M. Dyatlova, and M. I. Kabachnik, Bull.
          Acad. Sci. USSR, Div. Chem. Sci., 1968, 1150 (1211)

68MT   Y. Murakami and M. Takagi, J. Phys. Chem., 1968, 72, 116

68NL   O. Navratil and J. Liska, Coll. Czech. Chem. Comm., 1968, 33, 991

68NT   R. Nasanen, P. Tilus, and T. Uro, Suomen Kem., 1968, B41, 314

68O   W. F. O'Hara, Canad. J. Chem., 1968, 46, 1965

68OW   G. Ojelund and I. Wadso, Acta Chem. Scand., 1968, 22, 2691

68P   L. Przyborowski, Rocz. Chem., 1968, 42, 1181

68Pa   L. Przyborowski, Rocz. Chem., 1968, 42, 1383

68Pb   N. G. Podder, Curr. Sci. (India), 1968, 37, 48

68PG   R. Phillips and P. George, Biochim. Biophys. Acta, 1968, 162, 73

68PK   E. Peltonen and P. Kivalo, Suomen Kem., 1968, B41, 187

68PM   G. Popa, V. Magearu, and C. Luca, J. Electroanal. Chem., 1968, 17, 335

68PS   D. D. Perrin and V. S. Sharma, J. Chem. Soc. (A), 1968, 446

68RB   J. K. Romary, J. D. Barger, and J. E. Bunds, Inorg. Chem., 1968, 7, 1142

68RJ   L. Rasmussen and C. K. Jorgensen, Acta Chem. Scand., 1968, 22, 2313

68RS   N. P. Rudenko and A. I. Sevastyanov, Russ. J. Inorg. Chem., 1968, 13, 94 (184)

68S   H. Sigel, European J. Biochem., 1968, 3, 530

68Sa   H. Sigel, Angew. Chem., 1968, 80, 124; Angew. Chem. Internat. Edn.,
             1968, 7, 137

68Sb   M. I. Shtokalo, Russ. J. Inorg. Chem., 1968, 13, 392 (748)

68SA   G. K. Schweitzer and M. M. Anderson, J. Inorg. Nucl. Chem., 1968, 30, 1051

68T   L. I. Tikhonova, Russ. J. Inorg. Chem., 1968, 13, 1384 (2687)

68TE   B. A. Timimi and D. H. Everett, J. Chem. Soc. (B), 1968, 1380

68TF   M. Tanaka, S. Funahashi, and K. Shirai, Inorg. Chem., 1968, 7, 573

68TK   S. Takata, E. Kyuno, and R. Tsuchiya, Bull. Chem. Soc. Japan, 1968, 41, 2416

68TM   C. Tissier and C. Missegue, J. Chim. Phys., 1968, 65, 2037

68TS   T. D. Turnquist and E. B. Sandell, Anal. Chim. Acta, 1968, 42, 239

68VP   A. Vacca and P. Paoletti, J. Chem. Soc. (A), 1968, 2378

68W   D. R. Williams, J. Chem. Soc. (A), 1968, 2965

68ZB   L. Y. Zompa and R. F. Bogucki, J. Amer. Chem. Soc., 1968, 90, 4569

68ZK   A. Zuberbuhler and T. Kaden, Helv. Chim. Acta, 1968, 51, 1805

69BH   P. D. Bolton and F. M. Hall, J. Chem. Soc. (B), 1969, 259

69BHa   P. D. Bolton and F. M. Hall, J. Chem. Soc. (B), 1969, 1047

69BM   K. S. Bai and A. E. Martell, J. Inorg. Nucl. Chem., 1969, 31, 1697

69BMa   K. S. Bai and A. E. Martell, J. Amer. Chem. Soc., 1969, 91, 4412

69BN   E. A. Biryuk, V. A. Nazarenko, and L. N. Thu, Russ. J. Inorg. Chem.,
          1969, 14, 373 (714)

69BS   J. P. Belaich and J. C. Sari, Proc. Nat. Acad. Sci. U.S., 1969, 64, 763

69BZ   J. D. Barger, R. D. Zachariasen and J. K. Romary, J. Inorg. Nucl. Chem.,
          1969, 31, 1019

69CA   I. D. Chawla and A. C. Andrews, J. Inorg. Nucl. Chem., 1969, 31, 3809

69CI   J. J. Christensen, R. M. Izatt, D. P. Wrathall, and L. D. Hansen, J. Chem.
          Soc. (A), 1969, 1212

69CM   G. F. Condike and A. E. Martell, J. Inorg. Nucl. Chem., 1969, 31, 2455

69DM   N. M. Dyatlova, V. V. Medyntsev, T. M. Balashova, T. Ya. Medved, and N. I.
          Kabachnik, J. Gen. Chem. USSR, 1969, 39, 309 (329)

69EH   W. J. Eilbeck, F. Holmes, and T. W. Thomas, J. Chem. Soc. (A), 1969, 113

69ES   B. Evtimova, J. P. Scharff, and M. R. Paris, Bull. Soc. Chim. France, 1969, 81

69FK   D. Feinauer and C. Keller, Inorg. Nucl. Chem. Letters, 1969, 5, 625

69G   R. W. Green, Aust. J. Chem., 1969, 22, 721

69GK   M. I. Gelfman and N. A. Kustova, Russ. J. Inorg. Chem., 1969, 14, 110 (212)

69HG   P. R. Huber, R. Griesser, B. Prijs, and H. Sigel, European J. Biochem.,
          1969, 10, 238

69HO   J. O. Hill, G. Ojelund, and I. Wadso, J. Chem. Thermodynamics, 1969, 1, 111

69IM   E. D. Ivanova and P. K. Migal, Russ. J. Inorg. Chem., 1969, 14, 1723 (3269)

69J   R. Jarvinen, Suomen Kem., 1969, B42, 409

69KJ   A. Kircheiss, W. Jabs, and J. Bremer, Zeit. Chem. (Leipzig), 1969, 9, 116

69LA   B. E. Leach and R. J. Angelici, Inorg. Chem., 1969, 8, 907

69LV   P. Lumme and P. Virtanen, Suomen Kem., 1969, B42, 333

69LW   B. Lenarcik and Z. Warnke, Rocz. Chem., 1969, 43, 457

69MB   V. Mihailova and M. Bonnet, Bull. Soc. Chim. France, 1969, 4258

69MG   G. S. Manku, R. D. Gupta, A. N. Bhat, and B. D. Jain, Canad. J. Chem.,
          1969, 47, 3909

69MI   P. K. Migal and E. D. Ivanova, Russ. J. Inorg. Chem., 1969, 14, 1270 (2420)

69MM   M. S. Michailidis and R. B. Martin, J. Amer. Chem. Soc., 1969, 91, 4683

69MP   P. K. Migal and K. I. Ploae, Russ. J. Inorg. Chem., 1969, 14, 89 (171)

69NN   S. Nakatsuji, R. Nakajime, and T. Hara, Bull. Chem. Soc. Japan, 1969, 42, 3598

69NS   L. V. Nazarova, L. V. Suk, G. V. Budu, and G. V. Tatarova, Russ. J. Inorg. Chem.,
          1969, 14, 1254 (2389)

69NT   R. Nasanen, P. Tilus, E. Huttunen, and E. R. Turunen, Suomen Kem., 1969, B42, 390

69P   D. M. Palade, Russ. J. Inorg. Chem., 1969, 14, 231 (452)

69PP   J. Prasad and N. C. Peterson, Inorg. Chem., 1969, 8, 1622

69PPH   L. J. Porter, D. D. Perrin, and R. W. Hay, J. Chem. Soc. (A), 1969, 118

69PS   D. D. Perrin and V. S. Sharma, J. Chem. Soc. (A), 1969, 2060

69RJ   L. Rasmussen and C. K. Jorgensen, Inorg. Chim. Acta, 1969, 3, 543

69RJa   L. Rasmussen and C. K. Jorgensen, Inorg. Chim. Acta, 1969, 3, 547

69RT   D. B. Rorabacher, T. S. Turan, J. A. Defever, and W. G. Nickels, Inorg. Chem.,
          1969, 8, 1498

69RW   H. Reinert and R. Weiss, Z. Physiol. Chem., 1969, 350, 1310

69RWa   H. Reinert and R. Weiss, Z. Physiol. Chem., 1969, 350, 1321

69SR   A. I. Sevastyanov and N. P. Rudenko, Soviet Radiochem., 1969, 11, 246 (251)

69SV   R. T. Simpson and B. L. Vallee, Inorg. Chem., 1969, 8, 1185

69VD   L. C. Van Poucke, H. F. De Brabander, and Z. Eeckhaut, Bull. Soc. Chim. Belges,
          1969, 78, 131

69VS   V. M. Vdovenko and O. B. Stebunov, Soviet Radiochem., 1969, 11, 625 (635)

69WG   W. Wind and D. E. Goldberg, J. Inorg. Nucl. Chem., 1969, 31, 575

70AB   A. T. Advani, D. S. Barnes, and L. D. Pettit, J. Chem. Soc. (A), 1970, 2691

70ABa   B. K. Avinashi and S. K. Banerji, J. Indian Chem. Soc., 1970, 47, 1050

70ABb   B. K. Avinashi and S. K. Banerji, Chim. Anal., 1970, 52, 515

70BA   H. Buhler and G. Anderegg, Chimia (Switz.), 1970, 24, 433

70BB   K. Balachandran and S. K. Banerje, J. Prakt. Chem., 1970, 312, 266

70BL   G. Berthon and C. Luca, Anal. Chim. Acta, 1970, 51, 239

70BM   K. S. Bai and A. E. Martell, Inorg. Chem., 1970, 9, 1126

70BP   R. Barbucci, P. Paoletti, and A. Vacca, J. Chem. Soc. (A), 1970, 2202

70CA   I. D. Chawla and A. C. Andrews, J. Inorg. Nucl. Chem., 1970, 32, 91

70CR   J. J. Christensen, J. H. Rytting, and R. M. Izatt, J. Chem. Soc. (B), 1970, 1643

70CRa   J. J. Christensen, J. H. Rytting, and R. M. Izatt, Biochemistry, 1970, 9, 4907

70D   H. Diebler, Ber. Bunsen-Gesell. Phys. Chem., 1970, 74, 268

70DM   N. M. Dyatlova, T. Ya. Medved, M. V. Rudomino, and M. I. Kabachnik, Bull.
         Acad. Sci. USSR, Div. Chem. Sci., 1970, 767 (815)

70DT   G. S. Dokolina, Ya. I. Turyan, and O. N. Malyavinskaya, Russ. J. Phys. Chem.,
         1970, 44, 1679 (2942)

70E   D. J. Eatough, Anal. Chem., 1970, 42, 635

70EHP   W. J. Eilbeck, F. Holmes, and G. Phillips, J. Chem. Soc. (A), 1970, 689

70EHT   W. J. Eilbeck, F. Holmes, and T. W. Thomas, J. Chem. Soc. (A), 1970, 2062

70F   Y. Fujii, Nippon Kagaku Zasshi, 1970, 91, 671

70FE   W. L. Felty, C. G. Ekstrom, and D. L. Leussing, J. Amer. Chem. Soc., 1970, 92, 3006

70FK   Y. Fukuda, E. Kyuno, and R. Tsuchiya, Bull. Chem. Soc. Japan, 1970, 43, 745

70FR   G. Faraglia, F. J. C. Rossotti, and H. S. Rossotti, Inorg. Chim. Acta, 1970, 4, 488

70GD   D. P. Goel, Y. Dutt, and R. P. Singh, J. Inorg. Nucl. Chem., 1970, 32, 3119

70GO   I. Grenthe, H. Ots, and O. Ginstrup, Acta Chem. Scand., 1970, 24, 1067

70GS   R. Griesser and H. Sigel, Inorg. Chem., 1970, 9, 1238

70HH   J. L. Hall and T. W. Hall, Proc. W. Va. Acad. Sci., 1970, 42, 146

70HM   R. W. Hay and P. J. Morris, J. Chem. Soc. (B), 1970, 1577

70HZ   D. N. Hague and M. S. Zetter, Trans. Faraday Soc., 1970, 66, 1176

70KM   R. Karlicek, J. Majer, and J. Polakovicova, Chem. Zvesti, 1970, 24, 161

70KP   F. Kopecky, M. Pesak, and J. Celechovsky, Coll. Czech. Chem. Comm., 1970, 35, 576

70MA   V. M. Mylnikova and K. V. Astakhov, Russ. J. Phys. Chem., 1970, 44, 284 (512)

70MAa   V. M. Mylnikova and K. V. Astakhov, Russ. J. Phys. Chem., 1970, 44, 600 (1084)

70MAb   V. M. Mylnikova and K. V. Astakhov, Russ. J. Phys. Chem., 1970, 44, 1222 (2152)

70MAc   V. M. Mylnikova and K. V. Astakhov, Russ. J. Phys. Chem., 1970, 44, 1417 (2500)

70MB   J. L. Meyer and J. E. Bauman, Jr., J. Amer. Chem. Soc., 1970, 92, 4210

70ML   A. Milczarska and T. Lipiec, Rocz. Chem., 1970, 44, 1355

70MR   M. S. Mohan and G. A. Rechnitz, J. Amer. Chem. Soc., 1970, 92, 5839; Science,
         1970, 168, 1460; (see also N. C. Melchior, Science, 1971, 171, 1267)

70NB   L. V. Navarova and G. V. Buda, Russ. J. Inorg. Chem., 1970, 15, 1600 (3072)

70NK   R. Nasanen, M. Koskinen, P. Tilus, and A. Ilomaki, Suomen Kem., 1970, B43, 34

70NL   R. Nasanen and E. Lindell, Suomen Kem., 1970, B43, 358

70NT   R. Nasanen, P. Tilus, H. Jarvinen, and I. Komsi, Suomen Kem., 1970, B43, 154

70O   S. Oki, Anal. Chim. Acta, 1970, 49, 455

70P   L. Przyborowski, Rocz. Chem., 1970, 44, 1883

70PB   M. Paabo and R. G. Bates, J. Phys. Chem., 1970, 74, 702

70T   P. Tilus, Suomen Kem., 1970, B43, 501

70Ta   L. I. Tikhonova, Soviet Radiochem., 1970, 12, 483 (519)

70UP   V. V. Udovenko, G. B. Pomerants, and V. A. Krasnov, Russ. J. Inorg. Chem.,
        1970, 15, 1522 (2923)

70UR   V. V. Udovenko, L. G. Reiter, and N. I. Potaskalova, Russ. J. Inorg. Chem.,
        1970, 15, 49 (97)

70VS   W. Van de Poel and P. J. Slootmaekers, Bull. Soc. Chim. Belges, 1970, 79, 223

70WA   F. Wenk and G. Anderegg, Chimia (Switz.), 1970, 24, 427

70WB   D. C. Weatherford, E. J. Billo, J. P. Jones, and D. W. Margerum, Inorg. Chem.,
        1970, 9, 1557

70WK   E. W. Wilson, Jr., M. H. Kasperian, and R. B. Martin, J. Amer. Chem. Soc.,
        1970, 92, 5365

70WW   E. M. Wooley, R. W. Wilton, and L. G. Hepler, Canad. J. Chem., 1970, 48, 3249

70Z   A. Zuberbuhler, Helv. Chim. Acta, 1970 53, 669

71A   G. Anderegg, Helv. Chim. Acta, 1971, 54, 509

71AA   J. W. Allison and R. J. Angelici, Inorg. Chem., 1971, 10, 2233

71AB   B. K. Avinashi and S. K. Banerji, Indian J. Chem., 1971, 9, 496

71AV   P. J. Antikainen and A. Viitala, Suomen Kem., 1971, B44, 259

71BL   G. Berthon and C. Luca, Chim. Anal. (Paris), 1971, 53, 40

71BLa   C. Berthon and C. Luca, Chim. Anal. (Paris), 1971, 53, 501, 599, 611

71BS   L. Barnes and J. M. Sturtevant, Biochemistry, 1971, 10, 2120

71BSa   G. A. Bhat and R. S. Subrahmanya, J. Inorg. Nucl. Chem., 1971, 33, 3487

71CC   S. Cabani, G. Conti, and L. Lepori, Trans. Faraday Soc., 1971, 67, 1933

71CH   G. R. Cayley and D. N. Hague, Trans. Faraday Soc., 1971, 67, 786

71CHa   M. A. Cobb and D. N. Hague, Trans. Faraday Soc., 1971, 67, 3069

71EI   L. G. Egorova, I. I. Il'yashevich, N. V. Serebryakova, and G. N. Tyurenkova,
        J. Gen. Chem. USSR, 1971, 41, 655 (657)

71EK   V. I. Egorova and V. N. Kumok, J. Gen. Chem. USSR, 1971, 41, 1795 (1786)

71ES   L. G. Egorova, N. V. Serebryakova, and G. N. Tyureukova, J. Gen. Chem. USSR, 1971, 41, 1816 (1807)

71GD   D. Giron, G. Duc, and G. Thomas, Comp. Rend. Acad. Sci. Paris, Ser. C, 1971, 272, 1022

71GE   A. M. Goeminne and Z. Eeckhaut, Bull. Soc. Chim. Belges, 1971, 80, 605

71GR   R. W. Green and M. J. Rogerson, Aust. J. Chem., 1971, 24, 65

71GS   R. Griesser and H. Sigel, Inorg. Chem., 1971, 10, 2229

71HB   K. Houngbossa and G. Berthon, Comp, Rend. Acad. Sci. Paris, Ser. C, 1971, 273, 1247

71HBM   H. Hauer, E. J. Billo, and D. W. Margerum, J. Amer. Chem. Soc., 1971, 93, 4173

71HG   P. R. Huber, R. Griesser, and H. Sigel, Inorg. Chem., 1971, 10, 945

71HM   R. W. Hay and P. J. Morris, J. Chem. Soc. (A), 1971, 1518

71HMa   R. W. Hay and P. J. Morris, J. Chem. Soc. (A), 1971, 3562

71HP   G. R. Hedwig and H. K. J. Powell, Anal. Chem., 1971, 43, 1206

71HS   J. L. Hall, R. B. Simons, E. Morita, E. Joseph, and J. F. Gavlas, Anal. Chem., 1971, 43, 634

71HT   L. D. Hansen and D. J. Temer, Inorg. Chem., 1971, 10, 1439

71IG   H. M. N. H. Irving and P. J. Gee, Anal. Chim. Acta, 1971, 55, 315

71IP   I. I. Ilyashevich, V. N. Podchainova, N. V. Serebryakova, L. G. Egorova, and G. N. Tyurenkova, J. Gen. Chem. USSR, 1971, 41, 764 (758)

71K   T. Kaden, Helv. Chim. Acta, 1971, 54, 625

71KK   R. L. Karpel, K. Kustin, and M. A. Wolff, J. Phys. Chem., 1971, 75, 799

71KM   R. Karlicek and J. Majer, Coll. Czech. Chem. Comm., 1971, 36, 101

71KMF   F. Ya. Kulba, Yu. A. Makashev, and N. I. Fedyaev, Russ. J. Inorg. Chem., 1971, 16, 1254 (2352)

71KP   J. Kalinka and S. Petri, Rocz. Chem., 1971, 45, 1391

71KT   K. Kina and K. Toei, Bull. Chem. Soc. Japan, 1971, 44, 1289

71KZ   T. Kaden and A. Zuberbuhler, Helv. Chim. Acta, 1971, 54, 1361

71LL   B. E. Leach and D. L. Leussing, J. Amer. Chem. Soc., 1971, 93, 3377

71LP   P. Legittimo, F. Pantani, and G. Ciantelli, Gazz. Chim. Ital., 1971, 101, 465

71LS   P. Lingaiah and E. V. Sundaram, J. Indian Chem. Soc., 1971, 48, 961

71LW   B. Lenarcik, Z. Warnke, and J. Pioch, Rocz. Chem., 1971, 45, 727

71M   J. N. Mathur, Indian J. Chem., 1971, 9, 567

71Ma   G. S. Manku, J. Less-Common Metals, 1971, 24, 475

71MA   V. M. Mylnikova and K. V. Astakhov, Russ. J. Phys. Chem., 1971, 45, 183 (333)

71MAa   V. M. Mylnikova and K. V. Astakhov, Russ. J. Phys. Chem., 1971, 45, 186 (338)

71ML   J. G. Mason and I. Lipschitz, Talanta, 1971, 18, 1111

71MM   O. Makitie and S. Mirttinen, Suomen Kem., 1971, B44, 155

71MMa   R. J. Motekaitis, I. Murase, and A. E. Martell, J. Inorg. Nucl. Chem., 1971,
         33, 3353

71MMb   R. J. Motekaitis, I. Murase, and A. E. Martell, Inorg. Nucl. Chem. Letters,
         1971, 7, 1103

71MMW   J. P. Manners, K. G. Morallee, and R. J. P. Williams, J. Inorg. Nucl. Chem.,
         1971, 33, 2085

71OS   M. M. Osman, T. M. Salem, and M. S. El-Ezaby, J. Chem. Soc. (A), 1971, 1401

71OT   H. Ohtaki and N. Tanaka, J. Phys. Chem., 1971, 75, 90

71PB   P. Paoletti, R. Barbucci, A. Vacca, and A. Dei, J. Chem. Soc. (A), 1971, 310

71PD   P. Paoletti, A. Dei, and A. Vacca, J. Chem. Soc. (A), 1971, 2656

71PW   P. Paoletti, R. Walser, A. Vacca, and G. Schwarzenbach, Helv. Chim. Acta,
         1971, 54, 243

71RM   B. Rao and H. B. Mathur, J. Inorg. Nucl. Chem., 1971, 33, 809

71S   S. Sjoberg, Acta Chem. Scand., 1971 25, 2149

71SB   R. C. Sharma and P. K. Bhattacharya, J. Indian Chem. Soc., 1971, 48, 581

71SH   H. Sigel, P. R. Huber, and R. F. Pasternack, Inorg. Chem., 1971, 10, 2226

71SL   M. I. Shtokalo and V. V. Lukachina, Russ. J. Inorg. Chem., 1971, 16, 1334 (2502)

71SS   K. Srinivasan and R. S. Subrahmanya, J. Electroanal. Chem., 1971, 31, 233

71SSa   K. Srinivasan and R. S. Subrahmanya, J. Electroanal. Chem., 1971, 31, 245

71SSb   K. Srinivasan and R. S. Subrahmanya, J. Electroanal. Chem., 1971, 31, 257

71SW   G. Schwarzenbach and R. Walser, in Coordination Chemistry, Vol.1, A. E. Martell,
         ed., Van Nostrand Reinhold Co., New York, 1971, p. 450

71TK   M. M. Taqui Khan and C. R. Krishnamoorthy, J. Inorg. Nucl. Chem., 1971, 33, 1417

71U   J. Ulstrup, Acta Chem. Scand., 1971, 25, 3397

71V   W. Van de Poel, Bull. Soc. Chim. Belges, 1971, 80, 401

71WL   Z. Warnke and B. Lenarcik, Rocz. Chem., 1971, 45, 1385

71WN   W. Wozniak, J. Nicole, and G. Tridot, <u>Comp. Rend. Acad. Sci. Paris, Ser. C</u>, 1971, <u>272</u>, 635

71WS   O. A. Weber and V. Simeon, <u>J. Inorg. Nucl. Chem.</u>, 1971, <u>33</u>, 2097

71YM   O. Yamauchi, H. Miyata, and A. Nakahara, <u>Bull. Chem. Soc. Japan</u>, 1971, <u>44</u>, 2716

71ZK   A. Zuberbuhler, T. Kaden, and F. Koechlin, <u>Helv. Chim. Acta</u>, 1971, <u>54</u>, 1502

72AB   B. K. Avinashi and S. K. Banerji, <u>J. Indian Chem. Soc.</u>, 1972, <u>49</u>, 693

72AP   E. Arenare, P. Paoletti, A. Dei, and A. Vacca, <u>J. Chem. Soc. Dalton</u>, 1972, 736

72AU   W. Achilles and E. Uhlig, <u>Z. Anorg. Allg. Chem.</u>, 1972, <u>390</u>, 225

72B   J. Bjerrum, <u>Acta Chem. Scand.</u>, 1972, <u>26</u>, 2734

72Ba   Y. Bokra, <u>Bull. Soc. Chim. France</u>, 1972, 4483

72BB   Y. Bokra and G. Berthon, <u>J. Chim. Phys.</u>, 1972, <u>69</u>, 414

72BBa   K. Balachandran and S. K. Banerji, <u>J. Indian Chem. Soc.</u>, 1972, <u>49</u>, 543

72BBB   N. L. Babenko, A. I. Busev, and M. Sh. Blokh, <u>Russ. J. Inorg. Chem.</u>, 1972, <u>17</u>, 210 (402)

72BBL   A. P. Brunetti, E. J. Burke, M. C. Lim, and G. H. Nancollas, <u>J. Soln. Chem.</u>, 1972, <u>1</u>, 153

72BE   G. Berthon and O. Enea, <u>Thermochim. Acta</u>, 1972, <u>5</u>, 107

72BF   R. Barbucci, L. Fabbrizzi, and P. Paoletti, <u>J. Chem. Soc. Dalton</u>, 1972, 745

72BFa   R. Barbucci, L. Fabbrizzi, P. Paoletti, and A. Vacca, <u>J. Chem. Soc. Dalton</u>, 1972, 740

72BS   G. A. Bhat and R. S. Subrahmanya, <u>Inorg. Chim. Acta</u>, 1972, <u>6</u>, 403

72CA   K. E. Curtis and G. F. Atkinson, <u>Canad. J. Chem.</u>, 1972, <u>50</u>, 1649

72CH   M. A. Cobb and D. N. Hague, <u>J. Chem. Soc. Faraday I</u>, 1972, <u>68</u>, 932

72CHP   N. F. Curtis, G. R. Hedwig, and H. K. J. Powell, <u>Aust, J. Chem.</u>, 1972, <u>25</u>, 2025

72CM   S. C. Chang, J. K. H. Ma, J. T. Wang, and N. C. Li, <u>J. Coord. Chem.</u>, 1972, <u>2</u>, 31

72CP   Y. Couturier and C. Petitfaux, <u>Comp. Rend. Acad. Sci. Paris, Ser. C</u>, 1972, <u>275</u>, 953

72CPa   G. Carpeni and S. Poize, <u>J. Chim. Phys.</u>, 1972, <u>69</u>, 1592

72CS   J. J. Christensen, D. E. Smith, M. D. Slade, and R. M. Izatt, <u>Thermochim. Acta</u>, 1972, <u>5</u>, 35

72EB   O. Enea and G. Berthon, <u>Comp. Rend. Acad. Sci. Paris, Ser. C</u>, 1972, <u>274</u>, 1968

72EH   O. Enea, K. Houngbossa, and G. Berthon, <u>Electrochim. Acta</u>, 1972, <u>17</u>, 1585

72FB   L. Fabbrizzi, R. Barbucci, and P. Paoletti, <u>J. Chem. Soc.</u>, 1972, 1529

72FH   J. W. Fraser, G. R. Hedwig, H. K. J. Powell, and W. T. Robinson, Aust. J. Chem.,
         1972, 25, 747

72FS   C. M. Frey and J. E. Stuehr, J. Amer. Chem. Soc., 1972, 94, 8898

72GL   M. I. Gromova, M. N. Litvina, and V. M. Peshkova, J. Anal. Chem. USSR,
         1972, 27, 232 (270)

72GT   D. Giron-Forest and G. Thomas, Bull. Soc. Chim. France, 1972, 390

72HJ   J. L. Hall, E. Joseph, and M. B. Gum, J. Electroanal. Chem., 1972, 34, 529

72HM   R. W. Hay and P. J. Morris, J. Chem. Soc. Perkin II, 1972, 1021

72HMZ   D. N. Hague, S. R. Martin, and M. S. Zetter, J. Chem. Soc. Faraday I, 1972,
         68, 37

72J   J. B. Jensen, Acta Chem. Scand., 1972, 26, 4031

72JJ   Z. Jablonski, H. Jablonski, and W. Gorzelany, Rocz. Chem., 1972, 46, 365

72JW   R. F. Jameson and M. F. Wilson, J. Chem. Soc. Dalton, 1972, 2607

72KK   M. Koskinen and K. Kollin, Suomen Kem., 1972, B45, 114

72KM   R. Karlicek and J. Majer. Coll. Czech. Chem. Comm., 1972, 37, 151

72KMa   R. Karlicek and J. Majer, Coll. Czech. Chem. Comm., 1972, 37, 805

72KMF   F. Ya. Kulba, Yu. A. Makashev, and N. I. Fedyaev, Russ. J. Inorg. Chem.,
         1972, 17, 188 (361)

72KN   M. Koskinen and I. Nikkila, Suomen Kem., 1972, B45, 89

72KP   F. Kopecky, M. Pesak, and J. Celechovsky, Coll. Czech. Chem. Comm., 1972,
         37, 2739

72KS   F. Ya. Kulba, V. L. Stolyarov, and A. P. Zharkov, Russ. J. Inorg. Chem.,
         1972, 17, 57 (109); 197 (378)

72KV   R. G. Kallen, R. O. Viale, and L. K. Smith, J. Amer. Chem. Soc., 1972, 94, 576

72MH   E. M'Foundou, K. Houngbossa, and G. Berthon, Comp. Rend. Acad. Sci. Paris, Ser. C,
         1972, 274, 832

72MR   M. S. Mohan and G. A. Rechnitz, J. Amer. Chem. Soc., 1972, 94, 1714

72MS   J. M. Malin and R. E. Shepherd, J. Inorg. Nucl. Chem., 1972, 34, 3203

72ND   R. Nayan and A. K. Dey, Indian J. Chem., 1972, 10, 109

72NL   R. Nasanen and E. Lindell, Suomen Kem., 1972, B45, 18

72NM   R. Nakon and A. E. Martell, J. Amer. Chem. Soc., 1972, 94, 3026

72NMa   R. Nakon and A. E. Martell, J. Inorg. Nucl. Chem., 1972, 34, 1365

72NTE   R. Nasanen, P. Tilus, and E. Eskolin, Suomen Kem., 1972, B45, 87

72NTL   R. Nasanen, P. Tilus, E. Lindell, and E. Eskolin, Suomen Kem., 1972, B45, 111

72O   H. Ots, Acta Chem. Scand., 1972, 26, 3810

72S   A. B. Shalinets, Soviet Radiochem., 1972, 14, 279 (269)

72PB   C. S. G. Prasad, K. Balchandran, and S. K. Banerji, J. Indian Chem. Soc.,
          1972, 49, 1081

72PF   J. Pinart and J. Faucherre, Comp. Rend. Acad. Sci. Paris, Ser. C, 1972, 275, 1149

72PG   M. M. Palrecha and J. N. Gauer, J. Indian Chem. Soc., 1972, 49, 313

72PP   J. Pinart, C. Petitfaux, and J. Faucherre, Bull. Soc. Chim. France, 1972, 4534

72SGP   H. Sigel, R. Griesser, and B. Prijs, Z. Naturforsch. (B), 1972, 27, 353

72SW   H. Sigel, H. Wynberg, T. J. van Bergen, and K. Kahmann, Helv. Chim. Acta,
          1972, 55, 610

72TR   T. S. Turan and D. B. Rorabacher, Inorg. Chem., 1972, 11, 288

72TRa   M. M. Taqui Khan and P. R. Reddy, J. Inorg. Nucl. Chem., 1972, 34, 967

72TS   E. W. Tipping and H. A. Skinner, J. Chem. Soc. Faraday I, 1972, 68, 1764

72VE   L. C. Van Poucke and Z. Eeckhaut, Bull. Soc. Chim. Belges, 1972, 81, 363

72VT   L. C. Van Poucke, G. F. Thiers, and Z. Eeckhaut, Bull. Soc. Chim. Belges,
          1972, 81, 357

72WN   W. Wozniak, J. Nicole, and G. Tridot, Bull. Soc. Chim. France, 1972, 4445

72WNa   W. Wozniak, J. Nicole, and G. Tridot, Analusis, 1972, 1, 498

72ZB   S. Zimmer and R. Biltonen, J. Solution Chem., 1972, 1, 291

73AH   T. Arishima, K. Hamada, and S. Takamoto, Nippon Kagaku Kaishi, 1973, 1119

73AW   P. J. Antikainen and U. Witikainen, Acta Chem. Scand., 1973, 27, 2075

73B   J. Bjerrum, Acta Chem. Scand., 1973, 27, 970

73BB   M. J. Blais and G. Berthon, Bull. Soc. Chim. France, 1973, 2669

73BD   A. Braibanti, F. Dallavalle, E. Leporati, and G. Mori, J. Chem. Soc. Dalton,
          1973, 323

73BDa   A. Braibanti, F. Dallavalle, E. Leporati, and G. Mori, J. Chem. Soc. Dalton,
          1973, 2539

73BE   G. Berthon and O. Enea, Thermochim. Acta, 1973, 6, 57

73BEM   G. Berthon, O. Enea, and E. M'Foundou, Bull. Soc. Chim. France, 1973, 2967

73BF   R. Barbucci, L. Fabbrizzi, and P. Paoletti, Inorg. Chim. Acta, 1973, 7, 157

73BFa   R. Barbucci, L. Fabbrizzi, P. Paoletti, and A. Vacca, J. Chem. Soc. Dalton,
          1973, 1763

73BN   G. V. Budu and L. V. Nazarova, Russ. J. Inorg. Chem., 1973, 18, 807 (1531)

73CP   Y. Couturier and C. Petitfaux, Bull. Soc. Chim. France, 1973, 439

73CPa   Y. Couturier and C. Petitfaux, Bull. Soc. Chim. France, 1973, 445

73CV   H. S. Creyf and L. C. Van Poucke, J. Inorg. Nucl. Chem., 1973, 35, 3831

73DN   A. C. Dash and R. K. Nanda, J. Indian Chem. Soc., 1973, 50, 808

73FP   L. Fabbrizzi, P. Paoletti, M. C. Zobrist, and G. Schwarzenbach, Helv. Chim. Acta,
          1973, 56, 670

73GS   A. Gergely and I. Sovago, J. Inorg. Nucl. Chem., 1973, 35, 4355

73HB   K. Houngbossa, G. Berthon, and O. Enea, Thermochim. Acta, 1973, 6, 215

73HP   G. R. Hedwig and H. K. J. Powell, J. Chem. Soc. Dalton, 1973, 793

73HPa   G. R. Hedwig and H. K. J. Powell, J. Chem. Soc. Dalton, 1973, 1942

73HT   A. Hulanicki and M. Trojanowicz, Rocz. Chem., 1973, 47, 279

73JV   A. Jehlickova and F. Vlacil, Coll. Czech. Chem. Comm., 1973, 38, 3395

73K   M. Kunaszewska, Rocz. Chem., 1973, 47, 683

73KK   F. Karczynski, G. Kupryszewski, and H. Markuszewska, Rocz. Chem., 1973, 47, 1151

73LJ   P. Laget, P. Jallet, J. Wafflart, M. Moreau, H. Guerin, and J. Pieri,
          J. Chim. Phys., 1973, 70, 1285

73LP   C. L. Liotta, E. M. Perdue, and H. P. Hopkins, Jr., J. Amer. Chem. Soc.,
          1973, 95, 2439

73LS   P. Lingaiah and E. V. Sundaram, J. Indian Chem. Soc., 1973, 50, 479

73M   G. S. Manku, Bull. Chem. Soc. Japan, 1973, 46, 1704

73MB   P. K. Migal and E. S. Brunk, Russ. J. Inorg. Chem., 1973, 18, 497 (946)

73MM   R. J. Motekaitis, I. Murase, and A. E. Martell, to be published

73MMa   D. T. MacMillan, I. Murase, and A. E. Martell, to be published

73MS   M. L. Mittal, R. S. Saxena, and A. V. Pandey, J. Inorg. Nucl. Chem., 1973,
          35, 1691

73MW   J. K. H. Ma, J. T. Wang, and N. C. Li, J. Coord. Chem., 1973, 2, 281

73NA   R. Nakon and R. J. Angelici, Inorg. Chem., 1973, 12, 1269

73NAa   R. Nakon and R. J. Angelici, J. Amer. Chem. Soc., 1973, 75, 3170

73NB   L. V. Nazarova, G. V. Budu, and L. S. Matuzenko, Russ. J. Inorg. Chem.,
          1973, 18, 546 (1038)

73NK   R. Nasanen, M. Koskinen, P. Tilus, E. Lindell, J. Lihavainen, and M. Pinomaki,
          Suomen Kem., 1973, B46, 61

73NKa   R. Nasanen, M. Koskinen, P. Tilus, O. Klemola, E. Leinonen, and E. Wuolijoki, Suomen Kem., 1973, B46, 181

73OI   H. Ohtaki and Y. Ito, J. Coord. Chem., 1973, 3, 131

73P   C. Petitfaux, Ann. Chim., 1973, 8, 33

73PC   H. K. J. Powell and N. F. Curtis, Aust. J. Chem., 1973, 26, 977

73PF   P. Paoletti, L. Fabbrizzi, and R. Barbucci, Inorg. Chem., 1973, 12, 1861

73S   S. Sjoberg, Acta Chem. Scand., 1973, 27, 3721

73SB   R. C. Sharma and P. K. Bhattacharya, J. Indian Chem. Soc., 1973, 50, 232

73SBa   J. C. Sari and J. P. Belaich, J. Amer. Chem. Soc., 1973, 95, 7491

73SC   R. S. Saxena and U. S. Chaturvedi, Monat. Chem., 1973, 104, 1208

73SR   J. C. Sari, M. Ragot, and J. P. Belaich, Biochem. Biophys. Acta, 1973, 305, 1

73TK   M. M. Taqui Khan and C. R. Krishnamoorthy, J. Inorg. Nucl. Chem., 1973, 35, 1285

73TR   M. M. Taqui Khan and P. R. Reddy, J. Inorg. Nucl. Chem., 1973, 35, 2813

73U   S. G. W. Unruh, M. S. Thesis, Texas A&M Univ., 1973

73UP   V. V. Udovenko and G. B. Pomerants, Russ. J. Inorg. Chem., 1973, 18, 937

73YB   O. Yamauchi, H. Benno, and A. Nakahara, Bull. Chem. Soc. Japan, 1973, 46, 3458

73YN   O. Yamauchi, Y. Nakao, and A. Nakahara, Bull. Chem. Soc. Japan, 1973, 46, 3749

73ZT   K. Zwetanov, A. Tonev, and K. Kantschev, Monat. Chem., 1973, 104, 496

74A   G. Anderegg, Helv. Chim. Acta, 1974, 57, 1340

74B   Y. Bokra, Bull. Soc. Chim. France, 1974, 1269

74Ba   B. W. Budesinsky, Anal. Chim. Acta, 1974, 71, 343

74BEB   M. J. Blais, O. Enea, and G. Berthon, Thermochim. Acta, 1974, 8, 433

74BEH   G. Berthon, O. Enea, and K. Houngbossa, Thermochim. Acta, 1974, 9, 379

74CA   D. R. Crow and P. K. Aggarwal, Electrochim. Acta, 1974, 19, 309

74CP   Y. Couturier and C. Petitfaux, Bull. Soc. Chim. France, 1974, 855

74CPa   Y. Couturier and C. Petitfaux, Bull. Soc. Chim. France, 1974, 863

74GE   A. M. Goeminne and Z. Eeckhaut, J. Inorg. Nucl. Chem., 1974, 36, 357

74GS   B. Grgas-Kuznar, V. Simeon, and O. A. Weber, J. Inorg. Nucl. Chem., 1974, 36, 2151

74HM   D. N. Hague and S. R. Martin, J. Chem. Soc. Dalton, 1974, 254

74HP   G. R. Hedwig and H. K. J. Powell, J. Chem. Soc. Dalton, 1974, 47

74K   M. Komatsu, Bull. Chem. Soc. Japan, 1974, 47, 1636

74KZ   T. A. Kaden and A. D. Zuberbuhler, Helv. Chim. Acta, 1974, 57, 286

74MM   K. K. Mui, W. A. E. McBride, and E. Nieboer, Canad. J. Chem., 1974, 52, 1821

74MO   A. Massoumi, P. Overvoll, and F. J. Langmyhr, Anal. Chim. Acta, 1974, 68, 103

74MOa   A. Massoumi, P. Overvoll, and F. J. Langmyhr, Anal. Chim. Acta, 1974, 71, 205

74MW   G. K. R. Maker and D. R. Williams, J. Chem. Soc. Dalton, 1974, 1121

74MWa   G. K. R. Maker and D. R. Williams, J. Inorg. Nucl. Chem., 1974, 36, 1675

74NR   R. Nakon, P. R. Rechani, and R. J. Angelici, J. Amer. Chem. Soc., 1974, 96, 2117

74SM   R. M. Smith and A. E. Martell, unpublished data

74W   O. A. Weber, J. Inorg. Nucl. Chem., 1974, 36, 1341

$C_4H_{12}N_2$	40,40,41,42,52,54,83,88,120
$C_4H_{12}N_2S$	65
$C_4H_{12}N_2S_2$	66
$C_4H_{12}ON_2$	58,95
$C_4H_{12}O_3NP$	299,346
$C_4H_{13}N_3$	69,101
$C_4H_{13}O_6NP_2$	318
$C_4H_{14}O_4N_2P_2$	293
$C_4H_{14}O_6N_2P_2$	305
$C_5H_3N_4Cl$	344
C_5H_4NBr	340,340
C_5H_4NCl	340,340
$C_5H_4N_4$	344
$C_5H_4N_4S$	344
$C_5H_4ON_4$	345
$C_5H_4O_2N_4$	345
C_5H_5N	165
$C_5H_5N_5$	276
$C_5H_5N_5S$	344,344
C_5H_5ON	340,340,340
C_5H_5ONS	350
$C_5H_5ON_5$	345,353
$C_5H_5O_2N$	179,350
$C_5H_6N_2$	205,207,208
$C_5H_6N_6$	345
$C_5H_6ON_2$	206
$C_5H_6O_2N_2$	271
$C_5H_6O_3N_2$	342,342
C_5H_7NS	35
$C_5H_7N_3$	219
C_5H_7ON	21
$C_5H_8N_2$	148,150
$C_5H_8N_2S$	349
$C_5H_9N_3$	154,156,349,349
$C_5H_{10}O_3N_2$	30
$C_5H_{11}N$	5,73,337
$C_5H_{11}ON_2$	348
$C_5H_{11}O_2N$	333,333,333,336
$C_5H_{11}O_2NS$	334,334

$C_5H_{11}O_2N_3$	29,29
$C_5H_{12}N_2$	91,92,124
$C_5H_{12}ON_2$	14
$C_5H_{12}O_2N_2$	334
$C_5H_{13}N$	4,331,335
$C_5H_{13}ON$	17,337
$C_5H_{13}O_2N$	117,336,347
$C_5H_{13}O_3NS$	11
$C_5H_{14}N_2$	53,55,84,84,92,337
$C_5H_{14}ON_2$	96,97,348,348
$C_5H_{14}O_3NP$	317,346
$C_5H_{14}O_5NP$	317
$C_5H_{15}N_3$	103,103
$C_5H_{16}N_4$	71
$C_5H_{17}O_9N_2P_3$	346
$C_6H_4N_2$	177,177
$C_6H_4ON_4$	353
$C_6H_4O_2N_4$	353
C_6H_5ON	191
$C_6H_6N_4$	339
$C_6H_6ON_2$	184,196,197,198,340,340
$C_6H_6ON_4$	291,345,345
C_6H_7N	8,167,168,169
$C_6H_7N_2Cl$	50
$C_6H_7N_5$	344,344
C_6H_7ON	186,187,188,189,190,332
$C_6H_7ON_3$	192,193,194
$C_6H_7O_3NS$	12,13,331
$C_6H_7O_4NS$	332
$C_6H_8N_2$	49,147,206,209
$C_6H_8O_3N_2$	342
$C_6H_8O_4NP$	330
C_6H_9N	337
$C_6H_9N_3$	346
$C_6H_9O_2N$	333
$C_6H_{10}N_2$	150
$C_6H_{10}ON_4$	160
$C_6H_{11}N$	335
$C_6H_{11}N_3$	157,349

LIGAND NAME INDEX